Life Sciences Research and Development

AF743866

nova
science publishers

www.novapublishers.com

Life Sciences Research and Development

Properties and Applications of Alginate
Michael Y. Wilkerson (Editor)
2022. ISBN: 979-8-88697-371-6 (Softcover)
2022. ISBN: 979-8-88697-381-5 (eBook)

Dynamics and Interrelations between Nature, Science, and Society
Kaushalendra Kumar Jha, PhD,
Michael Campbell, PhD (Editors)
2022. ISBN: 979-8-88697-053-1 (Softcover)
2022. ISBN: 979-8-88697-130-9 (eBook)

Cyanobacteria and Their Importance
Rajeshwar P. Sinha, PhD (Editor)
2022. ISBN: 978-1-68507-934-5 (Hardcover)
2022. ISBN: 979-8-88697-071-5 (eBook)

Earthworm Engineering and Applications
Adarsh Pal Vig, PhD,
Surendra Singh Suthar, PhD,
Jaswinder Singh, PhD (Editors)
2022. ISBN: 978-1-68507-566-8 (Hardcover)
2022. ISBN: 978-1-68507-641-2 (eBook)

Earthworms and their Ecological Significance
Adarsh Pal Vig, PhD,
Surendra Singh Suthar, PhD,
Jaswinder Singh, PhD (Editors)
2022. ISBN: 978-1-68507-567-5 (Hardcover)
2022. ISBN: 978-1-68507-632-0 (eBook)

More information about this series can be found at
https://novapublishers.com/product-category/series/life-sciences-research-and-development/

Eylem Taskin, Celal Guven
and Salih Tunc Kaya
Editors

HMGB1

Functions, Inhibitors and Clinical Significance

nova
science publishers
www.novapublishers.com

DOI: 10.52305/KNYQ3299.

Library of Congress Cataloging-in-Publication Data

ISBN: 979-8-88697-408-9

Published by Nova Science Publishers, Inc. † New York

Contents

Preface

Damage-associated molecular patterns (DAMPs), a term also known as alarmins coined by Walter G. Land, Seong, and Matzinger, are endogenous danger molecules that are released from damaged or dying cells and activate the innate immune system by interacting with pattern recognition receptors (PRRs). DAMPs are physiologically located in the cell. However, it can be released from the cell when the cell is under stress conditions or microorganism invasion. In sterile inflammation, DAMPs are considered as "signal 0".

The best-known DAMPs include proteins such as heat shock proteins (HSP), histones, and cytokines, but some non-protein molecules can also act as "danger signals" such as ATP, extracellular DNA or RNA, and mitochondrial DNA. Also, one of the best well-known DAMPs is the high mobility group box-1 (HMGB1), the name coming due to its very fast moving in gel electrophoresis. Under physiologic conditions, HMGB1 is located at the nucleus. However, HMGB1 could be actively and passively secreted from the necrotic or dying cells. So, its function is diverse based on location. HMGB1 has played important roles in the regulation of transcription, replication, recombination, and DNA repair. However, the location of HMGB1 could be altered based on the redox status of cells, tissue type, and the stage of cell differentiation. Its function changes when the location alters. For example, HMGB1 has participated to mediate inflammation and immune responses at extracellular fluids as an alarmin to activate the innate immune system and promote tissue repair.

Furthermore, HMGB1 can promote autophagy by binding to BECN1 protein in the cytoplasm. Importantly, HMGB1 has been shown to contribute to the pathogenesis of various diseases including myocardial ischemia/reperfusion injury, epilepsy, diabetes, multiple sclerosis, cancer, as well as hepatic steatosis, and fatty liver disease. HMGB1 is also exposed to post-translational modifications, e.g., acetylation, methylation, phosphorylation, oxidation, glycosylation, and ADP-ribosylation. HMGB1 is a small protein around 25 kDa containing 215 amino acid residues expressed in almost all eukaryotic cell types. HMGB1 contains three cysteines located at positions 23 and 45 in the A box and 106 in the B box. In the book, HMGB1 and its functions are outlined with the clinical significance of its antagonist as well.

There are three sections in the book. The first section is named HMGB1 and Cancer, including two chapters. One of the chapters in the first section is focused on HMGB1 on cancer therapy and managing COVID-19 infection multiple sclerosis. The chapter sum up the active and/or their active seconder metabolites associated with HMGB1 of phytotherapy on cancer and COVID-19 treatments. Especially, the chapter focuses on the effect of herbal medicinal plants on cell proliferation for cancer treatment. Finally, the chapter gives us some suggestions for a potential oral tropical antiviral phytotherapy compound for COVID-19. The second chapter in the first section is the crosstalk between cancer and myocardial ischemia/reperfusion injury (MIR) through HMGB1 via ferroptotic cell death. The link between them might be inflammation through HMGB1 since inflammation is well documented to play an important role in the development of cancer and MIR. There is a new evaluation on the newly identified cell death pathway, ferroptosis, participating in the crosstalk between cancer and MIR based on inflammation by HMGB1.

The second section is HMGB1 and metabolic interactions, consisting of two chapters. The first chapter is HMGB1 and inflammation in adipose tissue, resulting in insulin resistance and type 2 diabetes. The chapter provides a comprehensive overview of the promotion and infiltration of immune cells in adipose leading to HMGB1 secretion triggering inflammation. Finally, in the chapter, some antagonist drugs are given to reverse the effect of HMGB1 on obesity and type-2 diabetes. The second chapter in the second section sums up recent data related to HMGB1 and liver injury, e.g., drug-induced liver injury, alcoholic liver disease, non-alcoholic fatty liver disease, viral hepatitis, sepsis, and acute chronic liver failure, hepatocellular death through oxidative stress, inflammatory signaling, and autophagy in hepatocytes. HMGB1 might play a double-edged sword role due to having a role in tissue repair and angiogenesis and triggering inflammation as well. So, the main effect of HMGB1 on the liver is oxidative stress and inflammation in acute periods.

The third section is about HMGB1 and neurodegenerative diseases. The section contains four chapters. The first chapter in the section evaluates HMGB1 and its antagonist in brain disorders, including epilepsy, headache, neuroimmunological disorders, neurodegenerative disorders, and stroke. The second chapter in the third section is about the role of HMGB1 on post-brain injury, including potential mechanisms and therapeutic opportunities as well. HMGB1 is well known to mediate neuro-inflammation and enhancement of endogenous brain repair. That is why the second chapter in the section outlines HMGB1 signaling in post-traumatic neuro-inflammation and explores prospects of therapeutic modulation to enhance recovery. The third chapter in the third section evaluates the interaction of HMGB1 and Multiple sclerosis via TLR4/NF-κB signaling pathway, leading to the release of proinflammatory cytokines causing an inflammatory response. Furthermore, the chapter contains some clues for developing effective treatment strategies in multiple sclerosis by blocking HMGB1-induced neuroinflammation. The last chapter in the third chapter aims to explain the effects of HMGB1 on Epilepsy. So, the chapter presents novel findings that the HMGB1 system could be targeted to prevent seizure generation and provide therapeutically relevant prognostic markers that can also predict a patient's response to therapy. Furthermore, it recaps epileptogenesis theory and examines the relationships between epilepsy and neuroinflammatory pathways using experimental and clinical data.

Eylem Taskin, Professor,
Adiyaman University, Turkey

Celal Guven, Associate Professor,
Adiyaman University, Turkey

Salih Tunc Kaya, Assistant Doctor,
Duzce University, Turkey

Section I. HMGB1 and Cancer

Chapter 1

HMGB1-Targeting Herbal Medicinal Plant Extracts for the Phytotherapy of Cancer and Potential Management of COVID-19

Tuba Reçber[1], PhD,
Ibrahim Celalettin Haznedaroglu[2,*], MD
and Mustafa Çelebier[1,†], PhD
[1]Hacettepe University, Faculty of Pharmacy,
Department of Analytical Chemistry, Ankara Turkey
[2]Hacettepe University, Faculty of Medicine,
Department of Hematology, Ankara, Turkey

Abstract

A group of non-histone nuclear proteins with high electrophoretic mobility was discovered almost 40 years ago and referred to as high-mobility group (HMG) proteins. HMGB, HMGN, and HMGA are members of HMG families. HMGB has a family of three nuclear proteins including HMGB1, HMGB2, and HMGB3. HMGB1 is a DNA binding protein responsible for the maintenance of nucleosome structure and regulation of gene transcription. Recent studies suggest that HMGB1 has critical roles in DNA recombination, repair, replication, and gene transcription. This situation makes it a multifunctional protein involved in a variety of cellular biological processes. The unique features of HMGB1 allow it to be suggested as a novel target for cancer therapy and managing COVID-19 infection. However, there is still not any regulated drug by any authorities directly targeting HMGB1. The present study is

* Corresponding Author's Email: ichaznedaroglu@gmail.com.
† Corresponding Author's Email: celebier@hacettepe.edu.tr.

In: HMGB1
Editors: Eylem Taskin, Celal Guven and Salih Tunc Kaya
ISBN: 979-8-88697-408-9

aimed to systematically present the reported herbal medicinal plants and/or their active seconder metabolites to target HMGB1. First, the potential role of targeting HMGB1 on cancer therapy and COVID-19 management was discussed and then the reported herbal medicinal plants for this feature were presented. The active components for these medicinal plants were evaluated and their roles in cancer therapy and antiviral features besides targeting HMGB1 were discussed. Thus, a potential non-toxic phytotherapy solution primary intending to target HMGB1 and simultaneously to impact cell proliferation was suggested for supporting conventional cancer therapy. In addition, the well-known antiviral herbal medicinal plant extracts were considered whether they already include some components having features targeting HMGB1, or not. This critical point of view allowed us to suggest potential oral topical antiviral herbal solutions to be used in the management of COVID-19.

Keywords: HMGB1, COVID-19, anticancer, antiviral, herbal medicinal plants, metabolomic pathways

Introduction

High mobility group box 1 (HMGB1) is a highly conserved multifunctional protein inside and outside cells that were isolated and characterized in the calf thymus in 1973. HMGB1 is named because of its electrophoretic mobility in polyacrylamide gels. HMGB1 contains two DNA-binding HMG-box domains (N-terminal A and central B) and an acidic C-terminal tail (Xue et al. 2021).

HMGB1, which has many biological functions inside and outside the cell, is mostly found in the nucleus of the cell. It helps maintain nuclear homeostasis and regulates DNA replication, transcription, recombination, and repair in the nucleus (Travers 2003). Outside the cell, HMGB1 functions as a very important member of Damage-associated molecular pattern molecules (DAMPs). Extracellular secretion of HMGB1 occurs as a result of its transfer to the cytoplasm under stressful conditions such as pathogen infection, oxidative damage, and metabolic crisis (VanPatten and Al-Abed 2017).

HMGB1 has an important role in many diseases, especially inflammatory diseases, and cancer. High levels of HMGB1 in serum can be associated with diseases and used as biomarkers to predict disease progression, including viral diseases (Chen, Long, et al. 2020).

The pandemic caused by the severe acute respiratory syndrome coronavirus-2 (SARS-CoV-2) was declared in March 2020, and since then a large number of confirmed cases and deaths have been reported globally by the World Health Organization. COVID-19 disease has exhibited a highly variable disease profile ranging from asymptomatic infection to pneumonia to life-threatening consequences (Zhou et al. 2021; Yang et al. 2020). It is thought that COVID-19 includes cytokine storm syndromes and fatal hyperinflammation, resulting from overactivation of the immune system. Clarifying the pathogenesis of sepsis caused by COVID-19 is very important for the development of new treatments and drugs. Elevated serum HMGB1 levels in severe COVID-19 patients are consistent with the outcome of cytokine storm and high mortality (Zhou et al. 2021; Chen, Long, et al. 2020). Exogenous HMGB1 promotes the expression of the SARS-CoV-2 entry receptor, ACE2 (Chen, Huang, et al. 2020; Adil et al. 2021). Therefore, serum HMGB1 levels have the potential to be an important marker in the prediction and treatment of clinical outcomes (Chen, Long, et al. 2020). In addition, autophagy is one of the mechanisms involved in COVID-19, and it has been recognized that HMGB1 regulates autophagy and may be a biomarker of acute lung injury (Sun and Tang 2014; Yang and Shen 2020). In summary, HMGB1 inhibitors are thought to be drug candidates for the treatment of COVID-19 disease (Zhou et al. 2021; Wyganowska-Swiatkowska et al. 2020). In addition to Covid-19, the importance of HMGB1 level in the diagnosis and treatment of cancer is a remarkable issue (Xue et al. 2021). It is known that HMGB1 supports cell survival and death by regulating physiological mechanisms such as inflammation, immunity, genome stability, proliferation, metastasis, metabolism, apoptosis, and autophagy. Many serum HMGB1 levels have been found to be high, such as breast cancer, lymphoma, melanoma, stomach cancer, liver cancer, cervical cancer, and osteosarcoma (Jube et al. 2012; Kang et al. 2013; Wang and Zhang 2020).

Since ancient times, herbal products have been one of the most important sources for obtaining active ingredients used in the discovery and development of drugs (Wu et al. 2015; Shawky, Nada, and Ibrahim 2020). In this study, we aimed to present a brief review of potential medicinal plants and herbal constituents that can be used for the treatment of COVID-19 and cancer disease by inhibiting HMGB1. These medicinal plants were searched for in literature if they had already been reported to be used as anticancer, antiviral, or antibacterial products. The combination of inhibiting HMGB1 besides their reported pharmacological properties may allow researchers to focus on the mechanism of the action by using multi-omics approaches. Thus, molecular mechanism research can be encouraged to focus on HMGB1 inhibition of some unique herbal origin chemical agents alone, and when they are in the extract to evaluate the synergic effects of these chemical agents with other reported constituents which may occur in the pharmacological activity.

Experiments

A Google Scholar search combined with PubMed and Web of Science allowed us to find reported HMGB1 targeting plants and herbal agents. These plants and herbal agents were listed. Another literature survey was planned and this time we focused on whether these reported HMGB1-targeting herbal therapeutic plants or therapeutic agents have anticancer, antiviral, and antibacterial features, or not. A brief discussion was performed using the already published reports. The pharmacological properties of these plants and agents were evaluated based on especially anticancer, antiviral, and antibacterial activities.

Results

Potential HMGB1-targeting herbal therapeutic plants and therapeutic agents were listed in Table 1 and Table 2, respectively. Some of these plants were discussed below.

Table 1. Potential HMGB1-targeting herbal therapeutic plants

Active molecule*	Source	Structure	Ref.
Acteoside (Verbascoside)	*Ligustrum purpurascen, Ligstrum purpurascens, Rehmannia glutinosa*	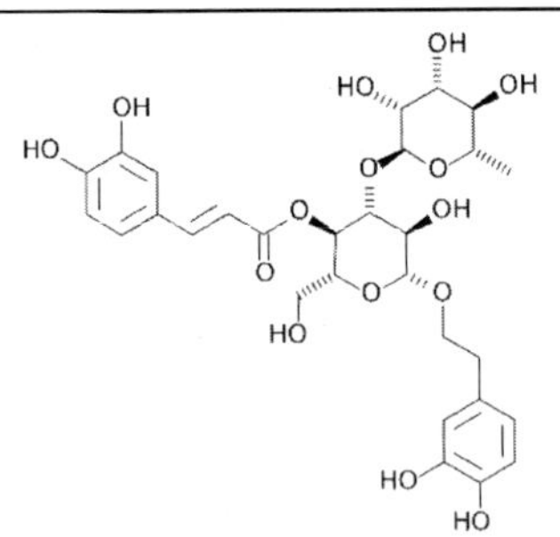	(Seo et al. 2013)
Angelica sinensis polysaccharide		-	(Wang et al. 2006)

Active molecule*	Source	Structure	Ref.
Apocynin	*Apocynum cannabinum*	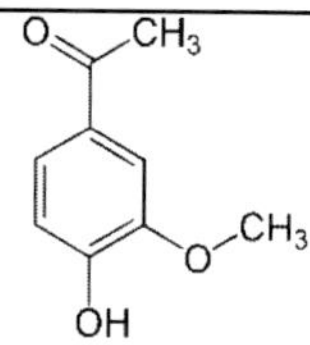	(Wang et al. 2015)
Astragalus mongolicus polysaccharide	*Astragalus Mongholicus* Bunge	-	(Zheng et al. 2013)
C19-diterpenoid	*Aconitum carmichaelii* Debeau	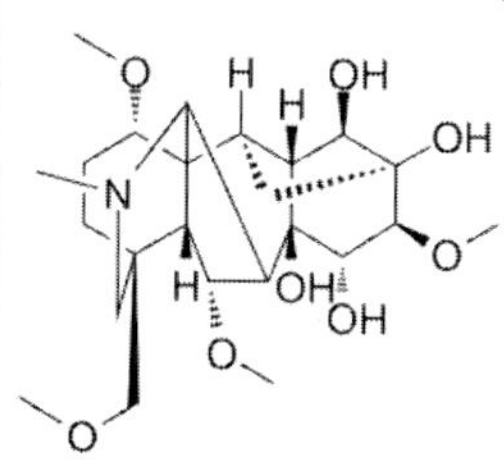	(Luo et al. 2016)
Caffeic acid	*Prunella vulgaris* L.	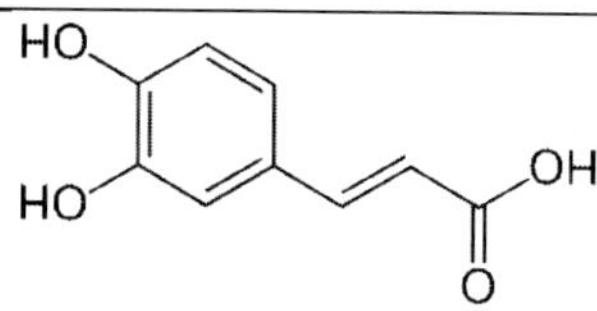	(Jun et al. 2012)
Chloroquine	*Cinchona bark*	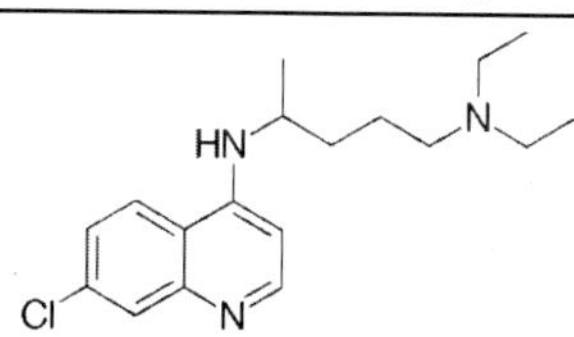	(Yang, Cao, et al. 2013)
Cucurbitacin E.	Cucurbitaceae	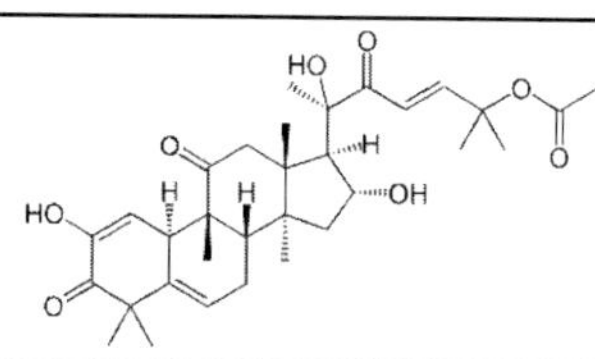	(Shang et al. 2019)
Curcumin	*Curcuma longa*	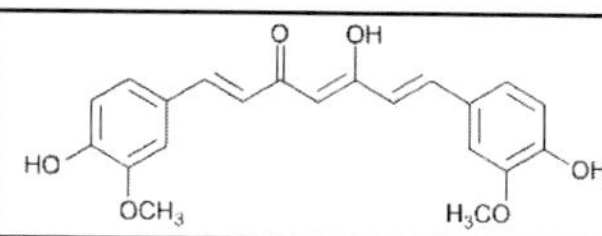	(Wang et al. 2012)
Danshensu	*Salvia miltiorrhiza* Bunge	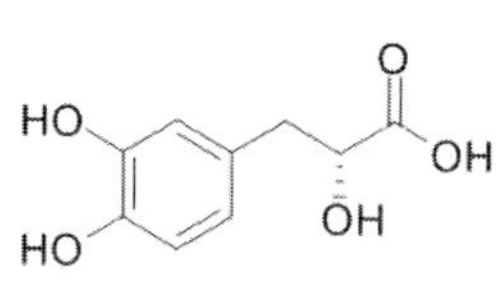	(Duan et al. 2016)

Table 1. (Continued)

Active molecule*	Source	Structure	Ref.
Dieckol	Ecklonia cava	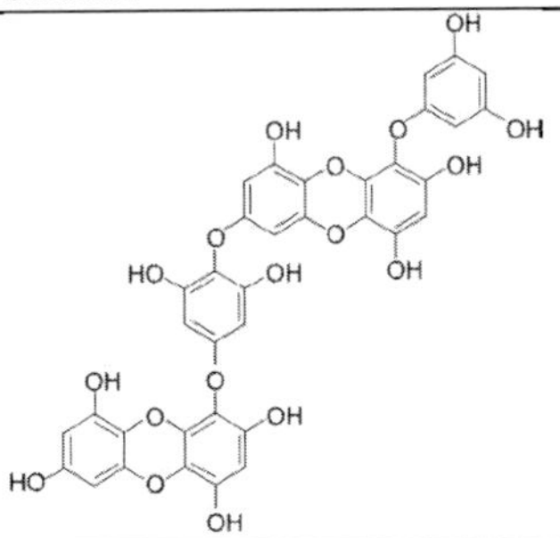	(Kim, Ku, et al. 2012)
Emodin-6-O-β-D-glucoside	*Reynoutria japonica*	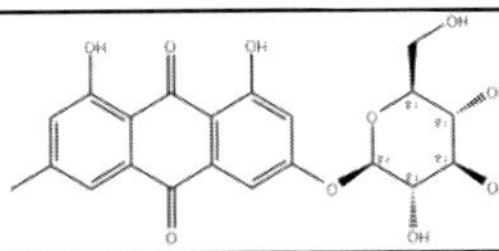	(Lee, Ku, et al. 2013a)
Epigallocatechin Gallate	*Camellia sinensis* (L.)	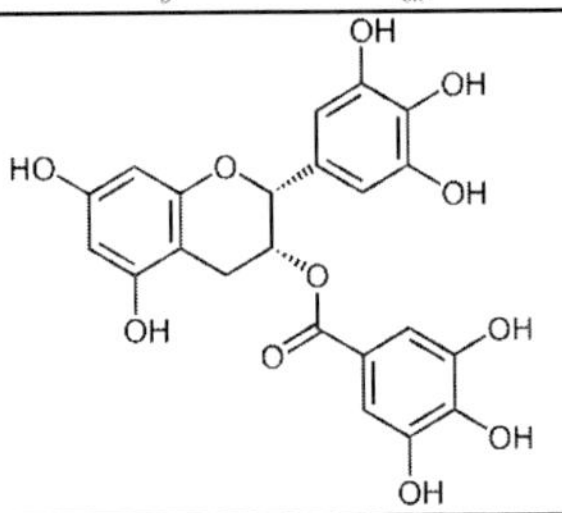	(Li et al. 2007)
Ferulic acid	*Angelica sinensis* (Oliv.) Diels	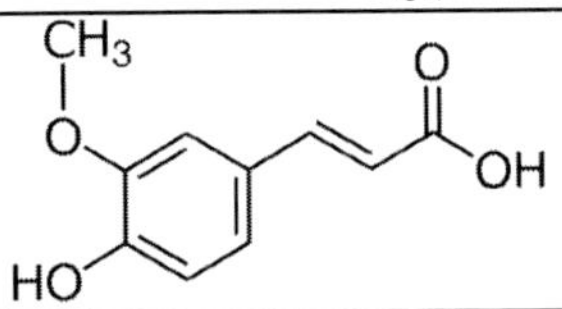	(Wang et al. 2006)
Forsythoside B	*Marrubium alysson*	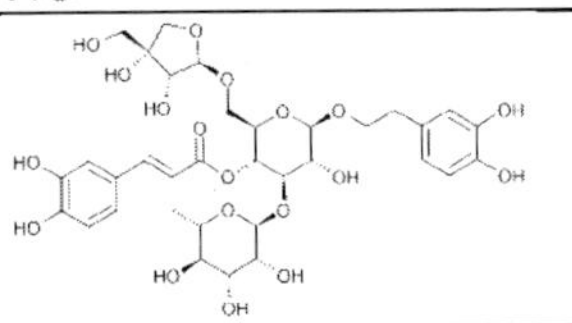	(Jiang, Zhang, and Zhu 2012)
Glycyrrhizic acid	*Glycyrrhiza glabra* L. (Licorice)	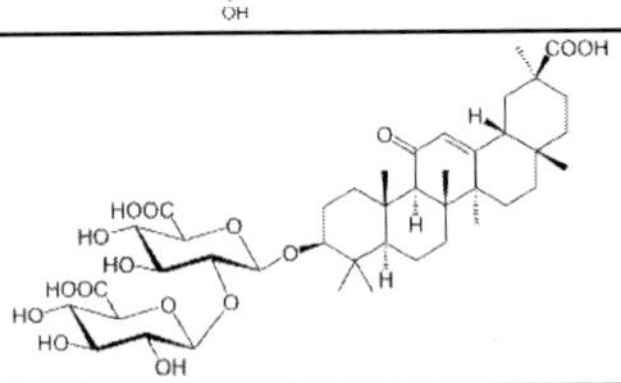	(Pang et al. 2016)

Active molecule*	Source	Structure	Ref.
Glycyrrhetinic acid		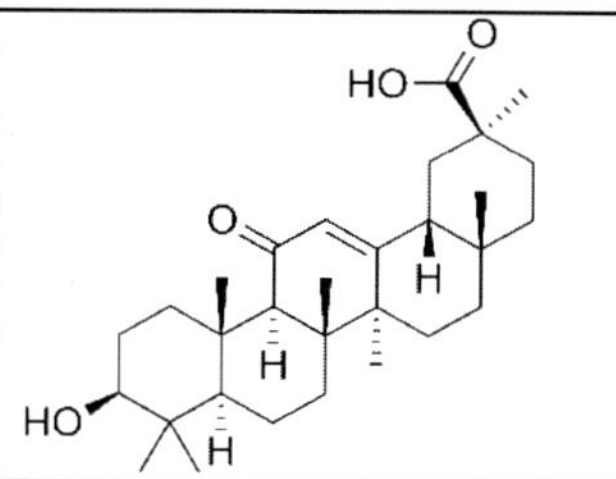	(Chen et al. 2017)
Isorhamnetin-3-O-galactoside (cacticine)	*Artemisia capillaris*	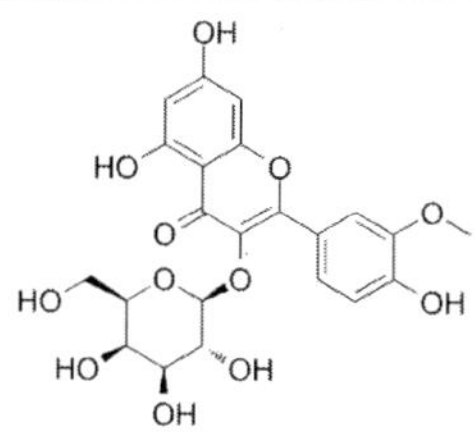	(Kim, Ku, and Bae 2013a)
Persicarin	*Persicaria hydropiper*	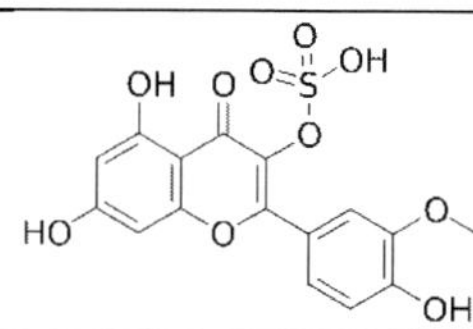	(Kim, Ku, and Bae 2013b)
Plumbagin	*Plumbago zeylanica*	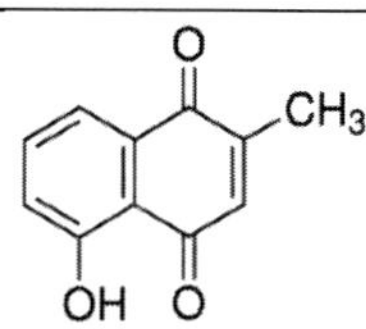	(Zaki et al. 2018)
Salidroside	*Rhodiola rosea* L.	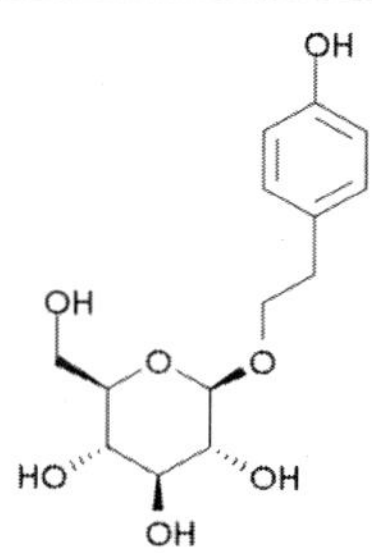	(Qi et al. 2017)
Scolymoside		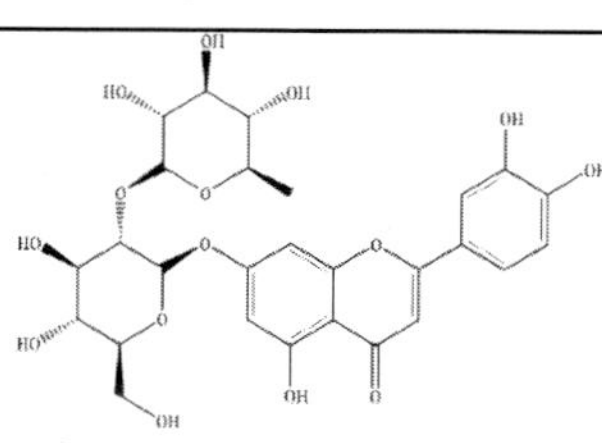	(Lee et al. 2015)

Table 1. (Continued)

Active molecule*	Source	Structure	Ref.
Shikonin	*Lithospermum erythrorhizon*	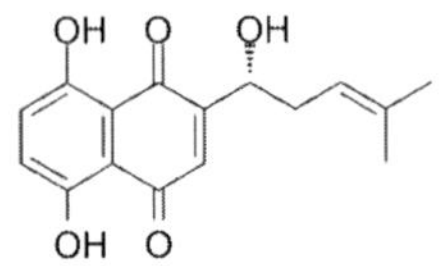	(Yang et al. 2014)
Tanshinone II	*Salvia miltiorrhiza*	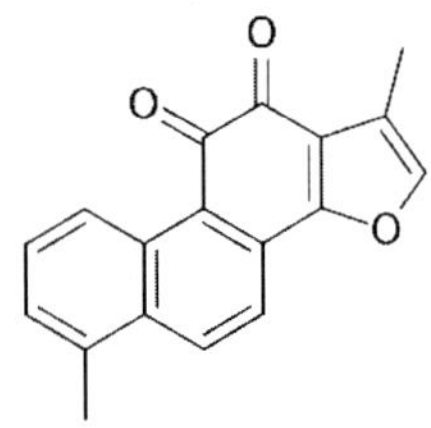	(Zhang et al. 2012)
Vicenin 2	*Cyclopia intermedia* E.Mey. (Honeybush)	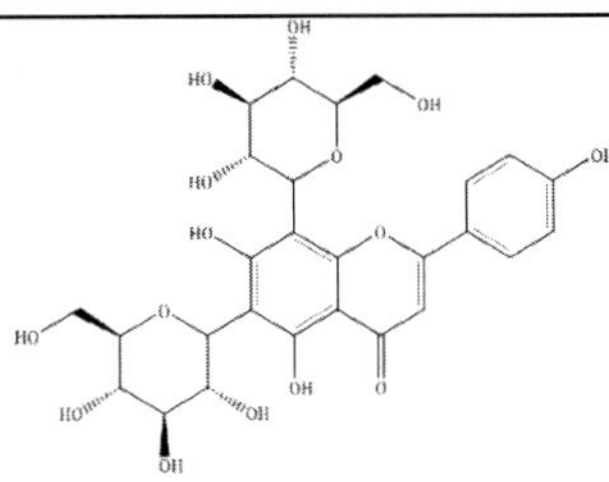	(Lee et al. 2015)
Vitexin	*Mung bean (Vigna Radiata)*	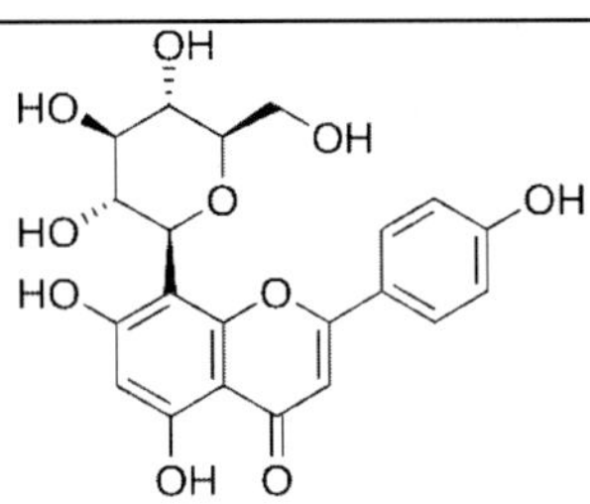	(Zhu et al. 2012)

*The molecules are listed in alphabetical order of their names.

Discussion

The HMGB1-targeting herbal therapeutic plants and agents listed in Table 1 and Table 2 were discussed below based on their reported anticancer, antibacterial, and antiviral features.

Table 2. Potential HMGB1-targeting herbal therapeutic agents

Active molecule*	Structure	Ref.
Baicalin	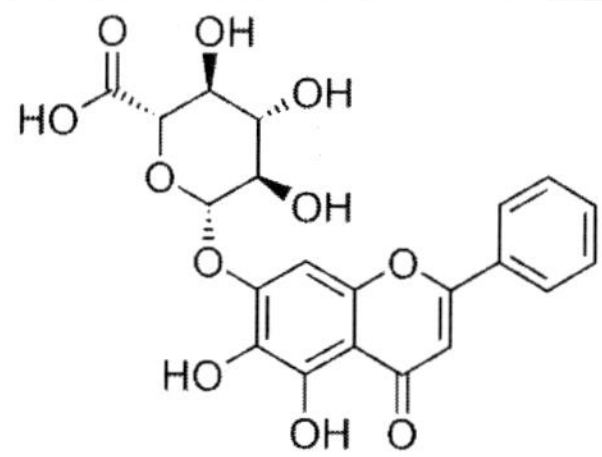	(Su et al. 2020; Chen, Liou, et al. 2014)
Berberine	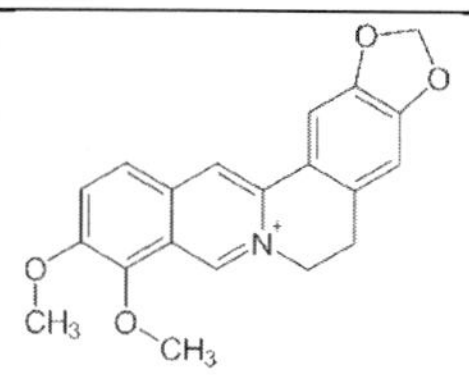	(Lee, Bae, et al. 2013)
Chlorogenic acid	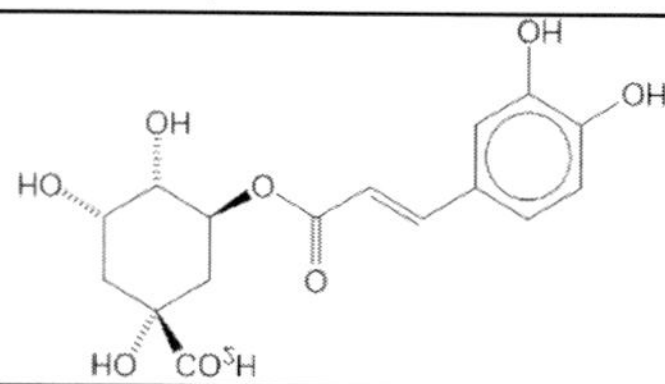	(Lee, Yoon, and Lee 2012)
Luteolin	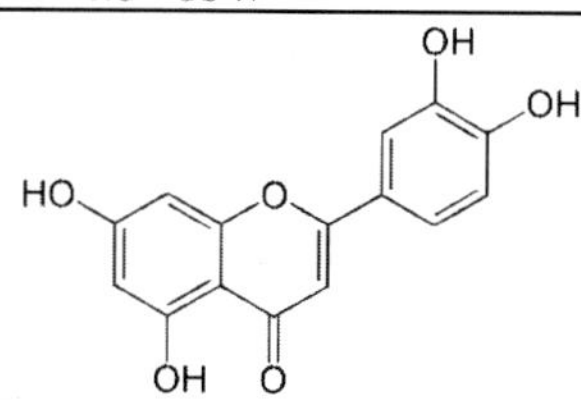	(Chen, Bi, et al. 2014)
Naringin	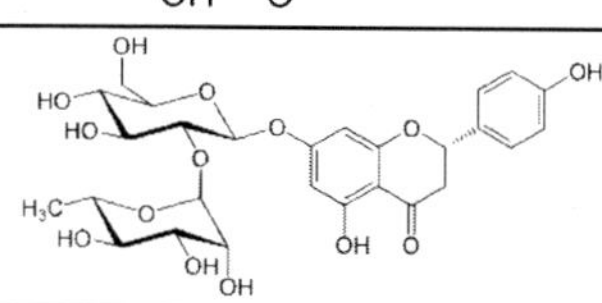	(Gil et al. 2016)
Quercetin	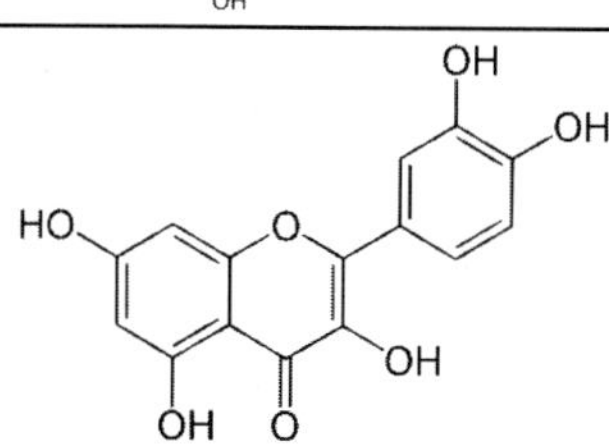	(Tang et al. 2009; Wang, Chen, et al. 2014)

Table 2. (Continued)

Active molecule*	Structure	Ref.
Resveratrol	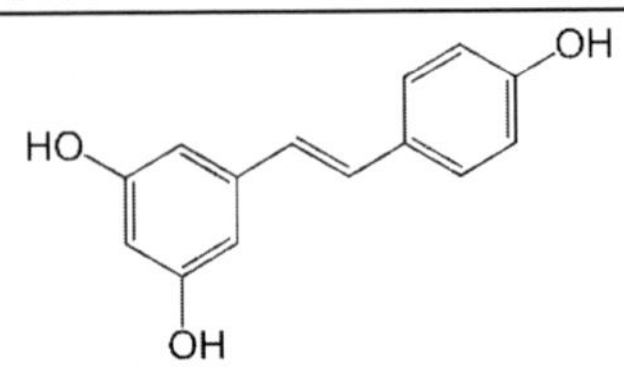	(Zainal et al. 2017)
Rosmarinic acid	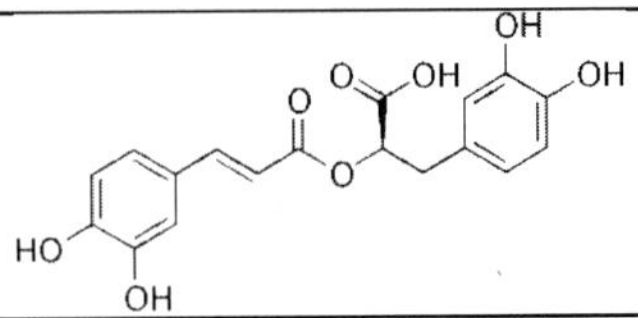	(Yang, Ku, et al. 2013)

*The molecules are listed in alphabetical order of their names.

Apocynum cannabinum

Apocynum, is a plant species from the Apocynaceae family that grows in North America, the southern half of Canada, and the United States, and is used by North American Indians as cannabis. Apocynin is a naturally occurring acetophenone, found in the roots of *Apocynum cannabinum* and *Picrorhiza kurroa*. Apocynin has been primarily reported as an NADPH oxidase (NOX) inhibitor and prevents translocation of its p47phox subunit to the plasma membrane, observed in neurodegeneration and hypertension (Savla, Laddha, and Kulkarni 2021). Prostate cancer is the second most frequent cancer diagnosis made in men and the fifth leading cause of death worldwide. Prostate cancer may be asymptomatic at the early stage and often has an indolent course that may require only active surveillance (Rawla 2019). Apocynin has been reported to inhibit cell proliferation of prostate cancer cell PLS10 via G1 (Reed 1997) arrest of the cell cycle *in vitro* (Suzuki et al. 2013). Klees et al. (Klees et al. 2006) reported that while apocynin itself is not effective, its derivatives inhibit migration of the breast cancer cell line MDAMB-435 at subtoxic concentrations and the migration of nonmalignant MCF10A breast cells is unaffected. These compounds also cause a significant rearrangement of the actin cytoskeleton, cell rounding, and decreased levels of active Rac1 and its related G protein Cdc42 (Stefanska and Pawliczak 2008). In a study published by Sammaiah et al., oxime esters of apocynin were reported to *in vitro* show antimicrobial and antioxidant activities (Sammaiah et al. 2015). Using apocynin as a preventive agent has also been reported to be protected from Helicobacter pylori-induced gastric carcinogenesis (Horemans et al. 2017). Human influenza viruses are respiratory pathogens that are easily transmitted from an infected patient to another susceptible individual (Deng et al. 2015). Vlahos et al. have reported that the production of reactive oxygen species (ROS), particularly superoxide, is an important host defense mechanism for killing invading pathogens, and NOX2 oxidase inhibitors, in combination with current therapeutics (i.e., antivirals and vaccines), could be useful for suppression of influenza A virus-induced lung disease. Thus, apocynin may be useful to manage influenza A (Deng et al. 2015).

Cinchona bark

Malaria, one of the oldest diseases known in the history of humanity, is the struggle of human beings to dominate nature on earth. Started to take effect. Malaria, which manifests itself in almost every place where agriculture is made, is also a social problem all around the World due to its results (Malaria 2005; Greenwood and Mutabingwa 2002). During the 18th century, albeit with some continuing ambiguity, malaria gradually became accepted as a defined set of intermittent fevers responding to 'therapeutic tests' using *Cinchona bark* or, from the 1820s, by using quinine (Gachelin et al. 2017). Quinine remains an important anti-malarial drug almost 400 years after its effectiveness was first documented. However, its continued use is challenged by its poor tolerability, poor compliance with complex dosing regimens, and the availability of more efficacious anti-malarial drugs (Achan et al. 2011). In the beginning of COVID-19 pandemic, it has been suggested to be used to prevent or treat coronavirus infection (Jomsky and Nicholas 2020; Latarissa et al. 2021; Bazotte et al. 2021; Große et al. 2021). Although it is a well-known antimicrobial (Kharal, Hussain, and Ali 2009; Akinwale et al. 2021; Antika, Triana, and Ernawati 2020), quinine has also been reported as a potential anticancer agent (Krishnaveni and Suresh 2015; Krishnaveni, Suresh, and Arunkumar 2016). In a study published by Krishnaveni and Suresh, it has been reported to perform growth inhibition on Hep-2 human laryngeal carcinoma cell lines (Krishnaveni and Suresh 2015). Another study showed its antiproliferative and apoptotic effect on KB oral cancer cells (Krishnaveni, Suresh, and Arunkumar 2016). Multi-drug resistance (MDR) in cancer therapy can be observed as a decrease in intracellular drug accumulation (decrease in drug entry into the cell or increase in excretion function), decrease in drug-target relationship, increase in detoxification function or change in drug distribution (Szakács et al. 2006). Quinine has been reported as an MDR inhibitor in some studies (Solary et al. 2003; Solary et al. 1992; Robert and Jarry 2003). Even though it is a basic compound easily found in markets as the tonic water (Herman, Jezorek, and Tang 2000; Lawson-Wood, Evans, and Seer Green 2019), its dietary levels may not cause theopathic effects (Donovan et al. 2003). However, its well-known toxicity for a long time (Bateman and Dyson 1986; Goldenberg and Wexler 1988) still makes it not to be utilized as an approved drug (AlKadi 2007) but instead using as an homeopatic agent in treatments (Khandelwal 2018; Grams 2019).

Salvia miltiorrhiza

Salvia miltiorrhiza is a well-known traditional Chinese herb, thet belongs to the family of Labiatae and is used in many parts of the world to treat various conditions due to their excellent medicinal values. It is a rich source of tanshinone I, tanshinone II, miltirone and salvianolic acid and a wide range of chemical compounds (Wang 2010). Diterpenes extracted from *Salvia miltiorrhiza* have been reported to show anti-inflammatory activity (Ngo et al. 2021; Lee, Yim, and Choi 2021), antibacterial activity (Lee, Yim, and Choi 2021; Kong et al. 2017) and antifungal activity (Lee and Kim 2016). In a study published by Gong et al. *Salvia miltiorrhiza* has been indicated as a potential 'anticancer' herb since cryptotanshinone (CT), tanshinone IIA (T2A) and tanshinone I (T1) inside inhibit human prostate cancer cell lines and in mice (Gong et al. 2011). In another study, a polysaccharide of *Salvia miltiorrhiza* provided an improvement in the immune function of rats having gastric cancer (Wang, Yang, et al. 2014). In a study published by Gao et al. totally 18 compounds where 4 of them have been reported as novel have been characterized through advanced analytical techniques including NMR and one of them suppressed secretion of nitric oxide (NO), tumor necrosis factor (TNF)-α and interleukin-6 (IL-6) (Gao, Sun, et al. 2016).

Reynoutria japonica

Reynoutria japonica is a plant used for treating allergic inflammation in Japan (Kimura, Okuda, and Kubo 1995) Emodin-6-O-β-D-glucoside (EG) is an active compound from *Reynoutria japonica*. It is a potential anti-inflammatory agent and provides barrier protective effects in human umbilical vein endothelial cells (Lee et al. 2014). However, recent studies provide the information that it has also the potential to inhibit HMGB1-induced inflammatory responses *in vitro* and *in vivo* (Lee, Ku, et al. 2013b). A recent study, published by Khalil et al. showed that 70% ethanol extract of the root portion of *Reynoutria japonica* was employed to be used against three strains (43504, 51, and 26695) of Helicobacter pylori and the results were highly satisfactory for isolated compounds 2-methoxy-6-acetyl-7-methyljuglone (1), 2-ethoxy-6-acetyl-7-methyljuglone (2), 2-methoxy-7-acetonyljuglone (3).

Artemisia capillaris

Artemisia is a hard herbaceous or shrubby plant genus that has 200-400 species from the Asteraceae family, usually growing in dry or semi-arid habitats in both the Northern and Southern hemispheres (Bora and Sharma 2011; Naghavi, Alaeimoghadam, and Ghafoori 2014). *Artemisia capillaris* is an herb having therapeutic efficacy especially, in liver diseases. It is potentially used as an alternative therapy option in Asia (Jang et al. 2015). A study published by He et al. confirmed that polysaccharides in this unique plant offer protective effect against oxidative injury in the liver (Jang et al. 2015). This protective property of *Artemisia capillaris* is mainly related to its antioxidative features (Kang and Lee 2013; Hong et al. 2007; Gao, Zhao, et al. 2016; Kim, Kang, et al. 2012). Kang and Lee reported that extraction methods, reflux extraction, autoclave extraction, low-temperature high-pressure extraction and ultrasonication extraction changed the extracted antioxidant substances (Kang and Lee 2013). Therefore, the effect of the extraction technique for *Artemisia capillaris* might be evaluated individually to check its therapeutic effects for diseases in further experiments. In 2007, a study published by Kim et al. presented that *Artemisia capillaris* extracts have cytotoxic effect against Hepa-1c1c7 and Sarcoma 180 cancer cells (Kim et al. 2007). A recent study published by Taleghani et al. confirmed that there had been 332 papers appeared when searching for the terms "Artemisia and cancer" until 2018 (Taleghani, Emami, and Tayarani-Najaran 2020). In 2018, Yan et al. reported that ethyl acetate fraction extracted from *Artemisia capillaris* employed against hepatocellular carcinoma cells and dose-dependently inhibited the growth and proliferation of two hepatocellular carcinoma cell lines (Yan et al. 2018).

Persicaria hydropiper

Persicaria hydropiper (L.) is known as Polygonum hydropiper and it is a popular medicinal plant used in traditional medicine. The plant is indigenous to the tropical northern hemisphere and temperate zone, including China, Bangladesh, India, and Japan. The main applications of this unique plant are to be used on hemorrhoids, antifertility, diarrhea, and dyspepsia (Nasir et al. 2021). Antioxidative features of its leaves were first introduced at the beginning of the 90s (Haraguchi, Hashimoto, and Yagi 1992). It was reported that the organic fractions of the title plant obtained from various parts of the plant showed antioxidant, antibacterial, antifungal, anthelmintic, cytotoxic, anti-inflammatory, anticancer, antirheumatic, anticholinesterase and neuro-protective activities (Huq, Jamal, and Stanslas 2014; Ayaz et al. 2015; Ayaz et al. 2016). It was reported that one of the main compounds isolated from the essential oil of *Polygonum hydropiper* L. showed antimicrobial activity both in bacteria and fungi. The lowest MIC for bacteria was observed against Enterococcus faecalis (31.25 μg/mL). Significant Minimum Inhibitory Concentration (MIC) for fungi was observed against *Scopulariopsis* sp (7.81 μg/mL), *Curvularia lunata* (7.81 μg/mL), *Epidermophyton floccosum* (7.81 μg/mL), *Trichophyton mentagrophytes* (16.62 μg/mL), *Trichophyton rubrum* (MTCC 296) (16.62 μg/mL), *Aspergillus niger* (31.25 μg/mL), *Botrytis cinerea* (31.25 μg/mL) *Magnaporthe grisea* (62.5 μg/mL), *Trichophyton simii* (125 μg/mL) and *Trichophyton rubrum* (clinical isolate) (125 μg/mL) as reported (Duraipandiyan, Indwar, and Ignacimuthu 2010).

Marrubium alysson

Marrubium alysson L. is commonly distributed in Egypt and its Arabic name is 'Hashisha Rabiah,' while the common English name is 'White horehound' (Alaa et al. 2016). *Marrubium alysson* includes Forsythoside B (Çaliş et al. 1992) which concentration-dependently down-regulated the levels of TNF-α, IL-6 and high-mobility group-box 1 protein (HMGB1) in lipopolysaccharide-stimulated RAW264.7 cells, inhibited the IκB kinase pathway and modulated nuclear factor- κB. (Jiang, Zhang, and Zhu 2012). A study published by Mohamed indicated that isolated flavonoids (diosmetin, luteolin, and quercetin) using chromatography from *Marrubium alysson* L.and identified using UV, 1H, 13C NMR spectroscopy and *Marrubium alysson* total extract showed *in vitro* anticancer activity against Ehrlich tumor cell lines and Human tumor cell lines U251 and MCF7 (brain tumor and breast carcinoma cell lines respectively) (Mohamed 2010). A study published by Hayet et al. indicated that the petroleum ether extract of *Marrubium alysson* had a MIC varied from 128 to 2000 μg/mL against different Enterobacteriaceae and antifungal activity against *Candida glabrata*, *Candida albicans*, *Candida parapsilosis* and *Candida kreusei* with a MIC of 256 μg/mL (Hayet et al. 2007). Another study published by Edziri et al. reported that petroleum ether, ethyl acetate, chloroform, butanol, and methanol extracts of the leaves of *Marrubium alysson* L tested against human cytomegalovirus (HCMV) strain AD-169 (ATCC Ref. VR 538) and coxsackie B virus type 3 (CoxB-3). The results were satisfactory for methanol, butanol, and ethyl acetate extracts for anti-Coxsaki B3 activity. The authors indicated that this antiviral activity may be related to flavonoids and tannins.

Ligustrum purpurascen, Ligstrum purpurascens, Rehmannia glutinosa

Ligustrum purpurascen, Ligstrum purpurascens, and *Rehmannia glutinosa* include Acteoside (Verbascoside) (Wu et al. 2014; Wong et al. 2001; Liu et al. 2020). Khalaf et al. discussed the antitumor activity of Acteoside in a published review (Khalaf, Jasim, and Ibrahim 2021). Khan et al. reported the "metabolic pathways" affected by the treatment of Acteoside on cancer (Khan et al. 2022). They mainly divided its potential effect into eight different sections: these were antioxidant effect, apoptosis effect, antiproliferative effect, antiangiogenesis effect, antimetastasis and antiinvasion effects, antitumor effect, synergistic effects with chemotherapeutic drugs, and anti-inflammatory activity.

Lithospermum erythrorhizon

Lithospermum erythrorhizon is a Chinese medicinal plant that is mainly used for wound healing (Kim, Kim, et al. 2012) and its formulation to be used as a cosmetic product presented moisturizing effects on human skin (Chang et al. 2008). Its features to be used against were an idea at the beginning of the 90s (Yamasaki et al. 1993). The main component of this plant was reported previously as red pigments having a composition of naphthoquinone compounds. Shikonin and its derivatives are compounds classified as naphthoquinone (Sankawa et al. 1977; Chen et al. 2002). Antiadenovirus activity of shikonin was reported in previous studies (Gao et al. 2011; Sun et al. 2018). Shikonin is one of the main components found in *Lithospermum erythrorhizon* (Hisa et al. 1998). Andujar et al. reviewed the anticancer activity of shikonin (Andujar et al. 2013). Wang et al. reported that the anticancer activity of shikonin was mainly related to P21 upregulation and apoptosis induction (Wang et al. 2020).

Mung bean (Vigna Radiata)

The usage of *Mung bean (Vigna Radiata)* seeds and sprouts is a well-known plant especially in Asia. The recent reports confirmed that bioactive compounds in mung bean play efficient role for the prevention of degenerative diseases (Ganesan and Xu 2018). Vitexin is one of the main components of *Mung bean* (Peng et al. 2008). Spector reviewed the relation between oxidative stress and three diseases, chronic lung disease, diabetes and Alzheimer's disease (Spector 2000). Jenner reported the effect of oxidative stress on Parkinson's disease (Jenner 2003). Lefer et al. reported the correlation between cardiac diseases and oxidative stress (Lefer and Granger 2000). Such studies indicated that antioxidative features of vitexin found in *Mung bean* may be used as a preventing agent against diseases related to oxidative stress and the consumption of *Mung bean* foods and association with the ratio of some diseases in different regions where *Mung bean* is a common product in diets might be investigated in this perspective (Dahiya et al. 2014). The antiviral activity of flavonoids and organic acids is a major area of pharmaceutical research (Li et al. 2002; Badshah et al. 2021). Therefore, usage of the supplements containing flavonoids is a trend in the last decade (Egert and Rimbach 2011). In a study published in 2002, the potent or moderate antiviral activity of orientin and vitexin against parainfluenza type 3 was shown (Li et al. 2002). In another study it was demonstrated that vitexin had antibacterial properties (Qiufeng et al. 2004). STAT3 signaling is a major intrinsic pathway for cancer inflammation because it is often activated in tumor-associated immune and inflammatory cells as well as malignant cells and is capable of inducing a large number of genes that are crucial for inflammation including IL-6, 10, 11, 17, 23, CXCL12, and COX-2 (Siveen et al. 2014). In a recent study, the inhibition of the constitutive activation of STAT3 signaling through vitexin was reported (Lee et al. 2020). In another study, PI3K/Akt/mTOR signaling in human non-small cell lung cancer A549 cells was affected by vitoxin treatment and apoptosis was induced (Liu et al. 2019). Human glioblastoma cells treated with vitexin demonstrated that G2/M-phase was arrested and apoptosis via Akt/mTOR signaling pathway was altered (Zhang et al. 2018). The impact of vitexin against U937 human leukemia cells was also evaluated (Lee et al. 2012). Such studies confirm that vitexin can be used as a potential compound against various types of cancer and the studies must be improved through multi-omics platforms.

Salvia miltiorrhiza Bunge

Salvia miltiorrhiza Bunge, generally called danshen, is a well-known herbal traditional medicine used in China. It has got cardiovascular and neuroprotective effects since it has got antioxidative, anti-inflammatory, and antifibrotic properties. The antiviral effect of this unique plant against Enterovirus 71 had already been published before (Wu et al. 2007). In a recent study, its potential against COVID-19 was evaluated (Petitjean et al. 2022). Chen et al. systematically reviewed the anticancer properties of *Salvia miltiorrhiza* Bunge (Chen, Guo, et al. 2014). Its major components are hydrophilic phenolic acids and lipophilic tanshinones (Chen, Guo, et al. 2014). Danshensu is a pure molecule derived from Danshen, which is the root of the herb *Salvia miltiorrhiza* (Zhang et al. 2019). Danshensu inhibits PI3K/AKT pathway in A549 and NCI-H1299 cells and oral cancer cell migration and invasion by modulating p38 signaling pathway (Wang et al. 2022; Kumar et al. 2020).

Rhodiola rosea L.

Rhodiola rosea L. was stated as a protective herbal medicinal plant having anti-inflammatory features to be used on cardiovascular diseases, neurodegenerative diseases, diabetes, sepsis, and cancer (Pu et al. 2020). Salidroside is one of the major components of *Rhodiola rosea* L and is reported as a potential agent against coxsackievirus B3 confirmed by *in vitro* and *in vivo* studies (Wang, Ding, et al. 2009). Dengue Virus Infection is a viral disease that is often transmitted by the bites of infected mosquitoes and is endemic in tropical regions. The causative virus belongs to the genus Flavivirus from the family *Flaviviridae* and has four serotypes: DEN-1, DEN-2, DEN-3 and DEN-4. Salidroside demonstrated anti-viral activity against dengue virus by upregulating host innate immune factors (Sharma, Mishra, and Ganju 2016). Latest studies showed that salidroside could inhibit the growth of cancer cells by modulating CDK4-cyclin D1 pathway for G1-phase arrest and/or modulating the Cdc2-cyclin B1 pathway for G2-phase arrest (Hu et al. 2010; Li, Xu, and Liu 2018).

***Prunella vulgaris* L.**

Caffeic acid is one of the main components of *Prunella vulgaris* L. (Psotová et al. 2003; Feng et al. 2010). It is a compound mainly known as the component of propolis (Russo, Longo, and Vanella 2002) which is a well-known supplement used all around the World especially, during the COVID-19 pandemic (Berretta et al. 2020; Ripari et al. 2021). Caffeic acid antiviral properties were reported for HIV (Bailly and Cotelle 2005), canine distemper virus (Wu et al. 2017), hepatitis B virus (Wang, Shi, et al. 2009), and herpes simplex virus (Ikeda et al. 2011). The anticancer property of caffeic acid was reported for several cancer types such as cervical cancer, lung cancer and prostate cancer (Ozturk et al. 2012; Kanimozhi and Prasad 2015; Min et al. 2018; Lin et al. 2012).

Curcuma longa

Turmeric is a spice derived from the rhizomes of *Curcuma longa*, a member of the ginger family *(Zingiberaceae).* Curcumin inside the *Curcuma longa* exhibit potent anti-inflammatory and antioxidant effects (Akram et al. 2010; Menon and Sudheer 2007). In a study published by Fadus et al, it was indicated that 877 compounds presented as drug all around the world between 1981 and 2002 and approximately 61% of these had origin from natural products (Fadus et al. 2017) and curcumin was one of these compounds having the potential to be used for various diseases since it was reported as anti-inflammatory and anti-neoplastic agent (Kohli et al. 2005; Aggarwal, Kumar, and Bharti 2003). What makes natural products become efficient is their non-toxic features of which therapeutic doses are below their toxic levels. Besides the anticancer properties of Curcumin, it also has the potential to be used as an antiviral compound (Mathew and Hsu 2018). It was also proposed as a COVID-19 treatment opportunity during the pandemic (Soni et al. 2020).

Plumbago zeylanica

Plumbago zeylanica has potential cytotoxic constituents (Nguyen et al. 2004). It is an Indian medicinal plant and it consists of plumbagin which is a plant-derived naphthoquinones obtained mainly from three families, including *Plumbaginaceae*, *Droseraceae*, and *Ebenaceae* (Panichayupakaranant and Ahmad 2016). Roy reviewed the anticancer properties of plumbagin (Roy 2021). Its antibacterial and antifungal properties were reported at the beginning of the 80s (Krishnaswamy and Purushothaman 1980) and these features were confirmed in recent studies in detail (Thakor and Janathia 2022). Plumbagin is a potent reactive oxygen species inducer (Nadhan et al. 2021). Thus, it may potentially be able to protect against cancer (Zhang, Ni, and Yang 2021; Srinivas et al. 2004; Tian et al. 2012) and demonstrate antiviral features (Panichayupakaranant and Ahmad 2016).

Aconitum carmichaelii Debeaux

Aconitum carmichaelii Debeaux (Ranunculaceae)'s processed lateral root has common medicinal usage in China and it is known as Fuzi. It is called "bushi" in Japan and "Kyeong-Po Buja" in Korea (Zhou et al. 2015). C19-Diterpenoid alkaloid arabinoside is one of the major components of the aqueous extract of the lateral root of this plant (Guo et al. 2018). Zhang et al. reported that C19-Diterpenoid was found to be effective on HepG2 hepatocellular carcinoma cells (Zhang et al. 2014). In another study, C19-Diterpenoid was able to overcome multidrug resistance to cancer therapy (Wada et al. 2022). Its antiviral properties were observed in some other studies published by various scientific groups (Xu et al. 2019; Li et al. 2022).

Astragalus Mongholicus Bunge

Astragalus membranaceus Bunge, called Huangqi in China, is herbal tonic water used in China to improve vital energy and to improve the growth of new tissues (Yu et al. 2005). Antioxidant, antiviral and antitumor properties of polysaccharides are a well-known issue (Jiao et al. 2016; Gonzalez, Alarcón, and Carrasco 1987) and a unique polysaccharide belonging to *Astragalus Mongholicus Bunge* was studied for in previously published researches for these medicinal purposes (Wang et al. 2018). Possible management of

COVID-19 infection by using *Astragalus* had already been discussed during the pandemic (Law et al. 2020; Ren et al. 2021). Thus, a multi-omic platform to evaluate the effect at the molecular level might be encouraged for further studies since the results were satisfactory for both anticancer and antiviral properties.

Cucurbitaceae E.

Cucurbitacin E. is a chemical found in *Cucurbitaceae* which is a plant family consisting of about 965 species in around 95 genera (Christenhusz and Byng 2016). It is effective against Bcap37 and MDA-MB-231 breast cancer cells by inducing G2/M phase arrest and cell apoptosis (Lan et al. 2013). Its effect on pancreatic cancer cell growth through STAT3 signaling is another issue previously published (Sun et al. 2010). Zhang et al. reported that Cucurbitacin E is an inhibitor of cofilin phosphorylation in human leukemia U937 cells (Nakashima et al. 2010).

Ecklonia cava

Ecklonia cava is an edible marine brown alga found in the ocean off Japan and Korea. It is used as an herbal medicine in the form of an extract called Seanol, a polyphenolic extract, and Ventol, a natural agent rich in fluorotannin (Wijesinghe and Jeon 2012). Dieckol is one of the major components of *Ecklonia cava* (Rajan et al. 2021) and it was mentioned as a potential antiviral agent against SARS-CoV-2 (Adegboyega, Johnson, and Omale 2021). In the COVID-19 pandemic, its potential usage was discussed in various studies (Sangtani et al. 2021; Kashyap, Nadhan, and Dhanasekaran 2021). The impact of this unique chemical on PI3K/AKT/mTOR signaling pathway was presented in previously published studies especially focused on cancer research (Zhang et al. 2021; Wang et al. 2019).

Angelica sinensis (Oliv.) Diels

Polysaccharides from *Angelica sinensis* which is used as a tonic in Chinese traditional medicine has biological activities such as hematopoiesis, immunomodulation, antitumor, antioxidant, radioprotection and hypoglycemic activity (Jin et al. 2012). Besides polysaccharides, this plant includes ferulic acid (Lu et al. 2005). Ferulic acid was observed to be protective against prostate cancer cells (Eroğlu et al. 2015). Its usage as a drug delivery system and its modification to be used in cancer therapy is recent scientific interests (Ezhuthupurakkal et al. 2018; Panwar et al. 2016; Zheng et al. 2019).

Cyclopia intermedia E.Mey. (Honeybush)

Today, the prevention of cancer development with chemical compounds (Chemoprevention) is becoming increasingly important as a cancer control method. Chemical compounds that prevent the development of cancer are found in the form of drugs or as compounds of natural origin in the diet. Numerous compounds in the environment and industry induce genotoxic effects in humans. The human body tries to prevent these toxic effects by detoxifying them through various enzyme systems. However, in case of increased exposure to such chemicals, carcinogenic and mutagenic results occur. Since different formations containing DNA occur in the cell because of exposure to carcinogens and mutagens, it is possible to determine the genotoxicity of mutagenic and carcinogenic compounds by using these formations. The genotoxic effect, defined as DNA damage, is accepted as a cancer-initiating mechanism. For this reason, the detection of DNA damage with various methods developed can help in determining the risk of future cancer events. It is suggested that the widespread use of antimutagens and anticarcinogens in daily life may be effective in preventing human cancer and genetic diseases (Ferguson 1994). Honeybush is indicated as a rich source of antimutagenic compounds (Kokotkiewicz and Luczkiewicz 2009). Therefore, some clinical experiments supported with metabolomics approaches might be useful to evaluate the antimutagenic properties of Honeybush in near future.

Camellia sinensis L.

The antimicrobial properties of *Camellia sinensis* L. are a well-known subject of scientific research (Hamilton-Miller 1995). The tea of *Camellia sinensis* L. has a high potential antioxidant activity (Gramza and Korczak 2005) since it constitutes polyphenolic compounds (Forrest and Bendall 1969). Polyphenols are potential anticancer agents (Fresco et al. 2006; Çelebier et al. 2015) and *Camellia sinensis* L. might be a good source of dietary polyphenols for chemoprevention of cancer (R Conde et al. 2015; Esghaei et al. 2018). Anticarcinogenic effects of epigallocatechin gallate were a subject of scientific interest at the beginning of the 90s (Fujiki et al. 1992). Recent studies correlated the previous findings and enlightened the molecular effect of epigallocatechin gallate on cancer therapy (Min and Kwon 2014) In the COVID-19 pandemic, its usage as a preventing dietary agent was suggested in various studies (Chowdhury and Barooah 2020; Tallei et al. 2021).

Glycyrrhiza glabra L. (Licorice)

Diammonium glycyrrhizinate is the ammonium salt of glycyrrhizic acid, found in *Glycyrrhiza glabra*. Glycyrrhizic acid presents an anti-inflammatory effect via inhibiting HMGB1 phosphorylation and secretion (Kim, Jin, et al. 2012). Glycyrrhizin is the major essential component of *Glycyrrhiza glabra* (Gantait et al. 2010). Its anticancer properties were reported in the literature (Wahab et al. 2021; Zheng, Han, and Wu 2019). Besides anticancer properties, its antiviral activity for COVID-19 infection was discussed during the pandemic (Boozari and Hosseinzadeh 2021; Çelebier and Haznedaroğlu İ 2021).

Conclusion

The pathobiological impact of HMGB1 on SARS infection has already been stated. The molecules both derived from herbal products and synthetics were suggested to target HMGB1 in the COVID-19 pandemic. The other situation that makes HMGB1 a unique protein is its potential role to be targeted in cancer therapy. As mentioned above 877 compounds were presented as drugs worldwide between 1981 and 2002 and approximately 61% of these compounds had origin from natural products. This situation encouraged us to perform a google scholar, PubMed, and Web of Science search to find the herbal medicinal products to be used in targeting HMGB1. The idea of this study was to discuss if an herbal medicinal plant is stated as an HMGB1 targeting agent, its anticancer, antibacterial, and antiviral features must be considered. Thus, this herbal medicinal plant can be proposed to be prepared as a medicinal product or dietary supplement to be used in cancer therapy or management of COVID-19 infection. Such products can perform synergic biochemical effects since they already have major components performing major pharmacological effects (anticancer, antibacterial, and antiviral) and some unique components targeting HMGB1. Above, the list of herbal medicinal plants and some active agents were given, and a brief overview of these plants was discussed. Such studies may promote further investigations using multi-omics platforms to understand the synergic effects of natural products against cancer and COVID-19 infection.

References

Achan, Jane, Ambrose O Talisuna, Annette Erhart, Adoke Yeka, James K Tibenderana, Frederick N Baliraine, Philip J Rosenthal, and Umberto D'Alessandro. 2011. "Quinine, an old anti-malarial drug in a modern world: role in the treatment of malaria." *Malaria Journal* 10 (1):1-12.

Adegboyega, Abayomi Emmanuel, Titilayo Omolara Johnson, and Simeon Omale. 2021. "Computational modeling of the pharmacological actions of some antiviral agents against SARS-CoV-2." In *Data Science for COVID-19*, 467-482. Elsevier.

Adil, Mir S, Arti Verma, Madhuri Rudraraju, S Priya Narayanan, and Payaningal R Somanath. 2021. "Akt-independent effects of triciribine on ACE2 expression in human lung epithelial cells: Potential benefits in restricting SARS-CoV2 infection." *Journal of Cellular Physiology* 236 (9):6597-6606.

Aggarwal, Bharat B, Anushree Kumar, and Alok C Bharti. 2003. "Anticancer potential of curcumin: preclinical and clinical studies." *Anticancer Research* 23 (1/A):363-398.

Akinwale, Oluremi Adejoke, Uyi Oluwatobi Emokpae, Opeyemi Mariam Adebogun, Morenike Olutumbi Adeoye-Isijola, and Olufunmiso Olusola Olajuyigbe. 2021. "*In vitro* effects of quinine on the antibacterial activity of erythromycin against bacteria of clinical relevance." *GSC Biological and Pharmaceutical Sciences* 14 (2):077-086.

Akram, Muhammad, Ahmed A Shahab-Uddin, KHAN Usmanghani, ABDUL Hannan, Ejaz Mohiuddin, and Muhammad Asif. 2010. "Curcuma longa and curcumin: a review article." *Rom J Biol Plant Biol* 55 (2):65-70.

Alaa, Raed, Mohammad M Abd-Alhaseeb, Eman S Habib, Amany K Ibrahim, and Safwat A Ahmed. 2016. "Screening of Marrubium alysson L. extract for pharmacological activity." *Journal of Chemical and Pharmaceutical Research* 8 (4):283-289.

AlKadi, Hussien O. 2007. "Antimalarial drug toxicity: a review." *Chemotherapy* 53 (6):385-391.

Andujar, I, MC Recio, RM Giner, and JL Ríos. 2013. "Traditional chinese medicine remedy to jury: the pharmacological basis for the use of shikonin as an anticancer therapy." *Current Medicinal Chemistry* 20 (23):2892-2898.

Antika, LD, D Triana, and T Ernawati. 2020. "Antimicrobial activity of quinine derivatives against human pathogenic bacteria." *IOP Conference Series: Earth and Environmental Science*.

Ayaz, Muhammad, Muhammad Junaid, Farhat Ullah, Abdul Sadiq, Fazal Subhan, Mir Azam Khan, Waqar Ahmad, Gowhar Ali, Muhammad Imran, and Sajjad Ahmad. 2016. "Molecularly characterized solvent extracts and saponins from Polygonum hydropiper L. show high anti-angiogenic, anti-tumor, brine shrimp, and fibroblast NIH/3T3 cell line cytotoxicity." *Frontiers in Pharmacology* 7:74.

Ayaz, Muhammad, Muhammad Junaid, Farhat Ullah, Abdul Sadiq, Mir Azam Khan, Waqar Ahmad, Muhammad Raza Shah, Muhammad Imran, and Sajjad Ahmad. 2015. "Comparative chemical profiling, cholinesterase inhibitions and anti-radicals properties of essential oils from Polygonum hydropiper L: A Preliminary anti-Alzheimer's study." *Lipids in Health and Disease* 14 (1):1-12.

Badshah, Syed Lal, Shah Faisal, Akhtar Muhammad, Benjamin Gabriel Poulson, Abdul Hamid Emwas, and Mariusz Jaremko. 2021. "Antiviral activities of flavonoids." *Biomedicine & Pharmacotherapy* 140:111596.

Bailly, Fabrice, and Philippe Cotelle. 2005. "Anti-HIV activities of natural antioxidant caffeic acid derivatives: toward an antiviral supplementation diet." *Current Medicinal Chemistry* 12 (15):1811-1818.

Bateman, DN, and EH Dyson. 1986. "Quinine toxicity." *Adverse Drug Reactions and Acute Poisoning Reviews* 5 (4):215-233.

Bazotte, Roberto Barbosa, Sandro Massao Hirabara, Tamires Afonso Duarte Serdan, Raquel Bragante Gritte, Talita Souza-Siqueira, Renata Gorjao, Laureane Nunes Masi, Marina Masetto Antunes, Vinicius Cruzat, and Tania Cristina Pithon-Curi. 2021. "4-Aminoquinoline compounds from the Spanish flu to COVID-19." *Biomedicine & Pharmacotherapy* 135:111138.

Berretta, Andresa Aparecida, Marcelo Augusto Duarte Silveira, José Manuel Cóndor Capcha, and David De Jong. 2020. "Propolis and its potential against SARS-CoV-2 infection mechanisms and COVID-19 disease: Running title: Propolis against SARS-CoV-2 infection and COVID-19." *Biomedicine & Pharmacotherapy* 131:110622.

Boozari, M., and H. Hosseinzadeh. 2021. "Natural products for COVID-19 prevention and treatment regarding to previous coronavirus infections and novel studies." *Phytother Res* 35 (2):864-876. https://doi.org/10.1002/ptr.6873.

Bora, Kundan Singh, and Anupam Sharma. 2011. "The genus Artemisia: a comprehensive review." *Pharmaceutical Biology* 49 (1):101-109.

Çaliş, İhsan, Mohammed Hosny, Taha Khalifa, and Peter Rüedi. 1992. "Phenylpropanoid glycosides from Marrubium alysson." *Phytochemistry* 31 (10):3624-3626.

Çelebier, M., and C. Haznedaroğlu İ. 2021. "Could Targeting HMGB1 be Useful for the Clinical Management of COVID-19 Infection?" *Comb Chem High Throughput Screen* 24 (4):587-590. https://doi.org/10.2174/1386207323999200728114927.

Çelebier, Mustafa Ç, Carolina Simo, Sacide Altınöz, and Alejandro Cifuentes. 2015. "Comparative proteomics to investigate the *in vitro* antiproliferative effect of dietary polyphenols against K562 leukemia cells/Gıda kaynaklı polifenollerin K562 lösemi hücreleri üzerine *in vitro* çoğalmayı önleyici etkisinin karşılaştırmalı proteomik çalışmalarla incelenmesi." *Turkish Journal of Biochemistry* 40 (2):95-104.

Chang, Man-Jau, Huey-Chun Huang, Hsien-Cheh Chang, and Tsong-Min Chang. 2008. "Cosmetic formulations containing Lithospermum erythrorhizon root extract show moisturizing effects on human skin." *Archives of Dermatological Research* 300 (6):317-323.

Chen, D, LM Bellussi, S Cocca, J Wang, GC Passali, X Hao, L Chen, and D Passali. 2017. "Glycyrrhetinic acid suppressed hmgb1 release by up-regulation of Sirt6 in nasal inflammation." *Journal of Biological Regulators Homeostatic Agents* 31 (2):269-277.

Chen, Dan, Aijing Bi, Xiaoliang Dong, Yi Jiang, Bing Rui, Jinjiao Liu, Zhimin Yin, and Lan Luo. 2014. "Luteolin exhibits anti-inflammatory effects by blocking the activity of heat shock protein 90 in macrophages." *Biochemical Biophysical Research Communications* 443 (1):326-332.

Chen, Huai-Min, Shu-Fen Liou, Jong-Hau Hsu, Tsan-Ju Chen, Tsung-Lin Cheng, Chaw-Chi Chiu, and Jwu-Lai Yeh. 2014. "Baicalein inhibits HMGB1 release and MMP-2/-9 expression in lipopolysaccharide-induced cardiac hypertrophy." *The American Journal of Chinese Medicine* 42 (04):785-797.

Chen, Liting, Xiaolu Long, Qian Xu, Jiaqi Tan, Gaoxiang Wang, Yang Cao, Jia Wei, Hui Luo, Hui Zhu, and Liang Huang. 2020. "Elevated serum levels of S100A8/A9 and HMGB1 at hospital admission are correlated with inferior clinical outcomes in COVID-19 patients." *Cellular & Molecular Immunology* 17 (9):992-994.

Chen, Ruochan, Yan Huang, Jun Quan, Jiao Liu, Haichao Wang, Timothy R Billiar, Michael T Lotze, Herbert J Zeh, Rui Kang, and Daolin Tang. 2020. "HMGB1 as a potential biomarker and therapeutic target for severe COVID-19." *Heliyon* 6 (12):e05672.

Chen, Xin, Lu Yang, Joost J Oppenheim, and OM Zack Howard. 2002. "Cellular pharmacology studies of shikonin derivatives." *Phytotherapy Research* 16 (3):199-209.

Chen, Xiuping, Jiajie Guo, Jiaolin Bao, Jinjian Lu, and Yitao Wang. 2014. "The anticancer properties of Salvia miltiorrhiza Bunge (Danshen): a systematic review." *Medicinal Research Reviews* 34 (4):768-794.

Chowdhury, Pritom, and Anoop Kumar Barooah. 2020. "Tea bioactive modulate innate immunity: In perception to COVID-19 pandemic." *Frontiers in Immunology* 11: 590716.

Christenhusz, Maarten JM, and James W Byng. 2016. "The number of known plants species in the world and its annual increase." *Phytotaxa* 261 (3):201–217-201–217.

Conde, R., Vanessa, Marco G Alves, Pedro F Oliveira, and Branca M Silva. 2015. "Tea (Camellia sinensis (L.)): a putative anticancer agent in bladder carcinoma?" *Anti-Cancer Agents in Medicinal Chemistry (Formerly Current Medicinal Chemistry-Anti-Cancer Agents)* 15 (1):26-36.

Dahiya, Pradeep Kumar, MJR Nout, Martinus A van Boekel, Neelam Khetarpaul, Raj Bala Grewal, and Anita Linnemann. 2014. "Nutritional characteristics of mung bean foods." *British Food Journal*.

Deng, Lei, Ki Joon Cho, Walter Fiers, and Xavier Saelens. 2015. "M2e-based universal influenza A vaccines." *Vaccines* 3 (1):105-136.

Donovan, Jennifer L, C Lindsay DeVane, David Boulton, Seetal Dodd, and John S Markowitz. 2003. "Dietary levels of quinine in tonic water do not inhibit CYP2D6 *in vivo*." *Food and Chemical Toxicology* 41 (8):1199-1201.

Duan, Shu-Peng, Li-Hong Zhu, Peng Li, Xin-Wen Song, Hong-Wei Wang, and Bao-Sheng Shen. 2016. "Effect and mechanism of danshensu on hepatitis B virus reverse transcriptase and antigen expression." *China Journal of Chinese Materia Medica* 41 (7):1297-1301.

Duraipandiyan, Veeramuthu, Francis Indwar, and Savarimuthu Ignacimuthu. 2010. "Antimicrobial activity of confertifolin from Polygonum hydropiper." *Pharmaceutical Biology* 48 (2):187-190.

Egert, Sarah, and Gerald Rimbach. 2011. "Which sources of flavonoids: complex diets or dietary supplements?" *Advances in Nutrition* 2 (1):8-14.

Eroğlu, Canan, Mücahit Seçme, Gülseren Bağcı, and Yavuz Dodurga. 2015. "Assessment of the anticancer mechanism of ferulic acid via cell cycle and apoptotic pathways in human prostate cancer cell lines." *Tumor Biology* 36 (12):9437-9446.

Esghaei, Maryam, Hadi Ghaffari, Bahman Rahimi Esboei, Zienab Ebrahimi Tapeh, Farah Bokharaei Salim, and Manijeh Motevalian. 2018. "Evaluation of anticancer activity of Camellia sinensis in the Caco-2 colorectal cancer cell line." *Asian Pacific Journal of Cancer Prevention: APJCP* 19 (6):1697.

Ezhuthupurakkal, Preedia Babu, Subastri Ariraman, Suyavaran Arumugam, Nithyananthan Subramaniyan, Suresh Kumar Muthuvel, Premkumar Kumpati, Bharathidasan Rajamani, and Thirunavukkarasu Chinnasamy. 2018. "Anticancer potential of ZnO nanoparticle-ferulic acid conjugate on Huh-7 and HepG2 cells and diethyl nitrosamine induced hepatocellular cancer on Wistar albino rat." *Nanomedicine: Nanotechnology, Biology and Medicine* 14 (2):415-428.

Fadus, Matthew C, Cecilia Lau, Jai Bikhchandani, and Henry T Lynch. 2017. "Curcumin: An age-old anti-inflammatory and anti-neoplastic agent." *Journal of Traditional and Complementary Medicine* 7 (3):339-346.

Feng, Liang, Xiaobin Jia, Mao-Mao Zhu, Yan Chen, and Feng Shi. 2010. "Antioxidant activities of total phenols of Prunella vulgaris L. *in vitro* and in tumor-bearing mice." *Molecules* 15 (12):9145-9156.

Ferguson, Lynnette R. 1994. "Antimutagens as cancer chemopreventive agents in the diet." *Mutation Research/Fundamental and Molecular Mechanisms of Mutagenesis* 307 (1):395-410.

Forrest, GI, and DS1184764 Bendall. 1969. "The distribution of polyphenols in the tea plant (Camellia sinensis L.)." *Biochemical Journal* 113 (5):741-755.

Fresco, P, FIGM Borges, C Diniz, and MPM Marques. 2006. "New insights on the anticancer properties of dietary polyphenols." *Medicinal Research Reviews* 26 (6):747-766.

Fujiki, Hirota, Seiji Yoshizawa, Takahiko Horiuchi, Masami Suganuma, Jun Yatsunami, Shinji Nishiwaki, Sachiko Okabe, Rie Nishiwaki-Matsushima, Takuo Okuda, and Takashi Sugimura. 1992. "Anticarcinogenic effects of (−)-epigallocatechin gallate." *Preventive Medicine* 21 (4):503-509.

Gachelin, Gabriel, Paul Garner, Eliana Ferroni, Ulrich Tröhler, and Iain Chalmers. 2017. "Evaluating Cinchona bark and quinine for treating and preventing malaria." *Journal of the Royal Society of Medicine* 110 (1):31-40.

Ganesan, Kumar, and Baojun Xu. 2018. "A critical review on phytochemical profile and health promoting effects of mung bean (Vigna radiata)." *Food Science and Human Wellness* 7 (1):11-33.

Gantait, Arunava, Subrata Pandit, Neelesh K Nema, and Pulok K Mukjerjee. 2010. "Quantification of glycyrrhizin in Glycyrrhiza glabra extract by validated HPTLC densitometry." *Journal of AOAC International* 93 (2):492-495.

Gao, Hong, Lei Liu, Zhang-yi Qu, Feng-xiang Wei, Shu-qiu Wang, Guang Chen, Le Qin, Fu-yang Jiang, Ying-chen Wang, and Lei Shang. 2011. "Anti-adenovirus activities of shikonin, a component of Chinese herbal medicine *in vitro*." *Biological and Pharmaceutical Bulletin* 34 (2):197-202.

Gao, Hongwei, Wen Sun, Jianping Zhao, Xiaxia Wu, Jin-Jian Lu, Xiuping Chen, Qiong-ming Xu, Ikhlas A Khan, and Shilin Yang. 2016. "Tanshinones and diethyl blechnics with anti-inflammatory and anti-cancer activities from Salvia miltiorrhiza Bunge (Danshen)." *Scientific Reports* 6 (1):1-13.

Gao, Qinghan, Xin Zhao, Lei Yin, Yuanbin Zhang, Bing Wang, Xiuli Wu, Xinhui Zhang, Xueyan Fu, and Weihong Sun. 2016. "The essential oil of Artemisia capillaris protects against CCl 4-induced liver injury *in vivo*." *Revista Brasileira de Farmacognosia* 26:369-374.

Gil, Minchan, Yun Kyu Kim, Sang Bum Hong, and Kyung Jin Lee. 2016. "Naringin decreases TNF-α and HMGB1 release from LPS-stimulated macrophages and improves survival in a CLP-induced sepsis mice." *PLoS One* 11 (10):e0164186.

Goldenberg, AM, and LF Wexler. 1988. "Quinine overdose: review of toxicity and treatment." *Clinical Cardiology* 11 (10):716-718.

Gong, Yi, Yanli Li, Yin Lu, Linglin Li, Hamid Abdolmaleky, George L Blackburn, and Jin-Rong Zhou. 2011. "Bioactive tanshinones in Salvia miltiorrhiza inhibit the growth of prostate cancer cells *in vitro* and in mice." *International Journal of Cancer* 129 (5):1042-1052.

Gonzalez, ME, Balbino Alarcón, and Luís Carrasco. 1987. "Polysaccharides as antiviral agents: antiviral activity of carrageenan." *Antimicrobial Agents and Chemotherapy* 31 (9):1388-1393.

Grams, Natalie. 2019. "Homeopathy—where is the science? A current inventory on a pre-scientific artifact." *EMBO Reports* 20 (3):e47761.

Gramza, Anna, and Jozef Korczak. 2005. "Tea constituents (Camellia sinensis L.) as antioxidants in lipid systems." *Trends in Food Science & Technology* 16 (8):351-358.

Greenwood, Brian, and Theonest Mutabingwa. 2002. "Malaria in 2002." *Nature* 415 (6872):670.

Große, Maximilian, Natalia Ruetalo, Mirjam Layer, Dan Hu, Ramona Businger, Sascha Rheber, Christian Setz, Pia Rauch, Janina Auth, and Maria Fröba. 2021. "Quinine inhibits infection of human cell lines with SARS-CoV-2." *Viruses* 13 (4):647.

Guo, Qinglan, Huan Xia, Xianhua Meng, Gaona Shi, Chengbo Xu, Chenggen Zhu, Tiantai Zhang, and Jiangong Shi. 2018. "C19-Diterpenoid alkaloid arabinosides from an aqueous extract of the lateral root of Aconitum carmichaelii and their analgesic activities." *Acta pharmaceutica sinica B* 8 (3):409-419.

Hamilton-Miller, JM. 1995. "Antimicrobial properties of tea (Camellia sinensis L.)." *Antimicrobial Agents and Chemotherapy* 39 (11):2375-2377.

Haraguchi, Hiroyuki, Kensuke Hashimoto, and Akira Yagi. 1992. "Antioxidative substances in leaves of Polygonum hydropiper." *Journal of Agricultural and Food Chemistry* 40 (8):1349-1351.

Hayet, Edziri, Ammar Samia, Groh Patrick, Mahjoub Mohamed Ali, Mastouri Maha, Gutmann Laurent, Zine Mighri, and Laouni Mahjoub. 2007. "Antimicrobial and cytotoxic activity of Marrubium alysson and Retama raetam grown in Tunisia." *Pakistan Journal of Biological Sciences: PJBS* 10 (10):1759-1762.

Herman, Harvey B, John R Jezorek, and Zhe Tang. 2000. "Analysis of Diet Tonic Water Using Capillary Electrophoresis. An Undergraduate Instrumental Analysis Experiment." *Journal of Chemical Education* 77 (6):743.

Hisa, Tomoyuki, Yusuke Kimura, Kenji Takada, Fujio Suzuki, and Masaharu Takigawa. 1998. "Shikonin, an ingredient of Lithospermum erythrorhizon, inhibits angiogenesis *in vivo* and *in vitro*." *Anticancer Research* 18 (2A):783-790.

Hong, Jung-Hee, Ji-Won Lee, Jung-Hyun Park, and In-Seon Lee. 2007. "Antioxidative and cytoprotective effects of Artemisia capillaris fractions." *Biofactors* 31 (1):43-53.

Horemans, Tessa, Gaëlle Boulet, Marian van Kerckhoven, Johannes Bogers, Sofie Thys, Chris Vervaet, Anouck Vervaeck, Peter Delputte, Louis Maes, and Paul Cos. 2017. "In-vivo evaluation of apocynin for prevention of Helicobacter pylori-induced gastric carcinogenesis." *European Journal of Cancer Prevention* 26 (1):10-16.

Hu, Xiaolan, Shuxin Lin, Daihua Yu, Shuifeng Qiu, Xianqi Zhang, and Ruhuan Mei. 2010. "A preliminary study: the anti-proliferation effect of salidroside on different human cancer cell lines." *Cell Biology and Toxicology* 26 (6):499-507.

Huq, AKM, Jamia Azdina Jamal, and Johnson Stanslas. 2014. "Ethnobotanical, phytochemical, pharmacological, and toxicological aspects of Persicaria hydropiper (L.) Delarbre." *Evidence-Based Complementary and Alternative Medicine* 2014.

Ikeda, Keiko, Kazuko Tsujimoto, Misao Uozaki, Mitsunori Nishide, Yukiko Suzuki, A Hajime Koyama, and Hisashi Yamasaki. 2011. "Inhibition of multiplication of herpes simplex virus by caffeic acid." *International Journal of Molecular Medicine* 28 (4):595-598.

Jang, Eungyeong, Bum-Joon Kim, Kyung-Tae Lee, Kyung-Soo Inn, and Jang-Hoon Lee. 2015. "A survey of therapeutic effects of Artemisia capillaris in liver diseases." *Evidence-Based Complementary and Alternative Medicine* 2015.

Jenner, Peter. 2003. "Oxidative stress in Parkinson's disease." *Annals of Neurology: Official Journal of the American Neurological Association and the Child Neurology Society* 53 (S3):S26-S38.

Jiang, Wang-Lin, Shu-Ping Zhang, and Hai-Bo Zhu. 2012. "Forsythoside B protects against experimental sepsis by modulating inflammatory factors." *Phytotherapy Research* 26 (7):981-987.

Jiao, Rui, Yingxia Liu, Hao Gao, Jia Xiao, and Kwok Fai So. 2016. "The anti-oxidant and antitumor properties of plant polysaccharides." *The American Journal of Chinese Medicine* 44 (03):463-488.

Jin, Mingliang, Ke Zhao, Qingsheng Huang, Chunlan Xu, and Peng Shang. 2012. "Isolation, structure and bioactivities of the polysaccharides from Angelica sinensis (Oliv.) Diels: A review." *Carbohydrate Polymers* 89 (3):713-722.

Jomsky, Mitchell G, and A Kerna Nicholas. 2020. "Could low-dose quinine prevent or treat coronavirus infection." *EC Pharmacol. Toxicol. Comment* 8:62-64.

Jube, Sandro, Zeyana S Rivera, Marco E Bianchi, Amy Powers, Ena Wang, Ian Pagano, Harvey I Pass, Giovanni Gaudino, Michele Carbone, and Haining Yang. 2012. "Cancer Cell Secretion of the DAMP Protein HMGB1 Supports Progression in Malignant MesotheliomaHMGB1 Secretion Supports Mesothelioma Growth." *Cancer Research* 72 (13):3290-3301.

Jun, Min Soo, Hee Sook Kim, Young Min Kim, Hye Jung Kim, Eun Jung Park, Jae Heun Lee, Kang Ro Lee, Yeong Shik Kim, and Ki Churl Chang. 2012. "Ethanol Extract of Prunella vulgaris var. lilacina Inhibits HMGB1 Release by Induction of Heme Oxygenase-1 in LPS-activated RAW 264.7 Cells and CLP-induced Septic Mice." *Phytotherapy Research* 26 (4):605-612.

Kang, Kyoung-Myoung, and Shin-Ho Lee. 2013. "Effects of Extraction Methods on the Antioxidative Activity of Artemisia sp." *Journal of the Korean Society of Food Science and Nutrition* 42 (8):1249-1254.

Kang, Rui, Qiuhong Zhang, Herbert J Zeh, Michael T Lotze, and Daolin Tang. 2013. "HMGB1 in Cancer: Good, Bad, or Both? HMGB1 in Cancer." *Clinical cancer research* 19 (15):4046-4057.

Kanimozhi, G, and NR Prasad. 2015. "Anticancer effect of caffeic acid on human cervical cancer cells." In *Coffee in Health and Disease Prevention*, 655-661. Elsevier.

Kashyap, Srishti, Revathy Nadhan, and Danny N Dhanasekaran. 2021. "Phytochemicals: potential lead compounds for COVID-19 therapeutics." *Journal of Food Bioactives* 15.

Khalaf, Hasan Alaa Aldeen, Ruaa Azziz Jasim, and Ismail Taha Ibrahim. 2021. "Verbascoside—A review of its antitumor activities." *Pharmacology & Pharmacy* 12 (6):109-126.

Khan, Rasel Ahmed, Rajib Hossain, Pranta Roy, Divya Jain, Abu Saim Mohammad Saikat, Anik Prasad Roy Shuvo, Muhammad Akram, Walaa Fikry Elbossaty, Ishaq N Khan, and Sakshi Painuli. 2022. "Anticancer effects of acteoside: Mechanistic insights and therapeutic status." *European Journal of Pharmacology* 916:174699.

Khandelwal, Mohit. 2018. "Implications of homeopathic medicines in case of malaria." *International Journal of Homoeopathic Sciences* 2 (1):17-19.

Kharal, Saleem Ahmed, Qurban Hussain, and Shujat Ali. 2009. "Quinine is bactericidal." *J Pak Med Assoc* 59 (4).

Kim, H, J Kim, J Park, SH Kim, Y Uchida, WM Holleran, and Y Cho. 2012. "Water extract of gromwell (Lithospermum erythrorhizon) enhances migration of human keratinocytes and dermal fibroblasts with increased lipid synthesis in an *in vitro* wound scratch model." *Skin Pharmacology and Physiology* 25 (2):57-64.

Kim, Hong-Tae, Ju-Wan Kim, Mee-Kyoung Lim, Tae-Won Jin, Sang-Geon Yeo, Kwang-Ho Jang, Tae-Ho Oh, and Keun-Woo Lee. 2007. "Cytotoxic effect of Artemisia capillaris extracts on the cancer cells on *in vitro*." *Journal of Veterinary Clinics* 24 (3):367-371.

Kim, Ra-Jeong, Min-Jung Kang, Cho-Rong Hwang, Woo-Jae Jung, and Jung-Hye Shin. 2012. "Antioxidant and cancer cell growth inhibition activity of five different varieties of Artemisia cultivars in Korea." *Journal of Life Science* 22 (6):844-851.

Kim, Seung-Woo, Yinchuan Jin, Joo-Hyun Shin, Il-Doo Kim, Hye-Kyung Lee, Sunghyouk Park, Pyung-Lim Han, and Ja-Kyeong Lee. 2012. "Glycyrrhizic acid affords robust neuroprotection in the postischemic brain via anti-inflammatory effect by inhibiting HMGB1 phosphorylation and secretion." *Neurobiology of Disease* 46 (1):147-156.

Kim, Tae Hoon, Sae-Kwang Ku, and Jong-Sup Bae. 2013a. "Anti-inflammatory activities of isorhamnetin-3-O-galactoside against HMGB1-induced inflammatory responses in both HUVECs and CLP-induced septic mice." *Journal of Cellular Biochemistry* 114 (2):336-345.

Kim, Tae Hoon, Sae-Kwang Ku, Taeho Lee, and Jong-Sup Bae. 2012. "Vascular barrier protective effects of phlorotannins on HMGB1-mediated proinflammatory responses *in vitro* and *in vivo*." *Food Chemical Toxicology* 50 (6):2188-2195.

Kim, Tae Hoon, Sae-Kwang Ku, and Jong-Sup Bae. 2013b. "Persicarin is anti-inflammatory mediator against HMGB1-induced inflammatory responses in HUVECs and in CLP-induced sepsis mice." *Journal of Cellular Physiology* 228 (4):696-703.

Kimura, Yoshiyuki, Hiromichi Okuda, and Michinori Kubo. 1995. "Effects of stilbenes isolated from medicinal plants on arachidonate metabolism and degranulation in human polymorphonuclear leukocytes." *Journal of Ethnopharmacology* 45 (2):131-139.

Klees, Robert F, Paul C De Marco, Roman M Salasznyk, Disha Ahuja, Michael Hogg, Sylvain Antoniotti, Lakshmi Kamath, Jonathan S Dordick, and George E Plopper. 2006. "Apocynin derivatives interrupt intracellular signaling resulting in decreased migration in breast cancer cells." *Journal of Biomedicine and Biotechnology* 2006.

Kohli, K, J Ali, MJ Ansari, and Z Raheman. 2005. "Curcumin: a natural antiinflammatory agent." *Indian Journal of Pharmacology* 37 (3):141.

Kokotkiewicz, A, and M Luczkiewicz. 2009. "Honeybush (Cyclopia sp.)–a rich source of compounds with high antimutagenic properties." *Fitoterapia* 80 (1):3-11.

Kong, Wei-Jun, Shan-Shan Zhang, Yan-Ling Zhao, Ming-Quan Wu, Ping Chen, Xiao-Ru Wu, Xin-Ping Ma, Wei-Ying Guo, and Mei-Hua Yang. 2017. "Combination of chemical fingerprint and bioactivity evaluation to explore the antibacterial components of Salvia miltiorrhizae." *Scientific Reports* 7 (1):1-12.

Krishnaswamy, Mohana, and KK Purushothaman. 1980. "Plumbagin: a study of its anticancer, antibacterial and antifungal properties." *Indian Journal of Experimental Biology* 18 (8):876-877.

Krishnaveni, M, and K Suresh. 2015. "Induction of apoptosis by quinine in human laryngeal carcinoma cell line." *International Journal of Current Research and Academic Review* 3 (3):169-178.

Krishnaveni, Marimuthu, Kathiresan Suresh, and Ramu Arunkumar. 2016. "Antiproliferative and apoptotic effects of quinine in human Hep-2 laryngeal cancer and KB oral cancer cell." *Bangladesh Journal of Pharmacology* 11 (3):593-602.

Kumar, V Bharath, Shu-Hui Lin, B Mahalakshmi, Yu-Sheng Lo, Chia-Chieh Lin, Yi-Ching Chuang, Ming-Ju Hsieh, and Mu-Kuan Chen. 2020. "Sodium Danshensu inhibits oral cancer cell migration and invasion by modulating p38 signaling pathway." *Frontiers in Endocrinology* 11:568436.

Lan, Tian, Linling Wang, Qian Xu, Weiguo Liu, Hongchuan Jin, Weimin Mao, Xian Wang, and Xiaojia Wang. 2013. "Growth inhibitory effect of Cucurbitacin E on breast cancer cells." *International Journal of Clinical and Experimental Pathology* 6 (9):1799.

Latarissa, Irma Rahayu, Melisa Intan Barliana, Anna Meiliana, and Keri Lestari. 2021. "Potential of Quinine Sulfate for COVID-19 Treatment and Its Safety Profile." *Clinical Pharmacology: Advances and Applications* 13:225.

Law, Siukan, Chuiman Lo, Jie Han, Albert Wingnang Leung, and Chuanshan Xu. 2020. "Traditional Chinese herb, Astragalus: possible for treatment and prevention of COVID-19?" *Herba Polonica* 66 (4).

Lawson-Wood, Kathryn, Kieran Evans, and UK Seer Green. 2019. *Determination of Quinine in Tonic Water Using Fluorescence Spectroscopy*.

Lee, Chan-Ho, Seong-Jin Yoon, and Sun-Mee Lee. 2012. "Chlorogenic acid attenuates high mobility group box 1 (HMGB1) and enhances host defense mechanisms in murine sepsis." *Molecular Medicine* 18 (11):1437-1448.

Lee, Chao-Ying, Yung-Shin Chien, Tai-Hui Chiu, Wen-Wen Huang, Chi-Cheng Lu, Jo-Hua Chiang, and Jai-Sing Yang. 2012. "Apoptosis triggered by vitexin in U937 human leukemia cells via a mitochondrial signaling pathway." *Oncology Reports* 28 (5):1883-1888.

Lee, Danbi, Jinbum Bae, Yoon Kyu Kim, Minchan Gil, Joo-Yong Lee, Chan-Sik Park, and Kyung Jin Lee. 2013. "Inhibitory effects of berberine on lipopolysaccharide-induced inducible nitric oxide synthase and the high-mobility group box 1 release in macrophages." *Biochemical Biophysical Research Communications* 431 (3):506-511.

Lee, Heung-Shick, and Younhee Kim. 2016. "Antifungal activity of Salvia miltiorrhiza against Candida albicans is associated with the alteration of membrane permeability and (1, 3)-β-D-glucan synthase activity." *Journal of Microbiology and Biotechnology* 26 (3):610-617.

Lee, Jong Hyun, Chakrabhavi Dhananjaya Mohan, Muthu K Shanmugam, Shobith Rangappa, Gautam Sethi, Kodappully Sivaraman Siveen, Arunachalam Chinnathambi, Tahani Awad Alahmadi, Sulaiman Ali Alharbi, and Salundi Basappa. 2020. "Vitexin abrogates invasion and survival of hepatocellular carcinoma cells through targeting STAT3 signaling pathway." *Biochimie* 175:58-68.

Lee, JungHyeok, Dongsool Yim, and SungSook Choi. 2021. "Antibacterial Activity and Anti-inflammatory Effect of Methanol Extracts of Saliva miltiorrhiza Against Oral Pathogenic Bacteria." *Korean Journal of Pharmacognosy* 52 (1):41-48.

Lee, Wonhwa, Eun-Kyung Yoon, Kyung-Min Kim, Dong Ho Park, and Jong-Sup Bae. 2015. "Antiseptic effect of vicenin-2 and scolymoside from Cyclopia subternata (honeybush) in response to HMGB1 as a late sepsis mediator *in vitro* and *in vivo*." *Canadian Journal of Physiology Pharmacology* 93 (8):709-720.

Lee, Wonhwa, Sae-Kwang Ku, Tae Hoon Kim, and Jong-Sup Bae. 2013a. "Emodin-6-O-β-d-glucoside inhibits HMGB1-induced inflammatory responses *in vitro* and *in vivo*." *Food Chemical Toxicology* 52:97-104.

Lee, Wonhwa, Sae-Kwang Ku, Tae Hoon Kim, and Jong-Sup Bae. 2013b. "Emodin-6-O-β-d-glucoside inhibits HMGB1-induced inflammatory responses *in vitro* and *in vivo*." *Food and Chemical Toxicology* 52:97-104.

Lee, Wonhwa, Sae-Kwang Ku, Doohyun Lee, Taeho Lee, and Jong-Sup Bae. 2014. "Emodin-6-O-β-d--glucoside inhibits high-glucose-induced vascular inflammation." *Inflammation* 37 (2):306-313.

Lefer, David J, and D Neil Granger. 2000. "Oxidative stress and cardiac disease." *The American Journal of Medicine* 109 (4):315-323.

Li, Liuying, Limin Zhang, Tingting Liao, Chuantao Zhang, Keling Chen, and Qun Huang. 2022. "Advances on pharmacology and toxicology of aconitine." *Fundamental & Clinical Pharmacology*.

Li, Tian, Kewei Xu, and Yifan Liu. 2018. "Anticancer effect of salidroside reduces viability through autophagy/PI3K/Akt and MMP-9 signaling pathways in human bladder cancer cells." *Oncology Letters* 16 (3):3162-3168.

Li, Wei, Mala Ashok, Jianhua Li, Huan Yang, Andrew E Sama, and Haichao Wang. 2007. "A major ingredient of green tea rescues mice from lethal sepsis partly by inhibiting HMGB1." *PLoS One* 2 (11):e1153.

Li, Yao-Lan, Shuang-Cheng Ma, Yi-Ting Yang, Shao-Ming Ye, and Paul Pui-Hay But. 2002. "Antiviral activities of flavonoids and organic acid from Trollius chinensis Bunge." *Journal of Ethnopharmacology* 79 (3):365-368.

Lin, HP, CY Lin, C Huo, LC Su, and CP Chuu. 2012. "Anticancer effect of caffeic acid phenethyl ester." *Pharmacologia* 3 (1):26-30.

Liu, Wei, Dong-Xue Yin, Tong Zhang, Qi Qiao, Yue-Qin Yang, and Wen-Lu Wang. 2020. "Phytochemical profiles and antioxidant activity of Rehmannia glutinosa from different production locations." *Chemistry & Biodiversity* 17 (8):e2000341.

Liu, Xiaoli, Qingfeng Jiang, Huaimin Liu, and Suxia Luo. 2019. "Vitexin induces apoptosis through mitochondrial pathway and PI3K/Akt/mTOR signaling in human non-small cell lung cancer A549 cells." *Biological Research* 52 (1):1-7.

Lu, Guang-Hua, Kelvin Chan, Kelvin Leung, Chi-Leung Chan, Zhong-Zhen Zhao, and Zhi-Hong Jiang. 2005. "Assay of free ferulic acid and total ferulic acid for quality assessment of Angelica sinensis." *Journal of Chromatography A* 1068 (2):209-219.

Luo, Jian-Xing, Yang Zhang, Xiao-Yu Hu, Guo Chen, Xi-Yun Liu, Hong-Ming Nie, Jing-Li Liu, and Da-chao Wen. 2016. "Aqueous extract from Aconitum carmichaelii Debeaux reduces liver injury in rats via regulation of HMGB1/TLR4/NF-KB/caspase-3 and PCNA signaling pathways." *Journal of Ethnopharmacology* 183:187-192.

Malaria, Roll Back. 2005. "World malaria report 2005." *World Health Organization and UNICEF*.

Mathew, Dony, and Wei-Li Hsu. 2018. "Antiviral potential of curcumin." *Journal of functional foods* 40:692-699.

Menon, Venugopal P, and Adluri Ram Sudheer. 2007. "Antioxidant and anti-inflammatory properties of curcumin." *The Molecular Targets and Therapeutic Uses of Curcumin in Health and Disease*:105-125.

Min, Jie, Hua Shen, Wang Xi, Qing Wang, Liang Yin, Yufeng Zhang, Yue Yu, Qian Yang, and Zhi-nong Wang. 2018. "Synergistic anticancer activity of combined use of caffeic acid with paclitaxel enhances apoptosis of non-small-cell lung cancer H1299 cells *in vivo* and *in vitro*." *Cellular Physiology and Biochemistry* 48 (4):1433-1442.

Min, Kyoung-jin, and Taeg Kyu Kwon. 2014. "Anticancer effects and molecular mechanisms of epigallocatechin-3-gallate." *Integrative Medicine Research* 3 (1):16-24.

Mohamed, NH. 2010. "Anticancer activity of Marrubium alysson L. and its phenolic constituents." *Drug Plants I*:185-193.

Nadhan, Revathy, Dipyaman Patra, Neethu Krishnan, Arathi Rajan, Srinivas Gopala, Dashnamoorthy Ravi, and Priya Srinivas. 2021. "Perspectives on mechanistic implications of ROS inducers for targeting viral infections." *European Journal of Pharmacology* 890:173621.

Naghavi, Mohammad Reza, Farzad Alaeimoghadam, and Hossein Ghafoori. 2014. "Artemisia species from Iran as valuable resources for medicinal uses." *World Acad. Sci. Eng. Technol* 11:1058-1064.

Nakashima, Souichi, Hisashi Matsuda, Yoshimi Oda, Seikou Nakamura, Masayuki Yamashita, and Masayuki Yoshikawa. 2010. "Cucurbitacin E as a new inhibitor of cofilin phosphorylation in human leukemia U937 cells." *Bioorganic & Medicinal Chemistry Letters* 20 (9):2994-2997.

Nasir, Abdul, Atif AK Khalil, Muhammad Z Bhatti, Ashfaq Ur Rehman, Jiayi Li, and Zahida Parveen. 2021. "Review on Pharmacological and Phytochemical Prospects of Traditional Medicinal Plant: Persicaria hydropiper (Smartweed)." *Current Topics in Medicinal Chemistry* 21 (12):1027-1036.

Ngo, Thi Men, Phuong Thao Tran, Le Son Hoang, Jeong-Hyung Lee, Byung Sun Min, and Jeong Ah Kim. 2021. "Diterpenoids isolated from the root of Salvia miltiorrhiza and their anti-inflammatory activity." *Natural Product Research* 35 (5):726-732.

Nguyen, Anh Tho, Hugues Malonne, Pierre Duez, Renée Vanhaelen-Fastre, Maurice Vanhaelen, and Jeanine Fontaine. 2004. "Cytotoxic constituents from Plumbago zeylanica." *Fitoterapia* 75 (5):500-504.

Ozturk, G, Z Ginis, S Akyol, G Erden, A Gurel, and O Akyol. 2012. "The anticancer mechanism of caffeic acid phenethyl ester (CAPE): review of melanomas, lung and prostate cancers." *Eur Rev Med Pharmacol Sci* 16 (15):2064-2068.

Pang, Honggang, Tinqin Huang, Jinning Song, Dandong Li, Yonglin Zhao, and Xudong Ma. 2016. "Inhibiting HMGB1 with glycyrrhizic acid protects brain injury after DAI via its anti-inflammatory effect." *Mediators of Inflammation* 2016.

Panichayupakaranant, Pharkphoom, and Md Iftekhar Ahmad. 2016. "Plumbagin and its role in chronic diseases." *Drug Discovery from Mother Nature*: 229-246.

Panwar, Richa, Asvene K Sharma, Mandeep Kaloti, Dharm Dutt, and Vikas Pruthi. 2016. "Characterization and anticancer potential of ferulic acid-loaded chitosan nanoparticles against ME-180 human cervical cancer cell lines." *Applied Nanoscience* 6 (6):803-813.

Peng, Xiaofang, Zongping Zheng, Ka-Wing Cheng, Fang Shan, Gui-Xing Ren, Feng Chen, and Mingfu Wang. 2008. "Inhibitory effect of mung bean extract and its constituents vitexin and isovitexin on the formation of advanced glycation endproducts." *Food Chemistry* 106 (2):475-481.

Petitjean, Simon JL, Marylène Lecocq, Camille Lelong, Robin Denis, Sylvie Defrère, Pierre-Antoine Mariage, David Alsteens, and Charles Pilette. 2022. "Salvia miltiorrhiza Bunge as a Potential Natural Compound against COVID-19." *Cells* 11 (8):1311.

Psotová, Jitka, Milan Kolář, Jaromír Soušek, Zdeněk Švagera, Jaroslav Vičar, and Jitka Ulrichová. 2003. "Biological activities of Prunella vulgaris extract." *Phytotherapy Research: An International Journal Devoted to Pharmacological and Toxicological Evaluation of Natural Product Derivatives* 17 (9):1082-1087.

Pu, Wei-ling, Meng-ying Zhang, Ru-yu Bai, Li-kang Sun, Wen-hua Li, Ying-li Yu, Yue Zhang, Lei Song, Zhao-xin Wang, and Yan-fei Peng. 2020. "Anti-inflammatory effects of Rhodiola rosea L.: A review." *Biomedicine & Pharmacotherapy* 121:109552.

Qi, Zhilin, Yao Zhang, Shimei Qi, Liefeng Ling, Lin Gui, Liang Yan, Jun Lv, and Qiang Li. 2017. "Salidroside inhibits HMGB1 acetylation and release through upregulation of SirT1 during inflammation." *Oxidative Medicine Cellular Longevity* 2017.

Qiufeng, Lin, Feng Shunqing, Cen Yingzhou, Yang Yiting, and Wang Lingyun. 2004. "Study on the antibacterial and antiviral activity compositions of Trollium chinensis Bunge." *Journal of Zhejiang University (science Edition)* 31 (4):412-415.

Rajan, Durairaj Karthick, Kannan Mohan, Shubing Zhang, and Abirami Ramu Ganesan. 2021. "Dieckol: a brown algal phlorotannin with biological potential." *Biomedicine & Pharmacotherapy* 142:111988.

Rawla, Prashanth. 2019. "Epidemiology of Prostate Cancer." *World Journal of Oncology* 10 (2):63-89. doi: 10.14740/wjon1191.

Reed, SI. 1997. "Control of the G1/S transition." *Cancer Surveys* 29:7-23.

Ren, Wei, Pan Liang, Yue Ma, Qin Sun, Qingrong Pu, Li Dong, Gang Luo, Maryam Mazhar, Jiali Liu, and Raoqiong Wang. 2021. "Research progress of traditional Chinese medicine against COVID-19." *Biomedicine & Pharmacotherapy* 137: 111310.

Ripari, Nicolas, Arthur Alves Sartori, Mariana da Silva Honorio, Fernanda Lopes Conte, Karen Ingrid Tasca, Karina Basso Santiago, and José Maurício Sforcin. 2021. "Propolis antiviral and immunomodulatory activity: a review and perspectives for COVID-19 treatment." *Journal of Pharmacy and Pharmacology* 73 (3):281-299.

Robert, Jacques, and Christian Jarry. 2003. "Multidrug resistance reversal agents." *Journal of medicinal chemistry* 46 (23):4805-4817.

Roy, Arpita. 2021. "Plumbagin: a potential anti-cancer compound." *Mini Reviews in Medicinal Chemistry* 21 (6):731-737.

Russo, A, R Longo, and A Vanella. 2002. "Antioxidant activity of propolis: role of caffeic acid phenethyl ester and galangin." *Fitoterapia* 73:S21-S29.

Sammaiah, Arukali, Shiva Shanker Kaki, Gorantla NVT Sai Manoj, Y Poornachandra, C Ganesh Kumar, and RBN Prasad. 2015. "Novel fatty acid esters of apocynin oxime exhibit antimicrobial and antioxidant activities." *European Journal of Lipid Science and Technology* 117 (5):692-700.

Sangtani, Rimjhim, Atreyee Ghosh, Hem C Jha, Hamendra Singh Parmar, and Kiran Bala. 2021. "Potential of algal metabolites for the development of broad-spectrum antiviral therapeutics: Possible implications in COVID-19 therapy." *Phytotherapy Research* 35 (5):2296-2316.

Sankawa, Ushio, Yutaka Ebizuka, Terutaka Miyazaki, Yasuo Isomura, Hideaki Otsuka, Shoji Shibata, Motoko Inomata, and Fumiko Fukuoka. 1977. "Antitumor activity of shikonin and its derivatives." *Chemical and Pharmaceutical Bulletin* 25 (9):2392-2395.

Savla, Shreya R, Ankit P Laddha, and Yogesh A Kulkarni. 2021. "Pharmacology of apocynin: A natural acetophenone." *Drug metabolism reviews* 53 (4):542-562.

Seo, Eun Sun, Bo Kang Oh, Jhang Ho Pak, Soon-Ho Yim, Sangilyandi Gurunathan, Young-Pil Kim, and Kyung Jin Lee. 2013. "Acteoside improves survival in cecal ligation and puncture-induced septic mice via blocking of high mobility group box 1 release." *Molecules Cells* 35 (4):348-354.

Shang, Jin, Weihua Liu, Chunyan Yin, Haiping Chu, and Meizhen Zhang. 2019. "Cucurbitacin E ameliorates lipopolysaccharide-evoked injury, inflammation and MUC5AC expression in bronchial epithelial cells by restraining the HMGB1-TLR4-NF-κB signaling." *Molecular Immunology* 114:571-577.

Sharma, Navita, KP Mishra, and Lilly Ganju. 2016. "Salidroside exhibits anti-dengue virus activity by upregulating host innate immune factors." *Archives of Virology* 161 (12):3331-3344.

Shawky, Eman, Ahmed A Nada, and Reham S Ibrahim. 2020. "Potential role of medicinal plants and their constituents in the mitigation of SARS-CoV-2: identifying related therapeutic targets using network pharmacology and molecular docking analyses." *RSC advances* 10 (47):27961-27983.

Siveen, Kodappully Sivaraman, Sakshi Sikka, Rohit Surana, Xiaoyun Dai, Jingwen Zhang, Alan Prem Kumar, Benny KH Tan, Gautam Sethi, and Anupam Bishayee. 2014. "Targeting the STAT3 signaling pathway in cancer: role of synthetic and natural inhibitors." *Biochimica et Biophysica Acta (BBA)-Reviews On Cancer* 1845 (2):136-154.

Solary, E, Denis Caillot, Bruno Chauffert, Rene-Olivier Casasnovas, Monique Dumas, Marc Maynadie, and Henri Guy. 1992. "Feasibility of using quinine, a potential multidrug resistance-reversing agent, in combination with mitoxantrone and cytarabine for the treatment of acute leukemia." *Journal of Clinical Oncology* 10 (11):1730-1736.

Solary, Eric, Bernard Drenou, Lydia Campos, Patricia de Crémoux, Francine Mugneret, Philippe Moreau, Bruno Lioure, Annie Falkenrodt, Brigitte Witz, and Marc Bernard. 2003. "Quinine as a multidrug resistance inhibitor: a phase 3 multicentric randomized study in adult de novo acute myelogenous leukemia." *Blood* 102 (4):1202-1210.

Soni, Vivek Kumar, Arundhati Mehta, Yashwant Kumar Ratre, Atul Kumar Tiwari, Ajay Amit, Rajat Pratap Singh, Subash Chandra Sonkar, Navaneet Chaturvedi, Dhananjay Shukla, and Naveen Kumar Vishvakarma. 2020. "Curcumin, a traditional spice component, can hold the promise against COVID-19?" *European Journal of Pharmacology* 886:173551.

Spector, Abraham. 2000. "Oxidative stress and disease." *Journal of Ocular Pharmacology and Therapeutics* 16 (2):193-201.

Srinivas, Priya, Gokul Gopinath, Asoke Banerji, Abhilash Dinakar, and Gopal Srinivas. 2004. "Plumbagin induces reactive oxygen species, which mediate apoptosis in human cervical cancer cells." *Molecular Carcinogenesis: Published in cooperation with the University of Texas MD Anderson Cancer Center* 40 (4):201-211.

Stefanska, J, and R Pawliczak. 2008. "Apocynin: molecular aptitudes." *Mediators of Inflammation* 2008.

Su, Haixia, Sheng Yao, Wenfeng Zhao, Minjun Li, Jia Liu, WeiJuan Shang, Hang Xie, Changqiang Ke, Meina Gao, and Kunqian Yu. 2020. "Discovery of baicalin and baicalein as novel, natural product inhibitors of SARS-CoV-2 3CL protease *in vitro*." *BioRxiv*.

Sun, Chunyan, Meixia Zhang, Xiaolei Shan, Xueying Zhou, Jiao Yang, Yanli Wang, Jesse Li-Ling, and Yihui Deng. 2010. "Inhibitory effect of cucurbitacin E on pancreatic cancer cells growth via STAT3 signaling." *Journal of Cancer Research and Clinical Oncology* 136 (4):603-610.

Sun, Wen, Li-Da Du, Shou-Bao Wang, Xiu-Ping Chen, and Guan-Hua Du. 2018. "Shikonin." In *Natural Small Molecule Drugs from Plants*, 731-736. Springer.

Sun, Xiaofang, and Daolin Tang. 2014. "HMGB1-dependent and-independent autophagy." *Autophagy* 10 (10):1873-1876.

Suzuki, Shugo, Pornsiri Pitchakarn, Shinya Sato, Tomoyuki Shirai, and Satoru Takahashi. 2013. "Apocynin, an NADPH oxidase inhibitor, suppresses progression of prostate cancer via Rac1 dephosphorylation." *Experimental and Toxicologic Pathology* 65 (7-8):1035-1041.

Szakács, Gergely, Jill K Paterson, Joseph A Ludwig, Catherine Booth-Genthe, and Michael M Gottesman. 2006. "Targeting multidrug resistance in cancer." *Nature Reviews Drug Discovery* 5 (3):219-234.

Taleghani, Akram, Seyed Ahmad Emami, and Zahra Tayarani-Najaran. 2020. "Artemisia: a promising plant for the treatment of cancer." *Bioorganic & Medicinal Chemistry* 28 (1):115180.

Tallei, Trina Ekawati, Nurdjannah Jane Niode, Rinaldi Idroes, BM Zidan, Saikat Mitra, Ismail Celik, Firzan Nainu, Duygu Ağagündüz, Talha Bin Emran, and Raffaele Capasso. 2021. "A comprehensive review of the potential use of green tea polyphenols in the management of COVID-19." *Evidence-based Complementary and Alternative Medicine* 2021.

Tang, Daolin, Rui Kang, Weimin Xiao, Huali Zhang, Michael T Lotze, Haichao Wang, and Xianzhong Xiao. 2009. "Quercetin prevents LPS-induced high-mobility group box 1 release and proinflammatory function." *American Journal of Respiratory Cell Molecular Biology* 41 (6):651-660.

Thakor, Niyatee, and Bhavyata Janathia. 2022. "Plumbagin: A potential candidate for future research and development." *Current Pharmaceutical Biotechnology*.

Tian, Linqiang, Delong Yin, YE Ren, Chen Gong, Anmin Chen, and Feng-Jin Guo. 2012. "Plumbagin induces apoptosis via the p53 pathway and generation of reactive oxygen species in human osteosarcoma cells." *Molecular Medicine Reports* 5 (1):126-132.

Travers, Andrew A. 2003. "Priming the nucleosome: a role for HMGB proteins?" *EMBO Reports* 4 (2):131-136.

VanPatten, Sonya, and Yousef Al-Abed. 2017. "High mobility group box-1 (HMGb1): current wisdom and advancement as a potential drug target: miniperspective." *Journal of Medicinal Chemistry* 61 (12):5093-5107.

Wada, Koji, Masuo Goto, Emika Ohkoshi, Kuo-Hsiung Lee, and Hiroshi Yamashita. 2022. "Bioactivity inspired C19-diterpenoid alkaloids for overcoming multidrug-resistant cancer." *Journal of Natural Medicines*:1-7.

Wahab, S., S. Annadurai, S. S. Abullais, G. Das, W. Ahmad, M. F. Ahmad, G. Kandasamy, R. Vasudevan, M. S. Ali, and M. Amir. 2021. "Glycyrrhiza glabra (Licorice): A Comprehensive Review on Its Phytochemistry, Biological Activities, Clinical Evidence and Toxicology." *Plants (Basel)* 10 (12). https://doi.org/10.3390/plants10 122751.

Wang, Bao-Qing. 2010. "Salvia miltiorrhiza: Chemical and pharmacological review of a medicinal plant." *Journal of Medicinal Plants Research* 4 (25):2813-2820.

Wang, Chao, Hao Nie, Kan Li, Yan-xiang Zhang, Fei Yang, Cheng-bin Li, Chang-fu Wang, and Quan Gong. 2012. "Curcumin inhibits HMGB1 releasing and attenuates concanavalin A-induced hepatitis in mice." *European Journal of Pharmacology* 697 (1-3):152-157.

Wang, Chun-Hong, Xiao-Feng Li, Li-Fang Jin, Yan Zhao, Geng-Jun Zhu, and Wei-Zhang Shen. 2019. "Dieckol inhibits non-small–cell lung cancer cell proliferation and migration by regulating the PI3K/AKT signaling pathway." *Journal of Biochemical and Molecular Toxicology* 33 (8):e22346.

Wang, Fangfang, Franklin Mayca Pozo, Danmei Tian, Xinran Geng, Xinsheng Yao, Youwei Zhang, and Jinshan Tang. 2020. "Shikonin inhibits cancer through P21 upregulation and apoptosis induction." *Frontiers in Pharmacology* 11:861.

Wang, Gui-Feng, Li-Ping Shi, Yu-Dan Ren, Qun-Fang Liu, Hou-Fu Liu, Ru-Jun Zhang, Zhuang Li, Feng-Hua Zhu, Pei-Lan He, and Wei Tang. 2009. "Anti-hepatitis B virus activity of chlorogenic acid, quinic acid and caffeic acid *in vivo* and *in vitro*." *Antiviral Research* 83 (2):186-190.

Wang, Haibo, Yuanyuan Ding, Jun Zhou, Xiaoli Sun, and Siwang Wang. 2009. "The *in vitro* and *in vivo* antiviral effects of salidroside from Rhodiola rosea L. against coxsackievirus B3." *Phytomedicine* 16 (2-3):146-155.

Wang, Haichao, Wei Li, Jianhua Li, Beatriz Rendon-Mitchell, Mahendar Ochani, Mala Ashok, Lihong Yang, Huan Yang, Kevin J Tracey, and Ping Wang. 2006. "The aqueous extract of a popular herbal nutrient supplement, Angelica sinensis, protects mice against lethal endotoxemia and sepsis." *The Journal of Nutrition* 136 (2):360-365.

Wang, Jia, Junying Jia, Li Song, Xue Gong, Jianping Xu, Min Yang, and Minhui Li. 2018. "Extraction, structure, and pharmacological activities of Astragalus polysaccharides." *Applied Sciences* 9 (1):122.

Wang, Lian, Jinming Chen, Bo Wang, Dingqian Wu, Hao Li, Huchen Lu, Haiwei Wu, and Ying Chai. 2014. "Protective effect of quercetin on lipopolysaccharide-induced acute lung injury in mice by inhibiting inflammatory cell influx." *Experimental Biology Medicine* 239 (12):1653-1662.

Wang, Miao, Weihua Tang, Nianjin Gong, and Peijun Liu. 2022. "Sodium Danshensu inhibits the progression of lung cancer by regulating PI3K/Akt signaling pathway." *Drug Development Research* 83 (1):88-96.

Wang, Nan, Jingyue Yang, Jianguo Lu, Qing Qiao, Tao Wu, Xilin Du, Guoqiang Bao, and Xianli He. 2014. "A polysaccharide from Salvia miltiorrhiza Bunge improves immune function in gastric cancer rats." *Carbohydrate Polymers* 111:47-55.

Wang, Shumin, and Yi Zhang. 2020. HMGB1 in inflammation and cancer. *BioMed Central*.

Wang, Xiaolan, Guangpin Chu, Zhihua Yang, Yinan Sun, Hanjing Zhou, Man Li, Jing Shi, Bo Tian, Chun Zhang, and Xianfang Meng. 2015. "Ethanol directly induced HMGB1 release through NOX2/NLRP1 inflammasome in neuronal cells." *Toxicology* 334:104-110.

Wijesinghe, WAJP, and You-Jin Jeon. 2012. "Exploiting biological activities of brown seaweed Ecklonia cava for potential industrial applications: a review." *International Journal of Food Sciences and Nutrition* 63 (2):225-235.

Wong, Ivan Yuen Fan, Yu Huang, Zhen-Dan He, Chi-Wai Lau, and Zhen-Yu Chen. 2001. "Relaxing effects of Ligstrum purpurascens extract and purified acteoside in rat aortic rings." *Planta Medica* 67 (04):317-321.

Wu, Andrew H, Li He, Wei Long, Qiuping Zhou, Shu Zhu, Ping Wang, Saijun Fan, and Haichao Wang. 2015. "Novel mechanisms of herbal therapies for inhibiting HMGB1 secretion or action." *Evidence-Based Complementary and Alternative Medicine* 2015.

Wu, Bin-Wen, Tai-Long Pan, Yann-Lii Leu, Yu-Kaung Chang, Pei-Ju Tai, Kwang-Huei Lin, and Jim-Tong Horng. 2007. "Antiviral effects of Salvia miltiorrhiza (Danshen) against enterovirus 71." *The American Journal of Chinese Medicine* 35 (01):153-168.

Wu, Xuli, Weiyi He, Haiping Zhang, Yao Li, Zhigang Liu, and Zhendan He. 2014. "Acteoside: a lipase inhibitor from the Chinese tea Ligustrum purpurascens kudingcha." *Food Chemistry* 142:306-310.

Wu, Zong-Mei, Zhen-Jiang Yu, Zhen-Qiang Cui, Lu-Yuan Peng, Hao-Ran Li, Chun-Lei Zhang, Hai-Qing Shen, Peng-Fei Yi, and Ben-Dong Fu. 2017. "*In vitro* antiviral efficacy of caffeic acid against canine distemper virus." *Microbial Pathogenesis* 110:240-244.

Wyganowska-Swiatkowska, Marzena, Michal Nohawica, Katarzyna Grocholewicz, and Gerard Nowak. 2020. "Influence of herbal medicines on HMGB1 release, SARS-CoV-2 viral attachment, acute respiratory failure, and sepsis. A literature review." *International Journal of Molecular Sciences* 21 (13):4639.

Xu, Weiming, Min Zhang, Hongwu Liu, Kun Wei, Ming He, Xiangyang Li, Deyu Hu, Song Yang, and Yuguo Zheng. 2019. "Antiviral activity of aconite alkaloids from Aconitum carmichaelii Debx." *Natural Product Research* 33 (10):1486-1490.

Xue, Jiaming, Joelle S Suarez, Michael Minaai, Shuangjing Li, Giovanni Gaudino, Harvey I Pass, Michele Carbone, and Haining Yang. 2021. "HMGB1 as a therapeutic target in disease." *Journal of Cellular Physiology* 236 (5):3406-3419.

Yamasaki, K, T Otake, H Mori, M Morimoto, N Ueba, Y Kurokawa, K Shiota, and T Yuge. 1993. "Screening test of crude drug extract on anti-HIV activity." *Yakugaku Zasshi: Journal of the Pharmaceutical Society of Japan* 113 (11):818-824.

Yan, Honghua, Kyung Hee Jung, Juyoung Kim, Marufa Rumman, Myung Sook Oh, and Soon-Sun Hong. 2018. "Artemisia capillaris extract AC68 induces apoptosis of hepatocellular carcinoma by blocking the PI3K/AKT pathway." *Biomedicine & Pharmacotherapy* 98:134-141.

Yang, Eun-Ju, Sae-Kwang Ku, Wonhwa Lee, Sangkyu Lee, Taeho Lee, Kyung-Sik Song, and Jong-Sup Bae. 2013. "Barrier protective effects of rosmarinic acid on HMGB1-induced inflammatory responses *in vitro* and *in vivo*." *Journal of Cellular Physiology* 228 (5):975-982.

Yang, Jing, YA Zheng, Xi Gou, Ke Pu, Zhaofeng Chen, Qinghong Guo, Rui Ji, Haojia Wang, Yuping Wang, and Yongning Zhou. 2020. "Prevalence of comorbidities and its effects in patients infected with SARS-CoV-2: a systematic review and meta-analysis." *International Journal of Infectious Diseases* 94:91-95.

Yang, Minghua, Lizhi Cao, Min Xie, Yan Yu, Rui Kang, Liangchun Yang, Mingyi Zhao, and Daolin Tang. 2013. "Chloroquine inhibits HMGB1 inflammatory signaling and protects mice from lethal sepsis." *Biochemical Pharmacology* 86 (3):410-418.

Yang, Naidi, and Han-Ming Shen. 2020. "Targeting the endocytic pathway and autophagy process as a novel therapeutic strategy in COVID-19." *International Journal of Biological Sciences* 16 (10):1724.

Yang, Ying, Jing Wang, Qiao Yang, Shanshan Wu, Zhenggang Yang, Haihong Zhu, Min Zheng, Weixia Liu, Wei Wu, and Jiliang He. 2014. "Shikonin inhibits the lipopolysaccharide-induced release of HMGB1 in RAW264. 7 cells via IFN and NF-κB signaling pathways." *International Immunopharmacology* 19 (1):81-87.

Yu, De-Hong, Yong-ming Bao, Chao-Liang Wei, and Li-Jia An. 2005. "Studies of chemical constituents and their antioxidant activities from Astragalus mongholicus Bunge." *Biomedical and Environmental Sciences* 18 (5):297.

Zainal, Nurhafiza, Chih-Peng Chang, Yi-Lin Cheng, Yan-Wei Wu, Robert Anderson, Shu-Wen Wan, Chia-Ling Chen, Tzong-Shiann Ho, Sazaly AbuBakar, and Yee-Shin Lin. 2017. "Resveratrol treatment reveals a novel role for HMGB1 in regulation of the type 1 interferon response in dengue virus infection." *Scientific Reports* 7 (1):1-12.

Zaki, Aya M, Dalia M El-Tanbouly, Rania M Abdelsalam, and Hala F Zaki. 2018. "Plumbagin ameliorates hepatic ischemia-reperfusion injury in rats: role of high mobility group box 1 in inflammation, oxidative stress and apoptosis." *Biomedicine Pharmacotherapy* 106:785-793.

Zhang, Guangning, Dongyuan Li, Hao Chen, Junchen Zhang, and Xingyi Jin. 2018. "Vitexin induces G2/M-phase arrest and apoptosis via Akt/mTOR signaling pathway in human glioblastoma cells." *Molecular Medicine Reports* 17 (3):4599-4604.

Zhang, Hui, Zengjun Guo, Ling Han, Xiaojuan You, and Ying Xu. 2014. "The antitumor effect and mechanism of taipeinine A, a new C19-diterpenoid alkaloid from Aconitum taipeicum, on the HepG2 human hepatocellular carcinoma cell line." *J BUON* 19 (3):705-12.

Zhang, Jinli, Qianqian Zhang, Guang Liu, and Ning Zhang. 2019. "Therapeutic potentials and mechanisms of the Chinese traditional medicine Danshensu." *European Journal of Pharmacology* 864:172710.

Zhang, Shouqiang, Hui Ren, Hanting Sun, and Songhua Cao. 2021. "Dieckol exerts anticancer activity in human osteosarcoma (MG-63) cells through the inhibition of PI3K/AKT/mTOR signaling pathway." *Saudi Journal of Biological Sciences* 28 (9):4908-4915.

Zhang, Yilin, Guantai Ni, and Hongying Yang. 2021. "Plumbagin attenuates high glucose-induced trophoblast cell apoptosis and insulin resistance via activating AKT/mTOR pathway." *Quality Assurance and Safety of Crops & Foods* 13 (3):102-108.

Zhang, Yusong, Wei Li, Shu Zhu, Arvin Jundoria, Jianhua Li, Huan Yang, Saijun Fan, Ping Wang, Kevin J Tracey, and Andrew E Sama. 2012. "Tanshinone IIA sodium sulfonate facilitates endocytic HMGB1 uptake." *Biochemical Pharmacology* 84 (11):1492-1500.

Zheng, C., L. Han, and S. Wu. 2019. "A metabolic investigation of anticancer effect of G. glabra root extract on nasopharyngeal carcinoma cell line, C666-1." *Mol Biol Rep* 46 (4):3857-3864. https://doi.org/10.1007/s11033-019-04828-1.

Zheng, Yinghao, Xinru You, Shuyu Guan, Jun Huang, Liying Wang, Jiayuan Zhang, and Jun Wu. 2019. "Poly (ferulic acid) with an anticancer effect as a drug nanocarrier for enhanced colon cancer therapy." *Advanced Functional Materials* 29 (15):1808646.

Zheng, Yun-Jiang, Bin Zhou, Zhi-Fang Song, Lei Li, Jun Wu, Ru-Yuan Zhang, and Yao-Qing Tang. 2013. "Study of Astragalus mongholicus polysaccharides on endothelial cells permeability induced by HMGB1." *Carbohydrate Polymers* 92 (1):934-941.

Zhou, Guohong, Liying Tang, Xidan Zhou, Ting Wang, Zhenzhen Kou, and Zhuju Wang. 2015. "A review on phytochemistry and pharmacological activities of the processed lateral root of Aconitum carmichaelii Debeaux." *Journal of Ethnopharmacology* 160:173-193.

Zhou, Yu-Wen, Yao Xie, Lian-Sha Tang, Dan Pu, Ya-Juan Zhu, Ji-Yan Liu, and Xue-Lei Ma. 2021. "Therapeutic targets and interventional strategies in COVID-19: mechanisms and clinical studies." *Signal Transduction and Targeted Therapy* 6 (1):1-25.

Zhu, Shu, Wei Li, Jianhua Li, Arvin Jundoria, Andrew E Sama, and Haichao Wang. 2012. "It is not just folklore: the aqueous extract of mung bean coat is protective against sepsis." *Evidence-Based Complementary Alternative Medicine* 2012.

Chapter 2

The Role of Ferroptotic Cell Death-Related HMGB1 in Inflammation Crosstalk with Cancer and Myocardial Ischemia/Reperfusion Injury

Salih Tunc Kaya[1], PhD, Celal Guven[2], PhD and Eylem Taskin[3,*], PhD

[1]Düzce University, Faculty of Arts and Science, Department of Biology, Düzce, Turkey
[2]Adiyaman University, Faculty of Medicine, Department of Biophysics, Adiyaman, Turkey
[3]Adiyaman University, Faculty of Medicine, Department of Physiology, Adiyaman, Turkey

Abstract

Inflammation is one of the critical defense pathways against pathogen and pathologic conditions, including cancer and myocardial ischemia/ reperfusion (MIR) injury. Therefore, inflammation has positive and negative impacts on tissues and organs. There are two types of inflammation, named as acute and chronic inflammation. The inflammation is also involved in the development of cancer and myocardial infarction. Damage-associated molecular patterns (DAMPs), such as High Mobility Group Box-1 (HMGB1), heat shock proteins (HSPs), ATP, and S100 proteins, are conservative danger molecules released from damaged or dying cells to trigger an immune response. However, they have essential roles in pathological inflammation to

* Corresponding Author's Email: eylemtaskin@yahoo.com.

In: HMGB1
Editors: Eylem Taskin, Celal Guven and Salih Tunc Kaya
ISBN: 979-8-88697-408-9

participate in self-defense in physiological and pathological circumstances. Recent evidence has emerged that inflammation may be a significant factor in the development and/or progression of cancer and myocardial infarction. Herein, we review current research on the role of HMGB1 in cancer and myocardial ischemia/reperfusion injury, focusing on inflammation. Furthermore, we have discussed the role of the posttranslational modification of HMGB1 in both illnesses. In addition, we focus on a new cell death pathway named ferroptosis, which may be involved in the development of both cancer and MIR through inflammation by HMGB1.

Keywords: cancer, ferroptosis, myocardial ischemia/reperfusion injury, inflammation, high mobility group box-1, posttranslational modification

Introduction

Cancer and myocardial ischemia/reperfusion (MIR) injury are still leading causes of morbidity and mortality worldwide, despite the existence of treatments. The link between cancer and MIR injury has received increased attention in recent years. In both cases, inflammation is a major contributor to their pathophysiology (Taruya et al. 2021; Kondapalli, Moslehi, and Bonaca 2020). Because of inflammation, physiological and pathophysiological changes can be observed. Several factors such as infection, tissue injury, and stress can result in inflammation. Thus, inflammation can be defined as an adaptive response to such changes in cells to maintain homeostasis (Medzhitov 2008). In cancer, inflammation is an important contributor to the development and progression of the disease (Mariani et al. 2022). Likewise, inflammation plays a significant role in the pathogenesis of MIR injury (Ong et al. 2018). Therefore, it is important to clarify the role of inflammation in the pathogenesis of cancer and MIR injury and to determine whether there is a link between them given the fact that inflammation is a common point to both, which is important to develop a deeper understanding of the relationships and the mechanisms underlying these pathologies.

There are a large number of published studies that imply the role of High Mobility Group Box-1 (HMGB 1) in inflammation (Lee et al. 2014; Tsung, Tohme, and Billiar 2014; Yu et al. 2015; Ye et al. 2019). HMGB1 is widely expressed by many cells, being an important transcription factor or transcription enhancer within the cell. It has been involved in several functions such as nucleosome shifting, DNA repair, recombination, telomere homeostasis, autophagy, and viral sensing (Luo et al. 2013). Besides, HMGB1 is released from stressed cells, and it serves as an alarm signal, cytokine, or growth factor to activate the immune response. HMGB1 is mainly found in the nucleus; however, it should be mentioned that it is also present in the cytoplasm and extracellular environment during cell activation and injury. Damaged cells and multiple cell types such as macrophages, endothelial cells, fibroblast, and cancer cells can actively secrete HMGB1 into the extracellular environment. On the other hand, HMGB1 can be passively released from necrotic cells under stress. There are several important roles of HMGB1 in cells. For example, HMGB1 can regulate ribosome number and activities. In addition, it plays a critical role in the regulation of autophagy and apoptosis in the cytoplasm. On the other side, extracellular HMGB1 also plays an important role in the differentiation and development of cells via activating immune responses. For example, it can become active in the secretion of inflammatory cytokines through Toll-Like Receptors 2/4 (TLR2/4) and Nuclear Factor kappa B (NF-κB) signaling (Lu et al. 2019).

Ferroptosis (FER) is a newly identified cell death pathway involved in the production of high lipid reactive oxygen species and iron accumulation (Yu et al. 2015). Cell death is generally expected to be different from other cell death pathways (Zhao, Huang, and Peng 2021). It has been suggested that FER is associated with ischemia/reperfusion, stroke, neurodegenerative disease, kidney disease, etc. (Fang et al. 2019; Li, Feng, et al. 2019; Ren et al. 2020). HMGB1 is released from ferroptotic cells (Efimova et al. 2020; Wen et al. 2019).

Overall, in this section, we discuss the role of HMGB1 in inflammation-associated cancer and MIR injury through ferroptosis. In addition, the interaction between them via HMGB1 will be focused on.

Inflammation

Inflammation is generally defined as a biological response to fight pathogens or heal damaged cells or injury. This response is a defense mechanism to protect the living organism. There has also been growing recognition of the vital link between inflammation and several diseases. In other words, inflammation plays a critical role in the mechanism underlying several diseases, such as cardiovascular diseases, cancer, depression, and diabetes (Libby 2007; Furman et al. 2019; Kronsten et al. 2022). In the inflammation, oxidative stress, tissue damage, and extracellular matrix remodeling underly physiological and pathophysiological changes in cells and/or organs (Libby 2007). Inflammation is one of the protective mechanisms against micro-organisms. However, the prolongation of inflammation could have a devastating effect on tissue, which causes illness, including cancer, cardiovascular diseases, diabetes, Alzheimer's disease, pulmonary diseases, and autoimmune diseases (Singh et al. 2019).

Inflammation can be divided into two types: acute inflammation and chronic or systemic inflammation. Acute and systemic inflammation has been well-reviewed in prior articles (Medzhitov 2008; Furman et al. 2019). Briefly, acute inflammation, on the one hand, is triggered by infection and cellular stress, which results in healing and/or tissue repairing. On the other hand, systemic inflammation was initiated by damage-associated molecular patterns (DAMPs), which bring about disruption in metabolism and damage to tissues. For example, chronic inflammation can induce plasticity of endothelial cells, which underpin the progression of fibrosis and cancer (Ciszewski et al. 2021). Therefore, we can say that inflammation has pros and cons for healthy.

The things that cause inflammation are so complex that it is somewhat difficult to classify. A myriad of causes of inflammation has been described. There are also several biomarkers of inflammation. Transcription factors, including NF-κB, nuclear factor of activated T cells (NFAT) and the activator protein 1 (AP-1), and cytokines such as tumor necrosis factor-α (TNF-α), interleukin-1 (IL-1), IL-6, IL-8 as well as enzymes or proteins cyclooxygenase-2 (COX-2), 5-lipoxygenase (5-LOX), matrix metallo-proteinases (MMP) are examples of biomarkers of inflammation (Roe 2021).

As mentioned before, inflammation is closely associated with many diseases. Among these, cancer and MIR injury have received great attention due to their prevalence and high mortality. A myriad of underlying mechanisms has been suggested in the pathophysiology of cancer and MIR injury, but among these, inflammation is the most common factor in both pathophysiology.

Next in the chapter, we will focus on the relation of inflammation with HMGB1 in both cancer and MIR injury and are getting deeper into cellular mechanisms. Little is known in the literature on the question of whether there is a link between inflammation-focused cancer and MIR injury. In addition, we have therefore discussed the role of ferroptotic cell death pathways in the aforementioned relationship.

Cancer and Inflammation

Cancer is a non-contagious disease, which has still led to higher mortality rates even though successful progress and improvement have been achieved in early diagnosis or screening, treatment, and prevention methods. The total number of cancer-related deaths worldwide in 2020 was estimated to be around 10 million (Ferlay et al. 2021). Moreover, it is predicted that the number of cancer cases will reach approximately 29 million in 2040 (Sung et al. 2021).

The role of inflammation in cancer has received increased attention. There is a growing body of literature that state whether there is a bridge between cancer and inflammation (Murata 2018; Greten and Grivennikov 2019; Kay et al. 2019; Singh et al. 2019; Piotrowski, Kulcenty, and Suchorska 2020; Song et al. 2020; Zhao, Wu, et al. 2021). Understanding the complexity of inflammation in cancer is vital to clarify the close relationship between them and to get more benefits from cancer treatment.

The first evidence that there may be a link between cancer and inflammation dates back to the 19th century. The existence of leukocytes in neoplastic tissues put forward the idea that there might be a relationship between cancer and inflammation (Balkwill and Mantovani 2001). As in many different types of diseases, inflammation is a common condition that has a significant impact on cancer. Also, inflammation is one of the most important contributors to cancer transformation, proliferation, survival, and metastasis as well as response to cure (Murata 2018; Zhao, Wu, et al. 2021). In other words, prolongation of inflammation adversely affects the process of cancer initiation, invitation, metastasis, and therapy.

As mentioned earlier, inflammation can be classified into two types: acute inflammation and chronic inflammation. Based on the types of inflammation, the fate of the cancer cell can show differences. Acute inflammation can induce apoptosis of cancer cells. On the other hand, chronic inflammation not only leads to a decline in the vulnerability of the cancer cell to chemotherapeutic agents or treatment but also evokes the progression of the cancer cell (Zhao, Wu, et al. 2021). Two pathways that underpin the relationship between inflammation and cancer have been classified. The first one is called extrinsic pathways or extrinsic tumor inflammation. The other is named intrinsic pathways or intrinsic tumor inflammation (Singh et al. 2019; Zhao, Wu, et al. 2021). The extrinsic pathways, connected to environmental factors, have been evoked by stress, infection, smoking, ultraviolet radiation, obesity, aflatoxin, and so on. In turn, it can result in the facilitation of tumor development or induce carcinogenesis under conditions of inflammation. For example, obesity may underlie the mechanism(s) for cancer initiation and progression of several cancer types such as breast, colorectal, liver, and pancreatic cancer via rising inflammation. Obesity evokes badly metabolic and endoplasmic reticulum stress, which in turn causes a surge in oxidative stress, NF-κB, and JNK (Kolb, Sutterwala, and Zhang 2016). On the other hand, the intrinsic pathways related to genetic factors or mutations initiating cancer result in a surge in the inflammatory response, which in turn triggers further inflammation and neoplasia. In other words, cancer and inflammation give rise to instability in the genetic material of cells.

Among the factors that cause cancer, 20-25% are thought to be associated with infection and inflammation (Murata 2018; Tewari et al. 2018). In 2018, scientists reported that nearly 2.2 million people faced cancer due to infection induced by mainly *Helicobacter pylori*, human papillomavirus hepatitis B virus, and hepatitis C virus (de Martel et al. 2020). Among all cancer-related death, 15-20% was stated likely to be associated with infection and inflammation progression (Del Prete et al. 2011; Piotrowski, Kulcenty, and Suchorska 2020).

Several regulators in cellular signaling pathways involved in inflammation have been well-reviewed in a current review article (Zhao, Wu, et al. 2021). Briefly, these include NF-kB, TLR pathways, mitogen-activated protein kinase (MAPK), cytokines such as IL, interferon (IFN), TNF-α, and so on. Therefore, in the next part of the chapter, we will focus on HMGB1-related pathways in inflammation taking part in cancer initiation, progression, and metastasis.

Phagocytic cells, mainly leukocytes and also epithelial cells (Murata 2018) can produce reactive oxygen species (ROS) and reactive nitrogen species (RNS) to fight infection (Singh et al. 2019). Macrophages have been stated to mainly dominate the cells around the chronic inflammation environment (Singh et al. 2019). However, these species give rise to damaged DNA in the proliferating cell (Coussens and Werb 2002). In addition, ROS and RNS might have a destructive effect based on the production of the mutagenic agent. For example, peroxynitrite could integrate with DNA, leading to a mutation in the proliferation of stromal and epithelial cells. Also, the production of TNF-α and macrophage migration inhibitory factor from macrophages and T-lymphocytes promotes DNA damage (Singh et al. 2019).

Myocardial Ischemia/Reperfusion Injury and Inflammation

MIR injury continues to remain one of the leading causes of mortality worldwide. MIR injury can be defined as two-sided damage. Myocardial tissues are firstly damaged by ischemia which is called the cessation of blood flow or dwindling in blood flow to myocardial tissues; then reperfusion, known as reflow of blood, rises myocardial damage via inducing the process of local inflammation and alleviating the production of ROS in the tissue, which then induced secondary injury. In addition to acute ischemia and reperfusion injury, long-term ischemia can result in apoptosis, necrosis, and autophagy (Wu, Liu, and Wang 2021). The general mechanisms underlying ischemia and reperfusion injury have been well-reviewed in a recent study (Wu et al. 2018). In the present part of the chapter, the crosstalk between inflammation and MIR injury is discussed with a focus on current literature.

As defined earlier, myocardial ischemia is defined as a serious condition observed because of decreased or cessation of coronary blood flow. After myocardial ischemia, reflow of coronary blood flow is defined as myocardial reperfusion, which further induces damage to myocardial cells. Reperfusion is the only way to rescue myocardial cells from infarction. However, it should be noted that the timing of reperfusion is crucial (Heusch 2020). As in cancer, a strong relationship between inflammation and MIR injury is present. Inflammation worsens cellular injury after MIR injury, especially following myocardial infarction (Frangogiannis 2012).

To date, inflammation is thought to influence MIR injury, which has been explored in several studies (Lucchesi 2001; Ong et al. 2018; Ardalan, Lloyd-Jones and Schanberg 2022). It has been proposed that focusing on inflammatory processes and the cellular response to inflammation has important insight into improving myocardial infarct size and left ventricular functions after MIR injury, which highlights the role of inflammation as a potential therapeutic target (Ong et al. 2018). Inflammation is one of the major contributors to reversible and irreversible alterations in myocardial cells in response to ischemia and reperfusion (Lucchesi 2001). In addition, it should be noted that following ischemia and reperfusion injury, an individual organ negatively affects distant organs by leading to the production and release of proinflammatory factors (Naito et al. 2020).

It has been known that chronic or systemic inflammation increases vulnerability to cardiovascular diseases (Ardalan, Lloyd-Jones, and Schanberg 2022), particularly acute myocardial infarction. Furthermore, systemic inflammation can also result in increased reperfusion injury, indicated by the fact that following MIR injury TNF-αcould raise local inflammation in the myocardial cell, thereby greatly amplifying reperfusion-induced damage (Blancke et al. 2005). In accordance, in a recent clinical study, it has been highlighted that larger infarct size and lowers left ventricular ejection fraction are related to higher levels of IL-6 sustained for four months, suggesting the detrimental role of prolonged inflammation in MIR injury (Tollefsen et al. 2021). Also, another significant study has indicated that following MIR injury, the levels of IL-1β, IL-6, IL-8, and TNF-α become higher (Ou et al. 2022). A recent study has stated that the extracellular vesicle released by MIR injury can be mediators of local and systemic inflammation, indicated by the transport of pro-inflammatory miR-155-5p to macrophages, which induces inflammatory response via JAK2/STAT1 pathways (Ge et al. 2021).

A Brief Overview of HMGB1

In the family of high mobility group proteins, three superfamilies are identified, which are HMGB, HMGN, and HMGA. Among them, HMGBs are well-known proteins in which four protein groups are present: HMGB1, HMGB2, HMGB3, and HMGB4 (Kang et al. 2014). In all cells, HMGB1 is the most abundant and has several functions in many biological processes, including cellular stress coordination, autophagy, apoptosis, etc. Importantly, HMGB1 can serve as a target in several diseases such as infection, inflammation, cancer, and so on (Kang et al. 2014). HMGB1 is found in the nucleus under normal physiological conditions, whereas it can be secreted into extracellular fluids under pathophysiological or stress conditions (Kwak et al. 2021; Yu, Tang, and Kang 2015). It has been known that HMGB1 can go between the nucleus and the cytoplasm. In nuclei, HMGB1 accumulates to bind chromatin. In addition, HMGB1 is also called a mobile protein in the nucleus because it can pass through organelles in the cytosol (Kang et al. 2014). It has been described that HMGB1 can be carried into the extracellular environment in two ways. The first one is that it is actively secreted from immunocompetent cells. The second one is that it is passively secreted from apoptotic cells. HMGB1 informs cells about danger via binding the receptor for advanced glycation end products (RAGE) and Toll-like receptors (TLRs) (Wang and Zhang 2020b). Due to its DNA binding and bending activities, nuclear HMGB1 acts as a DNA chaperone. That is why it regulates several key DNA events. In addition, HMGB1, present on the cell surface, functions as a regulator protein in the outgrowth of neurite, in the activation of platelet, cell differentiation as well as maturation and adhesion of erythroid. As mentioned before, immune cells can actively secrete HMGB1, and injured cells or dead cells can passively release HMGB1. From inflammation to tissue regeneration, HMGB1 has critical roles in several physiological and pathophysiological processes. HMGB1 also has several important extra-cellular activities and functions as DAMPs.

DAMP is a conservative molecule that plays a crucial role in cells. DAMPs, including HMGB1, HSPs, ATP, etc., are produced and released by damaged and/or dying cells to trigger tissue repair, regeneration, or cell death pathways Gong et al. 2020; Preissner, Fischer, and Deindl 2020; Richard et al. 2017). HMGB1 is a non-histone protein (Manivannan et al. 2021) and plays a role in immune system induction, giving an alarm signal due to its posttranslational modification because it is sensitive to cellular redox conditions due to containing three cysteine located in Cyt23., 45. and 105. positions. The posttranslational modification could alter its binding receptor. There are three posttranslational modification mechanisms of HMGB1 depending on the redox state of the cell. The first is all thiol HMGB1, which is released when all cysteine is reduced. In this form, it binds to RAGE to produce chemotactic function. The second is the disulfide HMGB1, which occurs when produced by the disulfide bridge between Cyt23. and 45. Cysteine. The receptors of disulfide HMGB1 are TLRs such as TLR4. As a result of binding to TLR4, the inflammatory process is triggered. The third is the sulfonyl HMGB1, produced when all three cysteines are in the hyperoxidized sulfonic acid state and exhibit no chemotactic or proinflammatory function (Agalave et al. 2021; Kwak et al. 2021).

The Functional Role of HMGB1 in MIR Injury and Inflammation

Myocardial ischemia/reperfusion is one of the fatal diseases that cause damage to heart tissue because of an interruption of coronary flow (Oozawa et al. 2008). The mechanism by which MIR injury induces inflammation remains unclear. However, it is noted that ischemia causes the release of endogenous danger signals to activate inflammation. Inflammation also induces necrosis of heart cells. One of the danger signals is HMGB1, which is involved in tissue repair and fighting microorganisms, including inflammation (Andrassy et al. 2008). HMGB1 is secreted from macrophages, monocytes, and necrotic cells into the extracellular fluid to initiate inflammation due to its cytokine-like functions (Gong et al. 2014). Moreover, cardiac mast cells participate in MIR injury and/or repair mechanism, probably because they are located around microvasculature beds, having a role in secretory mediators, such as chymase, histamine, and heparin (Xiong et al. 2020). We know that HMGB1 slicing protects the heart against MIR injury and cardiomyopathies; however, HMGB1 injection is detrimental to the heart by ameliorating contractility, apoptosis, and fibroblast activities (Raucci et al. 2019). TLRs, which are important proteins, play an essential role in the inflammatory response. HMGB1 released from tissue in response to stress factors is an endogenous ligand of TLRs. For example, the innate immune response is activated in the cardiac cell in response to ischemic insult, and then TLRs- mediated cellular signaling pathways are induced to produce chemokines and cytokines in myocardial cells (Liu, Wang, and Li 2016). In the inflammatory response, HMGB1 mainly interacts with TLR2, TLR4, and TLR9, which then induces NK-kB activation. In this way, cytokines and chemokines are produced for the inflammatory and immune responses (Kang et al. 2014). A recent study indicates that inhibition of the HMBG1/TLR4 signaling pathway can be protective against MIR injury via decreasing inflammation and apoptosis (Ding et al. 2021). In an experimental model of MIR damage, serum IL-6, TNF-α, IL-10, and HMGB1 were higher in rats, and higher expression of TLR4 was observed (Guo et al. 2020). In addition, it has been indicated that hypoxia/reoxygenation increases the levels of TNF-α, IL-1β, IL-6, and the expression of HMBG1 in cardiac cells (Chen et al. 2019). Similarly, the plasma level of IL-6, TNF-α, HMGB1, and expression of HMGB1, TLR4, and NF-κB was increased in MIR injury (Xia et al. 2018). In Kawasaki disease, HMGB1 release can induce abnormality in the coronary artery (Ueno et al. 2021).

As mentioned before, MIR injury is one of the pathophysiological conditions which is a serious life-threatening dysfunction (Sanchez-Hernandez et al. 2020). Reperfusion following ischemia induces the release of HMGB1 from tissues within one hour, and its level rises continuously for twenty-four hours (Oozawa et al. 2008). A positive correlation between HMGB1 and MIR injury is present (Beom et al. 2021). Therefore, several studies have reported that HMGB1 is associated with MIR injury (Raucci et al. 2019). However, it should be mentioned that oxidative posttranslational modifications of HMGB1 play different roles in the correlation. It was reported that disulfide HMGB1 gave rise to deactivating microglia resulting in attenuation of pain-like behavior in mice. In addition, there is no information on how long the posttranslational modification of HMGB1 is retained in the tissue. However, its posttranslational modification changes over time. Therefore, posttranslational modification has been reported to diversify binding receptor types, such as TLR2, TLR4, and RAGE (Agalave et al. 2021). After myocardial infarction, it was reported that injection of HMGB1 into the heart improved the regeneration capacity and functions (Limana et al. 2005). The therapeutic effect of HMGB1 may be related to its angiogenic effects in MIR injury (Takahashi et al. 2008) via enhancing VEGF or HIF-1α protein expression by activating the PI3K/Akt signaling pathway (Zhou et al. 2019; Yao et al. 2016). The effect of HMGB1 in attenuating tissue injury to maintain homeostasis is likely due to such highly balanced, double-edged effects (Takahashi et al. 2008). The fully reduced form of HMGB1 acts as a damage signal to activate chemotactic cells, such as monocytes and other leukocytes (Xue et al. 2020). When a heterocomplex formed between all-thiol HMGB1 and C-X-C motif chemokine 12 (CXCL12), cell migration is triggered as well as inflammatory cell recruitment by binding to the C-X-C chemokine receptor type four (CXCR4) (Yu, Tang, and Kang 2015; Xue et al. 2020; Pellegrini et al. 2019). It is noticed that the most abundant oxidative post-transitional modification of HMGB1 is the fully reduced form (Pellegrini et al. 2019). Sulfonyl HMGB1 is reported to reduce inflammation, acting as anti-inflammatory effects. Further oxidation of all cysteines to sulfonates is named sulfonyl HMGB1, which has neither chemokine nor cytokine-like activity (Pellegrini et al. 2019). In contrast, HMGB1 is reversibly transformed to disulfide, and all thiol form under the redox state conditions of the microenvironment, irreversibly transformed to sulfonyl form.

In the young, it is reported that inflammatory cardiomyopathy is predominant. In a study, it has been found that patients with myocarditis have a high level of HMGB1 when compared to healthy controls (Bangert et al. 2016). Therefore, HMGB1 may participate in inflammatory stimulation in MIR injury (Wu, Liu, and Wang 2021). Inflammation in MIR injury has been involved in neutrophils and vascular endothelium. After reperfusion, oxygen supply may trigger NF-κB, resulting in the activation of pro-inflammatory cytokines and chemokine genes, such as TNF-α, IL-1β, and IL-6 (Wu, Liu, and Wang 2021). NF-κB activation via RAGE-dependent manner has played an important role in some chronic inflammatory diseases. Moreover, RAGE and TLR2 might interact with each other through MyD88, a typical TLR adaptor protein. Elevation of immune effects of HMGB1 during necrosis is observed through RAGE for HMGB1-DNA interaction and TLR2 for HMGB1-nucleosome interaction as well (Bangert et al. 2016). In addition, the other receptor of HMGB1 might involve in the activation of NF-κB. Therefore, it is reported that the HMGB1/TLR2/TLR4 axis activates Jun N-terminal kinase (JNK) and NF-κB to initiate cardiac apoptosis during MIR injury when HMGB1 alone or in cooperation with TNF- α (Raucci et al. 2019). The importance of inflammatory cytokines is that they destroy the contractile function of the heart, leading to cardiac hypertrophy, fibrosis, and remodeling (Wu, Liu, and Wang 2021).

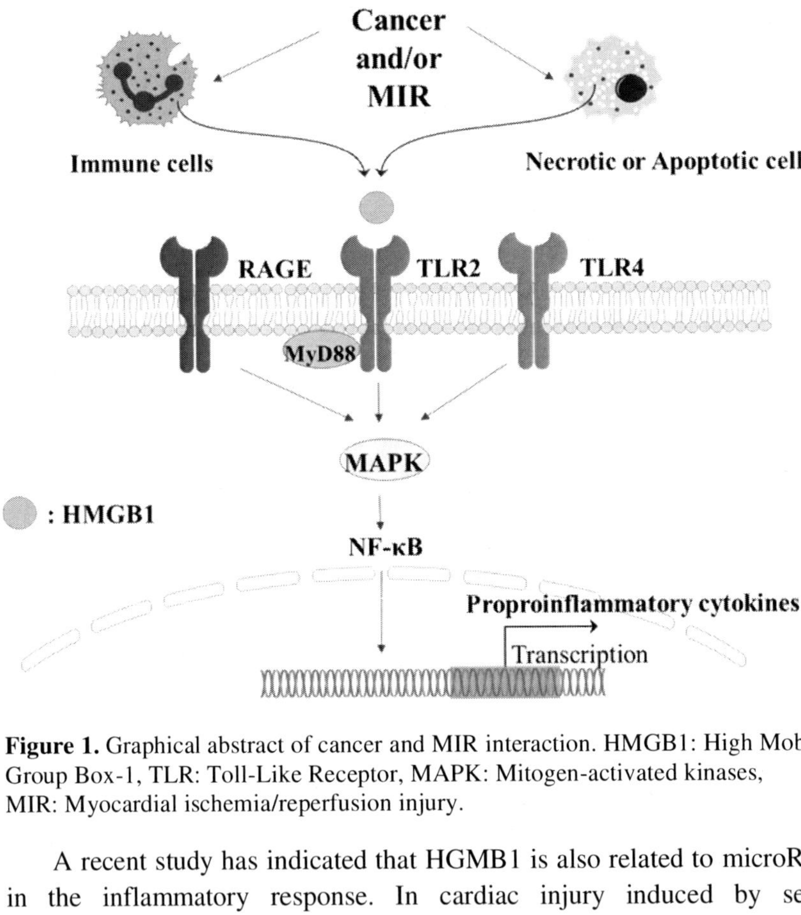

Figure 1. Graphical abstract of cancer and MIR interaction. HMGB1: High Mobility Group Box-1, TLR: Toll-Like Receptor, MAPK: Mitogen-activated kinases, MIR: Myocardial ischemia/reperfusion injury.

A recent study has indicated that HGMB1 is also related to microRNAs in the inflammatory response. In cardiac injury induced by sepsis, overexpression of miR-181b results in a decrease in HMGB1 protein expression, thereby improving inflammatory factors and apoptosis (Ling et al. 2021). Another recent study also highlights the role of HMGB1 in MIR as an inflammatory mediator. In the aforementioned study, inhibition of HMGB1/TLR4/NF-κB signaling pathways leads to improvement of autophagy and apoptosis in myocardial injury (Ding et al. 2021).

The Functional Role of HMGB1 in Cancer and Inflammation

HMGB1 has been reported to involve in the development of many cancers such as breast, cervical, liver, colorectal, lung, prostate, and ovarian cancers (Tripathi, Shrinet, and Kumar 2019). The functional role of HMGB1, especially extracellular HMGB1 in lung cancer, has been well-reviewed in a current study (Anderson et al. 2021). Based on the knowledge, it is reported that there is a link between HMGB1 and cancer. HMGB1 could be released from some cells under stress or dead conditions to alert them. Furthermore, cancer or necrotic cells exposed to chemotherapy also secrete HMGB1 due to its pro-inflammatory properties (Tang et al. 2010). It is stated that HMGB1 has a crucial role in the regulation of both innate and adaptive immune responses against infectious and non-infectious inflammatory diseases such as ischemia/reperfusion injury (IRI) (Wang and Zhang 2020a).

It is now well known that HMGB1, one of the redox-sensitive proteins, can alter the immunological potential of the cell due to its redox state. (Tang et al. 2010). HMGB1's posttranslational modification plays an essential role in its function and its receptors. For example, HMGB1 can change its disulfide structure, named disulfide-HMGB1 (DHMGB1), and bind to TLR4 to produce pro-inflammation cytokines. That is why DHMGB1 can exacerbate inflammation (Wang and Zhang 2020a). HMGB1 secreted from apoptotic cells stimulates macrophages. However, HMGB1 released from cancer cells is reported to have paradoxical dual effects. Both effects cause angiogenesis and stimulate anti-neoplastic T-cell responses (Tang et al. 2010).

A recent study reveals that extracellular HMGB1 plays a vital role in proliferation and metastasis (Zhu et al. 2021). In this study, it has been identified that HMGB1 released from apoptotic cells in the extracellular space in response to radiotherapy induces proliferation and metastasis in pancreatic cancer cells via activating phospho-GSK-3β (p-GSK-3β). On the other hand, in another study, an important observation has been obtained, indicating that intracellular HMGB1 can suppress tumorigenesis in the pancreas by inhibiting the inflammatory response and providing chromosomal stability (Kang et al. 2017).

HMGB1 is also an important molecule for apoptosis and autophagy, where HMGB1 is released into the extracellular space, inducing autophagy and augmenting apoptosis via RAGE in acute leukemia cells, evidenced by the fact that inhibition of RAGE results in opposite effects (Lai et al. 2021). Interaction of HMGB1 with RAGE reduced the apoptotic process via regulating the anti-apoptotic protein Bcl-2, whereas it promoted autophagy by activating ERK1/2 and attenuating phosphorylation of the mammalian target of rapamycin (mTOR) (Lai et al. 2021). In colon cancer, HMGB1 might be responsible for inflammatory process. Inhibition of HMGB1 resulted in the recovery of glutathione peroxidase 4, indicating the anti-inflammatory response. In the mentioned study, it was demonstrated that HMGB1 slicing was effective in the modulation of inflammation and oxidative stress via *NF-κB* (Yang et al. 2020). In an important study, it was shown that extracellular HMGB1 was increased in melanoma treated with Salmonella, resulting in the prevention of metastasis, which indicates that HMGB1 is critically important in cancer and inflammation (Pangilinan, Wu, and Lee 2021). In addition, it has been indicated that the HMGB1/RAGE axis can induce malignant transformation of normal cervical epithelial cells in inflammation induced by lipopolysaccharide (You et al. 2021). In another study, inhibition of HMGB1 was shown to be effective in reducing the number of tumors. In addition, it was shown that cell proliferation in cancer cells was increased via the ERK1/2 pathway when HMGB1 was released (Tan et al. 2020). In gastric cancer, HMGB1 induces malignant progression via activating PI3K/Akt/HIF-1α pathways; this is evidenced by the fact that treatment with Vitexin inhibits these pathways (Zhou et al. 2021).

HMGB1 has a role in inflammation in colon cancer via the *NF-κB* pathway. Moreover, posttranslational modification of HMGB1 via acetylation leads to oxidative stress and inflammation through *NF-κB* (Yang et al. 2020). In a recent study, it has been reported that HMGB1 is a good candidate for the cancer marker, especially serum hyper-acetylated HMGB1. RAGE has a role in cancer metastasis after HMGB1 binding (Sun et al. 2020). HMGB1 is also a ligand for TLRs and CD24 (Wu et al. 2013). HMGB1 is sensitive to cell redox situations. It has several modifications, such as acetylation, oxidation (Venereau et al. 2016), methylation, and phosphorylation (Wu et al. 2013). It has been reported that HMGB1 in renal cancer has been modified to mono-methylated at Lys 112. HMGB1 can act as a double-edged sword, such as the development of cancer through TLR2 or RAGE via ERK1/2 activation and triggering of anticancer pathways like T-cell responses via TLR4. Some posttranslational modifications of HMGB1, such as oxidation, hyper-acetylation, and phosphorylation, contribute to its cytoplasmic distribution in inflammatory and cancer cells. Acetylated and phosphorylated HMGB1 is reported to bind to TLR2 in the extracellular fluid, resulting in the development of brain cancer. Moreover, inhibition of the HMGB1/RAGE pathway has been suggested to suppress cancer progression due to its role in the inhibition of leukocyte recruitment through ERK1/2 activation (Wu et al. 2013; Diener et al. 2013).

Is There a Link between Cancer and Myocardial Ischemia and Reperfusion Injury?

To find the answer(s) to the question: is there a bridge linking cancer and myocardial ischemia/reperfusion injury? It is likely to come from an understanding of the cellular signaling pathways common in both diseases. And it is also not completely known whether a molecule released following MIR injury may take part in the initiation, progression, and metastasis of cancer or vice versa. Herein, it is important to focus on the cellular mechanism(s) of inflammation in cancer and MIR since inflammation is common to both. To the best of our knowledge, there is no study investigating the role of HMGB1 in the relationship between cancer and myocardial infarction in the literature up to now. Therefore, this part aims to draw attention to HMGB1.

In this sense, there are limited but valuable studies in the literature (Hasin et al. 2016; Malmborg et al. 2018; Li, Huang, et al. 2019; Velders, Hagstrom, and James 2020). The risk of cancer in patients with heart failure following myocardial infarction, markedly 1.5 years later, becomes two-fold higher than the patients without heart failure, and the percentage of facing cancer following the development of heart failure has been found to be 71%, regardless of sex and age (Hasin et al. 2016). Cancers related to the respiratory, digestive, and hematologic systems were common types in these patients. Another study has found that the risk of death and cancer, especially lung, prostate, and breast cancer, are linked to myocardial infarction, with a modest relation, and the risk becomes evident in the first year following myocardial infarction Malmborg et al. 2018). A systemic review and meta-analysis it has been aimed to investigate whether there is a relation between myocardial infarction and cancer (Li, Huang, et al. 2019). A surprising result has been obtained, such that a sex-specific difference is present in this concept, where the risk of cancer, notably lung cancer, increases following myocardial infarction in female patients but not in males. A recent cohort study has also indicated that the rate of cancer incidence has risen among the patients facing first myocardial infarction (Velders, Hagstrom, and James 2020).

It can be suggested that the relationship between cancer and myocardial infarction may stem from the that they are likely to share the same pathophysiological mechanisms, including inflammation, oxidative stress, obesity, etc. (Li, Huang, et al. 2019). However, it should not be ignored that there may be many other factors or molecules which we do not know or have not yet been clarified at the cellular level. Among them, it is strongly possible that inflammation can underly the link between cancer and myocardial infarction. As known, the immune system, especially the monocyte system and the damaged tissue repair system, has altered following myocardial infarction (Hasin et al. 2016). Due to its release from cells (macrophages, monocytes, NK cells, DCs, endothelial cells, and platelets) taking roles in immune systems, we speculate that HMGB1 is one of the most important candidates to construct a bridge to explain the relationship between cancer and myocardial infarction. Besides, HMGB1 is known to be secreted from necrotic and/or damaged cells (Yang et al. 2013). So, pharmacological inhibition of HMGB1 has been shown to reduce the severity of inflammatory illness in both sterile injury and infection. Redox modification of HMGB1 could change its activity, including inflammation as a role in DAMP.

The New Cell Death Pathways: Ferroptosis

Cell death pathways play an essential role in the normal physiological (Lei, Bai, and Sun 2019) and pathophysiological function of cells (Jones and Gores 1997; Elmore 2007; Galluzzi et al. 2018). Under stress conditions, such as extreme pH variations, cells release danger signaling molecules, e.g., HMGB1, to alert the cells or organism about a potential threat (Galluzzi et al. 2018). Cell death is divided into two classes: apoptosis and necrosis (Li et al. 2020). Ferroptosis (FER) is a newly identified cell death pathway. FER is documented to typically trigger iron accumulation and depletion of glutathione peroxidase 4 (GPx4). FER has been shown to develop cell alternations, such as cell volume shrinkage and increased mitochondrial membrane density (Yu et al. 2017). However, it should not be ruled out that FER and apoptosis may be related. For example, apoptosis could be converted to FER. Furthermore, FER may enhance the vulnerability to apoptotic cell death (Zhang et al. 2022). FER has a crucial role in regulating the growth and proliferation of some cancer, such as (Yu et al. 2017), development of some diseases, e.g., neurological diseases, acute kidney injury (Li et al. 2020), renal failure (Lei, Bai, and Sun 2019), ischemia/reperfusion (Li et al. 2020), liver fibrosis, stroke, type-I diabetes (Li et al. 2020), hematological system diseases (Lei, Bai, and Sun 2019). Iron is crucial for FER development because divalent iron triggers FER by lipid peroxidation production. Cell pathways are called Ferroptosis based on iron accumulation-dependent lipid peroxides (Yu et al. 2017; Zhou et al. 2020).

FER, the new programmed cell death, is also another way to be involved in the development of drug resistance and cancer therapy (Wu et al. 2020; Zhang et al. 2022). Enoyl coenzyme A hydratase is one of the enzymes playing a role in fatty acid β-oxidation. The enzyme decreases FER induced by nonalcoholic steatohepatitis (Liu et al. 2021). AMP-activated protein kinase participates in ferroptotic cell death by inhibition of BH4 (Duan et al. 2021). DAMPs are stated to involve in FER by initiating inflammation through RAGE to activate NF-κ (Yu et al. 2021). FER gives rise to the release of HMGB1 (Zhang et al. 2022). That is why the present chapter summarizes and reviews research on HMGB1 in the crosstalk between cancer and MIR through ferroptotic cell death.

The Role of HMGB1 in Interaction of Cancer and MIR by Ferroptosis

FER in cancer cells was first described in 2012 by Dixon et al. at the Stockwell Lab, induction by the small molecules erastin or Ras-selective lethal small molecule 3 (RSL3) (Xu et al. 2019; Chen et al. 2021; Zhao, Huang, and Peng 2021). There is little knowledge of how cancer and MIR interaction (Wen et al. 2019). The interaction might be through FER, one of the regulated cell deaths by triggering small molecules and drugs (Wen et al. 2019). FER has a role in some pathophysiological conditions, such as cancer, MIR, kidney diseases, stroke, and neurodegenerative diseases (Zhao, Huang, and Peng 2021). Moreover, only ischemia is noticed to lead to iron accumulation (Song et al. 2021). FER may be involved in sterile inflammation induced by DAMPs, e.g., HMGB1 (Li, Feng, et al. 2019; Mou et al. 2019). FER is also named as inflammatory cell death associated with the release of DAMPs, e.g., HMGB1 (Tang et al. 2021).

FER has been reported to be associated with iron metabolism via oxidative stress and lipid peroxidation. Moreover, it is suggested that HMGB1 is one of the key factors to trigger FER. For example, in one of the recent studies, it has been found when HMGB1 is blocked by antibodies or lacks advanced glycosylation end product-specific receptor (AGER), inflammation response during FER is limited. According to the results of the study, HMGB1 release plays an important role in FER triggered by autophagy. Also another important result of the study is that autophagy triggers posttranslational modification of HMGB1, such as the acetylation, in FER by inhibition of histone deacetylases (HDAC), especially HDAC1 and HDAC2. HMGB1 is a mediator of inflammation during FER by RAGE (Wen et al. 2019). Since HMGB1 is secreted by ferroptotic cells, initiation of innate immunity by binding its advanced glycosylation end-product-specific receptor (AGER/ RAGE) (Tang et al. 2021).

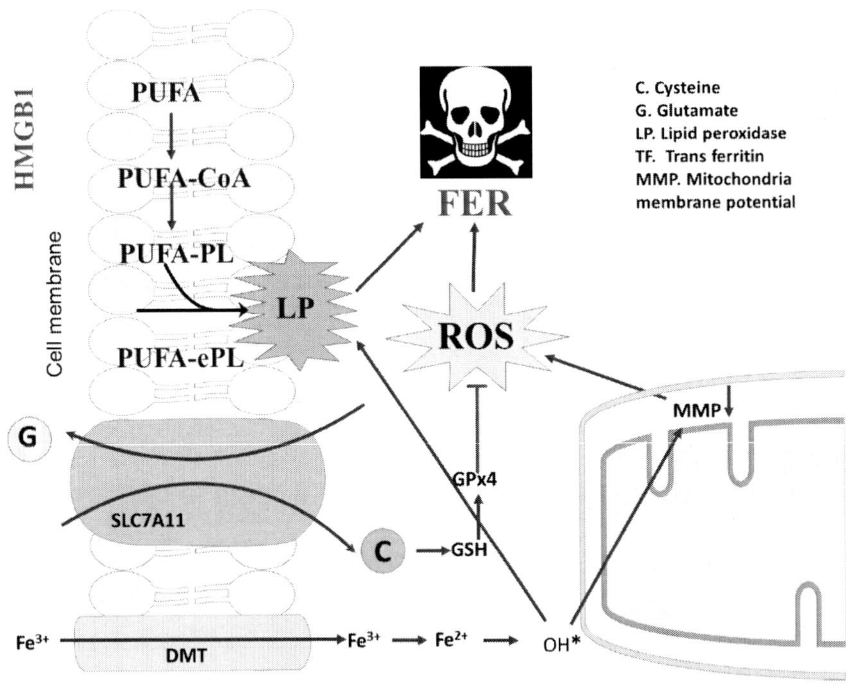

Figure 2. The ferroptotic cell death mechanism and pathways. C; Cysteine, G; Glutamate, LP; Lipid peroxidation, MMP; Mitochondria membrane potential, TF; Trans ferritin, SLC7A11. The figure was modified by Chen et al. 2021; Su et al. 2019.

FER is suggested to be based on the imbalance between antioxidant and oxidant systems. There are two main pathways to trigger FER, either endogenous or enzyme-regulated and exogenous or transporter-dependent. (Granata et al. 2022). So, the trigger of FER can be divided into two subgroups iron accumulation and alleviation of GPx4 (Chen et al. 2021). Both pathways are regulated by various metabolic compounds such as erastin, silpasdasine, sorafenib for extrinsic pathways and RSL3, ML210, ML162, FINO2, FIN54 for intrinsic pathways (Granata et al. 2022) and subcellular organelles, such as mitochondria, lysosomes, peroxisomes, Golgi apparatus, and endoplasmic reticulum (Chen et al. 2021). Oxidative stress is a key mechanism triggering FER (Wang et al. 2020). Probably that is why the low nicotinamide adenine dinucleotide phosphate (NADPH) is suggested to be a marker of FER (Tonnus and Linkermann 2016). So, plasma NADPH oxidase 1 (NOX1), long-chain acyl-CoA synthetase 4 (ACSL4), and prostaglandin-endoperoxide synthase 2 (PTGS2) are reported to be high in diabetic nephropathy patients, giving rise to trigger FER (Wu, Zhao, et al. 2021). The balance between oxidant and antioxidant is important for FER. Since free polyunsaturated fatty acids (PUFA) might be converted to PUFA-containing-phospholipid hydroperoxides (PUFA-OOH), which is meant to elevate ROS production, however, GPx4 can reverse PUFA-OOH to hydroxylated polyunsaturated fatty acid (PUFA-OH), a less toxic metabolite (Wang et al. 2020).

Cell or organelle membranes contain high levels of polyunsaturated fatty acids (PUFA), which is more sensitive to ROS injury due to easily oxidized to saturated fatty acids (SFAs) and monounsaturated fatty acid (MUFAs). Moreover, PUFA is well known for some compounds, such as arachidonic acid (AA) and adrenic acid (AdA), which are converted to proteoliposomes (PLs) on the cell membrane, especially phosphatidylethanolamine (PEs) by oxidation. That is why PEs is suggested to be a key intracellular factor to trigger FER by having AA and AdA. The PEs is a ligand for acyl-CoA synthetase long-chain family member 4 (ACSL4) and lysophosphatidylcholine acyltransferase 3 (LPCAT3) to cause lipid peroxidation by converting to PE-AA-OOH and PE-AdA-OOH. However, other enzymes, such as ACSL3 can retransform MUFAs to their acyl-CoA esters, thereby triggering FER. Lypoxgynase (LOX), an iron-containing enzyme, is the other enzyme that oxidizes membrane lipids, the production of esterified PUFAs. LOX is another iron-containing enzyme that his a role in the conversion of oxidation of membrane PLs. That is why it can trigger FER when the LOX gene is genetically or pharmacologically sliced, e.g., Erastin (Zhao, Huang, and Peng 2021) via activating VDAC, resulting in mitochondria hyperpolarization, thus leading to ROS, mitochondrial dysfunction, and cell death (Zhao, Zhou, et al. 2021). The hydro peroxidation of polyunsaturated fatty acids (PUFA-OOH) and phospholipids (PL-OOH) has been reported to decrease by GPx4 (Kagan et al. 2017) (Figure 2).

A previous study indicated that ferrostatin-1 attenuated cardiomyocytes death through FER and TLR4/Trif-dependent signaling in graft endothelial cells by declining pro-ferroptotic hydroperoxy-arachidonoyl-phosphatidylethanolamine and inhibiting neutrophil recruitment on heart transplantation (Li, Feng, et al. 2019). FER is also one of the cell deaths for cardiomyocytes on MIR, which is well known to be associated with excessive ROS production, calcium overload, and inflammation. Moreover, peroxidized phosphatidylethanolamine (PEox) is suggested to be a marker for FER in cardiomyocytes. Thus, FER can trigger endoplasmic reticulum stress-induced apoptosis by inhibiting the System Xc^{-}. In MIR injury, Erastin could trigger endoplasmic reticulum stress, leading to the initiation of FER (Zhao, Zhou, et al. 2021). It was found that the silence of HMGB1 cells significantly attenuated the expression of HSB1, known as HSP25 in mice and HSP27 in humans, but not other HSPs, such as HSP70 and HSP90. HSB1 is noticed to interact with the actin cytoskeleton, which has a role in cell shape, cellular locomotion, and the transport and morphology of intracellular vesicles and organelles, including mitochondria (Tang et al. 2011). Moreover, the effect of HMGB1 on mitochondria is mediated by HSP27 (Tang et al. 2011).

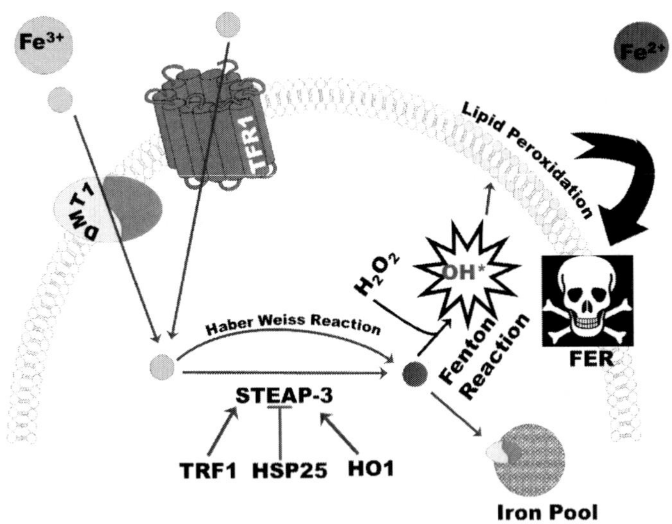

Figure 3. Iron homeostasis is related to Ferroptosis. The excess production of iron ions and ROS underly the signaling pathways of ferroptosis. DMT1. Divalent metal ion transporter, OH*. Hydroxyl radical, FER. Ferroptosis, TFR1. Transferrin receptor1, STEAP3. Metallo-reductase six-transmembrane epithelial antigen of prostate 3. The figure was modified by Wu, Zhao, et al. 2021; Zhao, Zhou, et al. 2021.

One of the previous studies reported that phosphorylated HSPB1 might be one of the potent candidates for cancer treatment due to inhibition of FER through attenuation of iron accumulation and ROS production (Sun et al. 2015). The cycling of Ferro (Fe^{2+}) -ferric (Fe^{3+}) iron plays a vital role in cell fate associated with FER (Zhang et al. 2022; Zhou et al. 2020) (Figure 1). Fe^{3+} can be transferred to the endosome by the transferrin receptor, then converted to Fe^{2+} by the Metallo-reductase six-transmembrane epithelial antigen of prostate 3 (STEAP3). Fe^{2+} is transferred to the cytosol by the divalent metal transporter-1 (DMT1). FER has been noticed to enhance DMT1. Acute MI is also reported to enhance DMT1. ROS also participates in the development of FER in MIR. Iron accumulation in MIR has been found in the scar area of the heart. (Wu, Iroegbu, et al. 2021). MIR has been reported to develop oxidative stress. That is why antioxidant therapy, i.e., glutathione, which is a key cellular antioxidant, has been shown to alleviate ROS production by MIR. FER is also triggered by ROS, but it is the lipid base of ROS that is catalyzed by 12/15-lipoxygenase (12/15-LOX). GPx4 removed lipid-ROS (Park et al. 2019) (Figure 3).

Conclusion

HMGB1 plays a role in the alert mechanism under stress and/or pathological conditions. It has several receptors. Moreover, it is sensitive to the redox status of cells. So, it can be modified, resulting in an alteration in its function and binding receptor types. In addition, HMGB1 participates in inflammation which underly the molecular mechanism of cancer and MIR injury. HMGB1 may be suggested as a marker to monitor the progress of cancer and MIR injury. FER might be a link between cancer and MIR through HMGB1. Several questions remain unanswered at present. Understanding how extracellular, nucleus and membrane HMGB1 regulate inflammation induced by FER in both cancer and MIR gives a chance to improve an effective strategy to treat cancer and cancer-associated MIR. However, further work is required to gain a better understanding of the FER mechanism underlying the association between cancer and myocardial ischemia/reperfusion injury in the aspect of HMGB1 as a DAMPs. Ferroptotic cell death-related HMGB1 and/or other DAMPs might be a good candidate strategy for the treatment of inflammation with cancer and MIR.

References

Agalave, N. M., R. Rudjito, A. B. Farinotti, P. E. Khoonsari, K. Sandor, Y. Nomura, T. A. Szabo-Pardi, C. M. Urbina, V. Palada, T. J. Price, H. Erlandsson Harris, M. D. Burton, K. Kultima, and C. I. Svensson. 2021. "Sex-dependent role of microglia in disulfide high mobility group box 1 protein-mediated mechanical hypersensitivity." *Pain* 162 (2):446-458. https://doi.org/10.1097/j.pain.0000000000002033.

Anderson, Brady, Mary Vue, Nya Gayluak, Sarah Jane Brown, Lynne T. Bemis, and Glenn E. Simmons. 2021. "Targeting HMGB1 in the Treatment of Non-Small Cell Lung Adenocarcinoma." *Onco* 1 (1):25-37.

Andrassy, M., H. C. Volz, J. C. Igwe, B. Funke, S. N. Eichberger, Z. Kaya, S. Buss, F. Autschbach, S. T. Pleger, I. K. Lukic, F. Bea, S. E. Hardt, P. M. Humpert, M. E. Bianchi, H. Mairbaurl, P. P. Nawroth, A. Remppis, H. A. Katus, and A. Bierhaus. 2008. "High-mobility group box-1 in ischemia-reperfusion injury of the heart." *Circulation* 117 (25):3216-26. https://doi.org/10.1161/CIRCULATIONAHA.108.769331.

Ardalan, K., D. M. Lloyd-Jones, and L. E. Schanberg. 2022. "Cardiovascular Health in Pediatric Rheumatologic Diseases." *Rheum Dis Clin North Am* 48 (1):157-181. https://doi.org/10.1016/j.rdc.2021.09.006.

Balkwill, F., and A. Mantovani. 2001. "Inflammation and cancer: back to Virchow?" *Lancet* 357 (9255):539-45. https://doi.org/10.1016/S0140-6736(00)04046-0.

Bangert, A., M. Andrassy, A. M. Muller, M. Bockstahler, A. Fischer, C. H. Volz, C. Leib, S. Goser, S. Korkmaz-Icoz, S. Zittrich, A. Jungmann, F. Lasitschka, G. Pfitzer, O. J. Muller, H. A. Katus, and Z. Kaya. 2016. "Critical role of RAGE and HMGB1 in inflammatory heart disease." *Proc Natl Acad Sci USA* 113 (2):E155-64. https://doi.org/10.1073/pnas.1522288113.

Beom, J. H., J. H. Kim, J. Seo, J. H. Lee, Y. E. Chung, H. S. Chung, S. P. Chung, C. H. Kim, and J. S. You. 2021. "Targeted temperature management at 33 degrees C or 36 induces equivalent myocardial protection by inhibiting HMGB1 release in myocardial ischemia/reperfusion injury." *PLoS One* 16 (1):e0246066. https://doi.org/10.1371/journal.pone.0246066.

Blancke, F., M. J. Claeys, P. Jorens, G. Vermeiren, J. Bosmans, F. L. Wuyts, and C. J. Vrints. 2005. "Systemic inflammation and reperfusion injury in patients with acute myocardial infarction." *Mediators Inflamm* 2005 (6):385-9. https://doi.org/10.1155/MI.2005.385.

Chen, J., Z. Jiang, X. Zhou, X. Sun, J. Cao, Y. Liu, and X. Wang. 2019. "Dexmedetomidine Preconditioning Protects Cardiomyocytes Against Hypoxia/Reoxygenation-Induced Necroptosis by Inhibiting HMGB1-Mediated Inflammation." *Cardiovasc Drugs Ther* 33 (1):45-54. https://doi.org/10.1007/s10557-019-06857-1.

Chen, X., R. Kang, G. Kroemer, and D. Tang. 2021. "Ferroptosis in infection, inflammation, and immunity." *J Exp Med* 218 (6). https://doi.org/10.1084/jem.20210518.

Ciszewski, W. M., M. E. Wawro, I. Sacewicz-Hofman, and K. Sobierajska. 2021. "Cytoskeleton Reorganization in EndMT-The Role in Cancer and Fibrotic Diseases." *Int J Mol Sci* 22 (21). https://doi.org/10.3390/ijms222111607.

Coussens, L. M., and Z. Werb. 2002. "Inflammation and cancer." *Nature* 420 (6917):860-7. https://doi.org/10.1038/nature01322.

de Martel, C., D. Georges, F. Bray, J. Ferlay, and G. M. Clifford. 2020. "Global burden of cancer attributable to infections in 2018: a worldwide incidence analysis." *Lancet Glob Health* 8 (2):e180-e190. https://doi.org/10.1016/S2214-109X(19)30488-7.

Del Prete, A., P. Allavena, G. Santoro, R. Fumarulo, M. M. Corsi, and A. Mantovani. 2011. "Molecular pathways in cancer-related inflammation." *Biochem Med (Zagreb)* 21 (3):264-75. https://doi.org/10.11613/bm.2011.036.

Diener, K. R., N. Al-Dasooqi, E. L. Lousberg, and J. D. Hayball. 2013. "The multifunctional alarmin HMGB1 with roles in the pathophysiology of sepsis and cancer." *Immunol Cell Biol* 91 (7):443-50. https://doi.org/10.1038/icb.2013.25.

Ding, H. S., J. Yang, J. Yang, X. Guo, Y. H. Tang, Y. Huang, Z. Chen, Z. X. Fan, and C. X. Huang. 2021. "Fluvastatin attenuated ischemia/reperfusion-induced autophagy and apoptosis in cardiomyocytes through down-regulation HMGB1/TLR4 signaling pathway." *Mol Biol Rep* 48 (5):3893-3901. https://doi.org/10.1007/s11033-021-06326-9.

Duan, J. Y., X. Lin, F. Xu, S. K. Shan, B. Guo, F. X. Li, Y. Wang, M. H. Zheng, Q. S. Xu, L. M. Lei, W. L. Ou-Yang, Y. Y. Wu, K. X. Tang, and L. Q. Yuan. 2021. "Ferroptosis and Its Potential Role in Metabolic Diseases: A Curse or Revitalization?" *Front Cell Dev Biol* 9:701788. https://doi.org/10.3389/fcell.2021.701788.

Efimova, I., E. Catanzaro, L. Van der Meeren, V. D. Turubanova, H. Hammad, T. A. Mishchenko, M. V. Vedunova, C. Fimognari, C. Bachert, F. Coppieters, S. Lefever, A. G. Skirtach, O. Krysko, and D. V. Krysko. 2020. "Vaccination with early ferroptotic cancer cells induces efficient antitumor immunity." *J Immunother Cancer* 8 (2). https://doi.org/10.1136/jitc-2020-001369.

Elmore, S. 2007. "Apoptosis: a review of programmed cell death." *Toxicol Pathol* 35 (4):495-516. https://doi.org/10.1080/01926230701320337.

Fang, X., H. Wang, D. Han, E. Xie, X. Yang, J. Wei, S. Gu, F. Gao, N. Zhu, X. Yin, Q. Cheng, P. Zhang, W. Dai, J. Chen, F. Yang, H. T. Yang, A. Linkermann, W. Gu, J. Min, and F. Wang. 2019. "Ferroptosis as a target for protection against cardiomyopathy." *Proc Natl Acad Sci USA* 116 (7):2672-2680. https://doi.org/10.1073/pnas.1821022116.

Ferlay, J., M. Colombet, I. Soerjomataram, D. M. Parkin, M. Pineros, A. Znaor, and F. Bray. 2021. "Cancer statistics for the year 2020: An overview." *Int J Cancer*. https://doi.org/10.1002/ijc.33588.

Frangogiannis, N. G. 2012. "Regulation of the inflammatory response in cardiac repair." *Circ Res* 110 (1):159-73. https://doi.org/10.1161/CIRCRESAHA.111.243162.

Furman, D., J. Campisi, E. Verdin, P. Carrera-Bastos, S. Targ, C. Franceschi, L. Ferrucci, D. W. Gilroy, A. Fasano, G. W. Miller, A. H. Miller, A. Mantovani, C. M. Weyand, N. Barzilai, J. J. Goronzy, T. A. Rando, R. B. Effros, A. Lucia, N. Kleinstreuer, and G. M. Slavich. 2019. "Chronic inflammation in the etiology of disease across the life span." *Nat Med* 25 (12):1822-1832. https://doi.org/10.1038/s41591-019-0675-0.

Galluzzi, L., I. Vitale, S. A. Aaronson, J. M. Abrams, D. Adam, P. Agostinis, E. S. Alnemri, L. Altucci, I. Amelio, D. W. Andrews, M. Annicchiarico-Petruzzelli, A. V. Antonov, E. Arama, E. H. Baehrecke, N. A. Barlev, et al. 2018. "Molecular mechanisms of cell death: recommendations of the Nomenclature Committee on Cell Death 2018." *Cell Death Differ* 25 (3):486-541. https://doi.org/10.1038/s41418-017-0012-4.

Ge, X., Q. Meng, L. Wei, J. Liu, M. Li, X. Liang, F. Lin, Y. Zhang, Y. Li, Z. Liu, H. Fan, and X. Zhou. 2021. "Myocardial ischemia-reperfusion induced cardiac extracellular vesicles harbour proinflammatory features and aggravate heart injury." *J Extracell Vesicles* 10 (4):e12072. https://doi.org/10.1002/jev2.12072.

Gong, G., L. Xiang, L. Yuan, L. Hu, W. Wu, L. Cai, L. Yin, and H. Dong. 2014. "Protective effect of glycyrrhizin, a direct HMGB1 inhibitor, on focal cerebral ischemia/reperfusion-induced inflammation, oxidative stress, and apoptosis in rats." *PLoS One* 9 (3):e89450. https://doi.org/10.1371/journal.pone.0089450.

Gong, T., L. Liu, W. Jiang, and R. Zhou. 2020. "DAMP-sensing receptors in sterile inflammation and inflammatory diseases." *Nat Rev Immunol* 20 (2):95-112. https://doi.org/10.1038/s41577-019-0215-7.

Granata, S., V. Votrico, F. Spadaccino, V. Catalano, G. S. Netti, E. Ranieri, G. Stallone, and G. Zaza. 2022. "Oxidative Stress and Ischemia/Reperfusion Injury in Kidney Transplantation: Focus on Ferroptosis, Mitophagy and New Antioxidants." *Antioxidants (Basel)* 11 (4). https://doi.org/10.3390/antiox11040769.

Greten, F. R., and S. I. Grivennikov. 2019. "Inflammation and Cancer: Triggers, Mechanisms, and Consequences." *Immunity* 51 (1):27-41. https://doi.org/10.1016/j.immuni.2019.06.025.

Guo, L. L., M. L. Guo, J. Yao, Y. Q. Weng, and X. Z. Zhang. 2020. "MicroRNA-421 improves ischemia/reperfusion injury via regulation toll-like receptor 4 pathway." *J Int Med Res* 48 (3):300060519871863. https://doi.org/10.1177/0300060519871863.

Hasin, T., Y. Gerber, S. A. Weston, R. Jiang, J. M. Killian, S. M. Manemann, J. R. Cerhan, and V. L. Roger. 2016. "Heart Failure After Myocardial Infarction Is Associated With Increased Risk of Cancer." *J Am Coll Cardiol* 68 (3):265-271. https://doi.org/10.1016/j.jacc.2016.04.053.

Heusch, G. 2020. "Myocardial ischaemia-reperfusion injury and cardioprotection in perspective." *Nat Rev Cardiol* 17 (12):773-789. https://doi.org/10.1038/s41569-020-0403-y.

Jones, B. A., and G. J. Gores. 1997. "Physiology and pathophysiology of apoptosis in epithelial cells of the liver, pancreas, and intestine." *Am J Physiol* 273 (6):G1174-88. https://doi.org/10.1152/ajpgi.1997.273.6.G1174.

Kagan, V. E., G. Mao, F. Qu, J. P. Angeli, S. Doll, C. S. Croix, H. H. Dar, B. Liu, V. A. Tyurin, V. B. Ritov, A. A. Kapralov, A. A. Amoscato, J. Jiang, T. Anthonymuthu, D. Mohammadyani, Q. Yang, B. Proneth, J. Klein-Seetharaman, S. Watkins, I. Bahar, J. Greenberger, R. K. Mallampalli, B. R. Stockwell, Y. Y. Tyurina, M. Conrad, and H. Bayir. 2017. "Oxidized arachidonic and adrenic PEs navigate cells to ferroptosis." *Nat Chem Biol* 13 (1):81-90. https://doi.org/10.1038/nchembio.2238.

Kang, R., R. Chen, Q. Zhang, W. Hou, S. Wu, L. Cao, J. Huang, Y. Yu, X. G. Fan, Z. Yan, X. Sun, H. Wang, Q. Wang, A. Tsung, T. R. Billiar, H. J. Zeh, 3rd, M. T. Lotze, and D. Tang. 2014. "HMGB1 in health and disease." *Mol Aspects Med* 40:1-116. https://doi.org/10.1016/j.mam.2014.05.001.

Kang, R., Y. Xie, Q. Zhang, W. Hou, Q. Jiang, S. Zhu, J. Liu, D. Zeng, H. Wang, D. L. Bartlett, T. R. Billiar, H. J. Zeh, 3rd, M. T. Lotze, and D. Tang. 2017. "Intracellular

HMGB1 as a novel tumor suppressor of pancreatic cancer." *Cell Res* 27 (7):916-932. https://doi.org/10.1038/cr.2017.51.

Kay, J., E. Thadhani, L. Samson, and B. Engelward. 2019. "Inflammation-induced DNA damage, mutations and cancer." *DNA Repair (Amst)* 83:102673. https://doi.org/10.1016/j.dnarep.2019.102673.

Kolb, R., F. S. Sutterwala, and W. Zhang. 2016. "Obesity and cancer: inflammation bridges the two." *Curr Opin Pharmacol* 29:77-89. https://doi.org/10.1016/j.coph.2016.07.005.

Kondapalli, L., J. Moslehi, and M. P. Bonaca. 2020. "Inflammation begets inflammation: cancer and acute MI." *Eur Heart J* 41 (23):2194-2196. https://doi.org/10.1093/eurheartj/ehz951.

Kronsten, V. T., T. H. Tranah, C. Pariante, and D. L. Shawcross. 2022. "Gut-derived systemic inflammation as a driver of depression in chronic liver disease." *J Hepatol* 76 (3):665-680. https://doi.org/10.1016/j.jhep.2021.11.008.

Kwak, M. S., W. J. Rhee, Y. J. Lee, H. S. Kim, Y. H. Kim, M. K. Kwon, and J. S. Shin. 2021. "Reactive oxygen species induce Cys106-mediated anti-parallel HMGB1 dimerization that protects against DNA damage." *Redox Biol* 40:101858. https://doi.org/10.1016/j.redox.2021.101858.

Lai, W., X. Li, Q. Kong, H. Chen, Y. Li, L. H. Xu, and J. Fang. 2021. "Extracellular HMGB1 interacts with RAGE and promotes chemoresistance in acute leukemia cells." *Cancer Cell Int* 21 (1):700. https://doi.org/10.1186/s12935-021-02387-9.

Lee, S. A., M. S. Kwak, S. Kim, and J. S. Shin. 2014. "The role of high mobility group box 1 in innate immunity." *Yonsei Med J* 55 (5):1165-76. https://doi.org/10.3349/ymj.2014.55.5.1165.

Lei, P., T. Bai, and Y. Sun. 2019. "Mechanisms of Ferroptosis and Relations With Regulated Cell Death: A Review." *Front Physiol* 10:139. https://doi.org/10.3389/fphys.2019.00139.

Li, J., F. Cao, H. L. Yin, Z. J. Huang, Z. T. Lin, N. Mao, B. Sun, and G. Wang. 2020. "Ferroptosis: past, present and future." *Cell Death Dis* 11 (2):88. https://doi.org/10.1038/s41419-020-2298-2.

Li, N., Z. Huang, Y. Zhang, H. Sun, J. Wang, and J. Zhao. 2019. "Increased cancer risk after myocardial infarction: fact or fiction? A systemic review and meta-analysis." *Cancer Manag Res* 11:1959-1968. https://doi.org/10.2147/CMAR.S193658.

Li, W., G. Feng, J. M. Gauthier, I. Lokshina, R. Higashikubo, S. Evans, X. Liu, A. Hassan, S. Tanaka, M. Cicka, H. M. Hsiao, D. Ruiz-Perez, A. Bredemeyer, R. W. Gross, D. L. Mann, Y. Y. Tyurina, A. E. Gelman, V. E. Kagan, A. Linkermann, K. J. Lavine, and D. Kreisel. 2019. "Ferroptotic cell death and TLR4/Trif signaling initiate neutrophil recruitment after heart transplantation." *J Clin Invest* 129 (6):2293-2304. https://doi.org/10.1172/JCI126428.

Libby, P. 2007. "Inflammatory mechanisms: the molecular basis of inflammation and disease." *Nutr Rev* 65 (12 Pt 2):S140-6. https://doi.org/10.1111/j.1753-4887.2007.tb00352.x.

Limana, F., A. Germani, A. Zacheo, J. Kajstura, A. Di Carlo, G. Borsellino, O. Leoni, R. Palumbo, L. Battistini, R. Rastaldo, S. Muller, G. Pompilio, P. Anversa, M. E. Bianchi, and M. C. Capogrossi. 2005. "Exogenous high-mobility group box 1 protein induces myocardial regeneration after infarction via enhanced cardiac C-kit+ cell proliferation and differentiation." *Circ Res* 97 (8):e73-83. https://doi.org/10.1161/01.RES.0000186276.06104.04.

Ling, L., L. Zhi, H. Wang, Y. Deng, and C. Gu. 2021. "MicroRNA-181b Inhibits Inflammatory Response and Reduces Myocardial Injury in Sepsis by Downregulating HMGB1." *Inflammation* 44 (4):1263-1273. https://doi.org/10.1007/s10753-020-01411-w.

Liu, B., W. Yi, X. Mao, L. Yang, and C. Rao. 2021. "Enoyl coenzyme A hydratase 1 alleviates nonalcoholic steatohepatitis in mice by suppressing hepatic ferroptosis." *Am J Physiol Endocrinol Metab* 320 (5):E925-E937. https://doi.org/10.1152/ajpendo.00614.2020.

Liu, J., H. Wang, and J. Li. 2016. "Inflammation and Inflammatory Cells in Myocardial Infarction and Reperfusion Injury: A Double-Edged Sword." *Clin Med Insights Cardiol* 10:79-84. https://doi.org/10.4137/CMC.S33164.

Lu, H., Z. Zhang, P. A. Barnie, and Z. Su. 2019. "Dual faced HMGB1 plays multiple roles in cardiomyocyte senescence and cardiac inflammatory injury." *Cytokine Growth Factor Rev* 47:74-82. https://doi.org/10.1016/j.cytogfr.2019.05.009.

Lucchesi, Benedict R. 2001. "Myocardial Reperfusion Injury—Role of Free Radicals and Mediators of Inflammation." In *Heart Physiology and Pathophysiology*, edited by Nicholas Sperelakis, Yoshihisa Kurachi, Andre Terzic and Michael V. Cohen, 1181-1210. San Diego: Academic Press.

Luo, Y., Y. Chihara, K. Fujimoto, T. Sasahira, M. Kuwada, R. Fujiwara, K. Fujii, H. Ohmori, and H. Kuniyasu. 2013. "High mobility group box 1 released from necrotic cells enhances regrowth and metastasis of cancer cells that have survived chemotherapy." *Eur J Cancer* 49 (3):741-51. https://doi.org/10.1016/j.ejca.2012.09.016.

Malmborg, M., C. B. Christiansen, M. D. Schmiegelow, C. Torp-Pedersen, G. Gislason, and M. Schou. 2018. "Incidence of new onset cancer in patients with a myocardial infarction - a nationwide cohort study." *BMC Cardiovasc Disord* 18 (1):198. https://doi.org/10.1186/s12872-018-0932-z.

Manivannan, S., O. Marei, O. Elalfy, and M. Zaben. 2021. "Neurogenesis after traumatic brain injury - The complex role of HMGB1 and neuroinflammation." *Neuropharmacology* 183:108400. https://doi.org/10.1016/j.neuropharm.2020.108400.

Mariani, P., D. Russo, M. Maisto, G. Troiano, V. C. A. Caponio, M. Annunziata, and L. Laino. 2022. "Pre-treatment neutrophil-to-lymphocyte ratio is an independent prognostic factor in head and neck squamous cell carcinoma: Meta-analysis and trial sequential analysis." *J Oral Pathol Med* 51 (1):39-51. https://doi.org/10.1111/jop.13264.

Medzhitov, R. 2008. "Origin and physiological roles of inflammation." *Nature* 454 (7203):428-35. https://doi.org/10.1038/nature07201.

Mou, Y., J. Wang, J. Wu, D. He, C. Zhang, C. Duan, and B. Li. 2019. "Ferroptosis, a new form of cell death: opportunities and challenges in cancer." *J Hematol Oncol* 12 (1):34. https://doi.org/10.1186/s13045-019-0720-y.

Murata, M. 2018. "Inflammation and cancer." *Environ Health Prev Med* 23 (1):50. https://doi.org/10.1186/s12199-018-0740-1.

Naito, H., T. Nojima, N. Fujisaki, K. Tsukahara, H. Yamamoto, T. Yamada, T. Aokage, T. Yumoto, T. Osako, and A. Nakao. 2020. "Therapeutic strategies for ischemia reperfusion injury in emergency medicine." *Acute Med Surg* 7 (1):e501. https://doi.org/10.1002/ams2.501.

Ong, S. B., S. Hernandez-Resendiz, G. E. Crespo-Avilan, R. T. Mukhametshina, X. Y. Kwek, H. A. Cabrera-Fuentes, and D. J. Hausenloy. 2018. "Inflammation following acute myocardial infarction: Multiple players, dynamic roles, and novel therapeutic opportunities." *Pharmacol Ther* 186:73-87. https://doi.org/10.1016/j.pharmthera.2018.01.001.

Oozawa, S., S. Mori, T. Kanke, H. Takahashi, K. Liu, Y. Tomono, M. Asanuma, I. Miyazaki, M. Nishibori, and S. Sano. 2008. "Effects of HMGB1 on ischemia-reperfusion injury in the rat heart." *Circ J* 72 (7):1178-84. https://doi.org/10.1253/circj.72.1178.

Ou, D., D. Ni, R. Li, X. Jiang, X. Chen, and H. Li. 2022. "Galectin-1 alleviates myocardial ischemia-reperfusion injury by reducing the inflammation and apoptosis of cardiomyocytes." *Exp Ther Med* 23 (2):143. https://doi.org/10.3892/etm.2021.11066.

Pangilinan, C. R., L. H. Wu, and C. H. Lee. 2021. "Salmonella Impacts Tumor-Induced Macrophage Polarization, and Inhibits SNAI1-Mediated Metastasis in Melanoma." *Cancers (Basel)* 13 (12). https://doi.org/10.3390/cancers13122894.

Park, T. J., J. H. Park, G. S. Lee, J. Y. Lee, J. H. Shin, M. W. Kim, Y. S. Kim, J. Y. Kim, K. J. Oh, B. S. Han, W. K. Kim, Y. Ahn, J. H. Moon, J. Song, K. H. Bae, D. H. Kim, E. W. Lee, and S. C. Lee. 2019. "Quantitative proteomic analyses reveal that GPX4 downregulation during myocardial infarction contributes to ferroptosis in cardiomyocytes." *Cell Death Dis* 10 (11):835. https://doi.org/10.1038/s41419-019-2061-8.

Pellegrini, L., E. Foglio, E. Pontemezzo, A. Germani, M. A. Russo, and F. Limana. 2019. "HMGB1 and repair: focus on the heart." *Pharmacol Ther* 196:160-182. https://doi.org/10.1016/j.pharmthera.2018.12.005.

Piotrowski, I., K. Kulcenty, and W. Suchorska. 2020. "Interplay between inflammation and cancer." *Rep Pract Oncol Radiother* 25 (3):422-427. https://doi.org/10.1016/j.rpor.2020.04.004.

Preissner, K. T., S. Fischer, and E. Deindl. 2020. "Extracellular RNA as a Versatile DAMP and Alarm Signal That Influences Leukocyte Recruitment in Inflammation and Infection." *Front Cell Dev Biol* 8:619221. https://doi.org/10.3389/fcell.2020.619221.

Raucci, A., S. Di Maggio, F. Scavello, A. D'Ambrosio, M. E. Bianchi, and M. C. Capogrossi. 2019. "The Janus face of HMGB1 in heart disease: a necessary update." *Cell Mol Life Sci* 76 (2):211-229. https://doi.org/10.1007/s00018-018-2930-9.

Ren, J. X., X. Sun, X. L. Yan, Z. N. Guo, and Y. Yang. 2020. "Ferroptosis in Neurological Diseases." *Front Cell Neurosci* 14:218. https://doi.org/10.3389/fncel.2020.00218.

Richard, S. A., Y. Jiang, L. H. Xiang, S. Zhou, J. Wang, Z. Su, and H. Xu. 2017. "Posttranslational modifications of high mobility group box 1 and cancer." *Am J Transl Res* 9 (12):5181-5196.

Roe, K. 2021. "An inflammation classification system using cytokine parameters." *Scand J Immunol* 93 (2):e12970. https://doi.org/10.1111/sji.12970.

Sanchez-Hernandez, C. D., L. A. Torres-Alarcon, A. Gonzalez-Cortes, and A. N. Peon. 2020. "Ischemia/Reperfusion Injury: Pathophysiology, Current Clinical Management, and Potential Preventive Approaches." *Mediators Inflamm* 2020:8405370. https://doi.org/10.1155/2020/8405370.

Singh, N., D. Baby, J. P. Rajguru, P. B. Patil, S. S. Thakkannavar, and V. B. Pujari. 2019. "Inflammation and cancer." *Ann Afr Med* 18 (3):121-126. https://doi.org/10.4103/aam.aam_56_18.

Song, X. D., Y. N. Wang, A. L. Zhang, and B. Liu. 2020. "Advances in research on the interaction between inflammation and cancer." *J Int Med Res* 48 (4):300060519895347. https://doi.org/10.1177/0300060519895347.

Song, Y., B. Wang, X. Zhu, J. Hu, J. Sun, J. Xuan, and Z. Ge. 2021. "Human umbilical cord blood-derived MSCs exosome attenuate myocardial injury by inhibiting ferroptosis in acute myocardial infarction mice." *Cell Biol Toxicol* 37 (1):51-64. https://doi.org/10.1007/s10565-020-09530-8.

Su, L. J., J. H. Zhang, H. Gomez, R. Murugan, X. Hong, D. Xu, F. Jiang, and Z. Y. Peng. 2019. "Reactive Oxygen Species-Induced Lipid Peroxidation in Apoptosis, Autophagy, and Ferroptosis." *Oxid Med Cell Longev* 2019:5080843. https://doi.org/10.1155/2019/5080843.

Sun, X., Z. Ou, M. Xie, R. Kang, Y. Fan, X. Niu, H. Wang, L. Cao, and D. Tang. 2015. "HSPB1 as a novel regulator of ferroptotic cancer cell death." *Oncogene* 34 (45):5617-25. https://doi.org/10.1038/onc.2015.32.

Sun, Y. D., H. Zhang, Y. Q. Chen, C. X. Wu, J. B. Zhang, H. R. Xu, J. Z. Liu, and J. J. Han. 2020. "HMGB1, the Next Predictor of Transcatheter Arterial Chemo-embolization for Liver Metastasis of Colorectal Cancer?" *Front Oncol* 10:572418. https://doi.org/10.3389/fonc.2020.572418.

Sung, H., J. Ferlay, R. L. Siegel, M. Laversanne, I. Soerjomataram, A. Jemal, and F. Bray. 2021. "Global Cancer Statistics 2020: GLOBOCAN Estimates of Incidence and Mortality Worldwide for 36 Cancers in 185 Countries." *CA Cancer J Clin* 71 (3):209-249. https://doi.org/10.3322/caac.21660.

Takahashi, K., S. Fukushima, K. Yamahara, K. Yashiro, Y. Shintani, S. R. Coppen, H. K. Salem, S. W. Brouilette, M. H. Yacoub, and K. Suzuki. 2008. "Modulated inflammation by injection of high-mobility group box 1 recovers post-infarction chronically failing heart." *Circulation* 118 (14 Suppl):S106-14. https://doi.org/10.1161/CIRCULATIONAHA.107.757443.

Tan, G., C. Huang, J. Chen, and F. Zhi. 2020. "HMGB1 released from GSDME-mediated pyroptotic epithelial cells participates in the tumorigenesis of colitis-associated colorectal cancer through the ERK1/2 pathway." *J Hematol Oncol* 13 (1):149. https://doi.org/10.1186/s13045-020-00985-0.

Tang, D., R. Kang, K. M. Livesey, G. Kroemer, T. R. Billiar, B. Van Houten, H. J. Zeh, 3rd, and M. T. Lotze. 2011. "High-mobility group box 1 is essential for mitochondrial quality control." *Cell Metab* 13 (6):701-11. https://doi.org/10.1016/j.cmet.2011.04.008.

Tang, D., R. Kang, H. J. Zeh, 3rd, and M. T. Lotze. 2010. "High-mobility group box 1 and cancer." *Biochim Biophys Acta* 1799 (1-2):131-40. https://doi.org/10.1016/j.bbagrm.2009.11.014.

Tang, D., X. Chen, R. Kang, and G. Kroemer. 2021. "Ferroptosis: molecular mechanisms and health implications." *Cell Res* 31 (2):107-125. https://doi.org/10.1038/s41422-020-00441-1.

Taruya, A., Y. Nakajima, A. Tanaka, M. Kashiwagi, T. Tanimoto, A. Kuroi, Y. Shiono, K. Shimamura, T. Kubo, H. Sougawa, T. Masuno, Y. Ozaki, K. Satogami, S. Ota, Y. Katayama, Y. Ino, H. Hoshiya, and T. Akasaka. 2021. "Cancer-related vulnerable lesions in patients with stable coronary artery disease." *Int J Cardiol* 335:1-6. https://doi.org/10.1016/j.ijcard.2021.03.050.

Tewari, A. K., J. A. Stockert, S. S. Yadav, K. K. Yadav, and I. Khan. 2018. "Inflammation and Prostate Cancer." *Adv Exp Med Biol* 1095:41-65. https://doi.org/10.1007/978-3-319-95693-0_3.

Tollefsen, I. M., C. Shetelig, I. Seljeflot, J. Eritsland, P. Hoffmann, and G. O. Andersen. 2021. "High levels of interleukin-6 are associated with final infarct size and adverse clinical events in patients with STEMI." *Open Heart* 8 (2). https://doi.org/10.1136/openhrt-2021-001869.

Tonnus, W., and A. Linkermann. 2016. ""Death is my Heir"--Ferroptosis Connects Cancer Pharmacogenomics and Ischemia-Reperfusion Injury." *Cell Chem Biol* 23 (2):202-203. https://doi.org/10.1016/j.chembiol.2016.02.005.

Tripathi, A., K. Shrinet, and A. Kumar. 2019. "HMGB1 protein as a novel target for cancer." *Toxicol Rep* 6:253-261. https://doi.org/10.1016/j.toxrep.2019.03.002.

Tsung, A., S. Tohme, and T. R. Billiar. 2014. "High-mobility group box-1 in sterile inflammation." *J Intern Med* 276 (5):425-43. https://doi.org/10.1111/joim.12276.

Ueno, K., Y. Nomura, Y. Morita, and Y. Kawano. 2021. "Prednisolone Suppresses the Extracellular Release of HMGB-1 and Associated Inflammatory Pathways in Kawasaki Disease." *Front Immunol* 12:640315. https://doi.org/10.3389/fimmu.2021.640315.

Velders, M. A., E. Hagstrom, and S. K. James. 2020. "Temporal Trends in the Prevalence of Cancer and Its Impact on Outcome in Patients With First Myocardial Infarction: A Nationwide Study." *J Am Heart Assoc* 9 (4):e014383. https://doi.org/10.1161/JAHA.119.014383.

Venereau, E., F. De Leo, R. Mezzapelle, G. Careccia, G. Musco, and M. E. Bianchi. 2016. "HMGB1 as biomarker and drug target." *Pharmacol Res* 111:534-544. https://doi.org/10.1016/j.phrs.2016.06.031.

Wang, S., and Y. Zhang. 2020a. "HMGB1 in inflammation and cancer." *J Hematol Oncol* 13 (1):116. https://doi.org/10.1186/s13045-020-00950-x.

Wang, Shumin, and Yi Zhang. 2020b. "HMGB1 in inflammation and cancer." *Journal of Hematology & Oncology* 13 (1):116. https://doi.org/10.1186/s13045-020-00950-x.

Wang, Y., Z. Wei, K. Pan, J. Li, and Q. Chen. 2020. "The function and mechanism of ferroptosis in cancer." *Apoptosis* 25 (11-12):786-798. https://doi.org/10.1007/s10495-020-01638-w.

Wen, Q., J. Liu, R. Kang, B. Zhou, and D. Tang. 2019. "The release and activity of HMGB1 in ferroptosis." *Biochem Biophys Res Commun* 510 (2):278-283. https://doi.org/10.1016/j.bbrc.2019.01.090.

Wu, F., Z. H. Zhao, S. T. Ding, H. H. Wu, and J. J. Lu. 2013. "High mobility group box 1 protein is methylated and transported to cytoplasm in clear cell renal cell carcinoma."

Asian Pac J Cancer Prev 14 (10):5789-95. https://doi.org/10.7314/apjcp.2013.14.10.5789.

Wu, M. Y., G. T. Yiang, W. T. Liao, A. P. Tsai, Y. L. Cheng, P. W. Cheng, C. Y. Li, and C. J. Li. 2018. "Current Mechanistic Concepts in Ischemia and Reperfusion Injury." *Cell Physiol Biochem* 46 (4):1650-1667. https://doi.org/10.1159/000489241.

Wu, X., C. D. Iroegbu, J. Peng, J. Guo, J. Yang, and C. Fan. 2021. "Cell Death and Exosomes Regulation After Myocardial Infarction and Ischemia-Reperfusion." *Front Cell Dev Biol* 9:673677. https://doi.org/10.3389/fcell.2021.673677.

Wu, Y., C. Yu, M. Luo, C. Cen, J. Qiu, S. Zhang, and K. Hu. 2020. "Ferroptosis in Cancer Treatment: Another Way to Rome." *Front Oncol* 10:571127. https://doi.org/10.3389/fonc.2020.571127.

Wu, Y., H. Liu, and X. Wang. 2021. "Cardioprotection of pharmacological postconditioning on myocardial ischemia/reperfusion injury." *Life Sci* 264:118628. https://doi.org/10.1016/j.lfs.2020.118628.

Wu, Y., Y. Zhao, H. Z. Yang, Y. J. Wang, and Y. Chen. 2021. "HMGB1 regulates ferroptosis through Nrf2 pathway in mesangial cells in response to high glucose." *Biosci Rep* 41 (2). https://doi.org/10.1042/BSR20202924.

Xia, J., J. Y. Xue, K. Chen, Z. Y. Lin, W. J. Wu, and Q. F. Zhao. 2018. "[The role of high-mobility group box protein 1 in the signaling pathways of myocardial ischemia-reperfusion injury in rats]." *Zhonghua Yi Xue Za Zhi* 98 (40):3268-3273. https://doi.org/10.3760/cma.j.issn.0376-2491.2018.40.011.

Xiong, W., M. Yao, R. Zhou, Y. Qu, Y. Yang, Z. Wang, N. Song, H. Chen, and J. Qian. 2020. "Oxytocin ameliorates ischemia/reperfusion-induced injury by inhibiting mast cell degranulation and inflammation in the rat heart." *Biomed Pharmacother* 128:110358. https://doi.org/10.1016/j.biopha.2020.110358.

Xu, T., W. Ding, X. Ji, X. Ao, Y. Liu, W. Yu, and J. Wang. 2019. "Molecular mechanisms of ferroptosis and its role in cancer therapy." *J Cell Mol Med* 23 (8):4900-4912. https://doi.org/10.1111/jcmm.14511.

Xue, J., J. S. Suarez, M. Minaai, S. Li, G. Gaudino, H. I. Pass, M. Carbone, and H. Yang. 2020. "HMGB1 as a therapeutic target in disease." *J Cell Physiol*. https://doi.org/10.1002/jcp.30125.

Yang, H., D. J. Antoine, U. Andersson, and K. J. Tracey. 2013. "The many faces of HMGB1: molecular structure-functional activity in inflammation, apoptosis, and chemotaxis." *J Leukoc Biol* 93 (6):865-73. https://doi.org/10.1189/jlb.1212662.

Yang, Y., L. Yang, S. Jiang, T. Yang, J. Lan, Y. Lei, H. Tan, and K. Pan. 2020. "HMGB1 mediates lipopolysaccharide-induced inflammation via interacting with GPX4 in colon cancer cells." *Cancer Cell Int* 20:205. https://doi.org/10.1186/s12935-020-01289-6.

Yao, H. C., M. Zhou, Y. H. Zhou, L. H. Wang, D. Y. Zhang, Q. F. Han, T. Liu, L. Wu, K. L. Tian, and M. Zhang. 2016. "Intravenous high mobility group box 1 upregulates the expression of HIF-1alpha in the myocardium via a protein kinase B-dependent pathway in rats following acute myocardial ischemia." *Mol Med Rep* 13 (2):1211-9. https://doi.org/10.3892/mmr.2015.4648.

Ye, Y., Z. Zeng, T. Jin, H. Zhang, X. Xiong, and L. Gu. 2019. "The Role of High Mobility Group Box 1 in Ischemic Stroke." *Front Cell Neurosci* 13:127. https://doi.org/10.3389/fncel.2019.00127.

You, L., H. Cui, F. Zhao, H. Sun, H. Zhong, G. Zhou, and X. Chen. 2021. "Inhibition of HMGB1/RAGE axis suppressed the lipopolysaccharide (LPS)-induced vicious transformation of cervical epithelial cells." *Bioengineered* 12 (1):4995-5003. https://doi.org/10.1080/21655979.2021.1957750.

Yu, H., P. Guo, X. Xie, Y. Wang, and G. Chen. 2017. "Ferroptosis, a new form of cell death, and its relationships with tumourous diseases." *J Cell Mol Med* 21 (4):648-657. https://doi.org/10.1111/jcmm.13008.

Yu, Y., D. Tang, and R. Kang. 2015. "Oxidative stress-mediated HMGB1 biology." *Front Physiol* 6:93. https://doi.org/10.3389/fphys.2015.00093.

Yu, Y., L. Yang, J. Lv, X. Huang, J. Yi, C. Pei, and Y. Shao. 2015. "The role of high mobility group box 1 (HMGB-1) in the diabetic retinopathy inflammation and apoptosis." *Int J Clin Exp Pathol* 8 (6):6807-13.

Yu, Y., Y. Yan, F. Niu, Y. Wang, X. Chen, G. Su, Y. Liu, X. Zhao, L. Qian, P. Liu, and Y. Xiong. 2021. "Ferroptosis: a cell death connecting oxidative stress, inflammation and cardiovascular diseases." *Cell Death Discov* 7 (1):193. https://doi.org/10.1038/s41420-021-00579-w.

Zhang, Y., L. Xin, M. Xiang, C. Shang, Y. Wang, Y. Wang, X. Cui, and Y. Lu. 2022. "The molecular mechanisms of ferroptosis and its role in cardiovascular disease." *Biomed Pharmacother* 145:112423. https://doi.org/10.1016/j.biopha.2021.112423.

Zhao, H., L. Wu, G. Yan, Y. Chen, M. Zhou, Y. Wu, and Y. Li. 2021. "Inflammation and tumor progression: signaling pathways and targeted intervention." *Signal Transduct Target Ther* 6 (1):263. https://doi.org/10.1038/s41392-021-00658-5.

Zhao, W. K., Y. Zhou, T. T. Xu, and Q. Wu. 2021. "Ferroptosis: Opportunities and Challenges in Myocardial Ischemia-Reperfusion Injury." *Oxid Med Cell Longev* 2021:9929687. https://doi.org/10.1155/2021/9929687.

Zhao, Y., Z. Huang, and H. Peng. 2021. "Molecular Mechanisms of Ferroptosis and Its Roles in Hematologic Malignancies." *Front Oncol* 11:743006. https://doi.org/10.3389/fonc.2021.743006.

Zhou, P., Z. H. Zheng, T. Wan, J. Wu, C. W. Liao, and X. J. Sun. 2021. "Vitexin Inhibits Gastric Cancer Growth and Metastasis through HMGB1-mediated Inactivation of the PI3K/AKT/mTOR/HIF-1alpha Signaling Pathway." *J Gastric Cancer* 21 (4):439-456. https://doi.org/10.5230/jgc.2021.21.e40.

Zhou, S. Y., G. Z. Cui, X. L. Yan, X. Wang, Y. Qu, Z. N. Guo, and H. Jin. 2020. "Mechanism of Ferroptosis and Its Relationships With Other Types of Programmed Cell Death: Insights for Potential Interventions After Intracerebral Hemorrhage." *Front Neurosci* 14:589042. https://doi.org/10.3389/fnins.2020.589042.

Zhou, Y. H., Q. F. Han, L. Gao, Y. Sun, Z. W. Tang, M. Wang, W. Wang, and H. C. Yao. 2019. "HMGB1 Protects the Heart Against Ischemia-Reperfusion Injury via PI3K/AkT Pathway-Mediated Upregulation of VEGF Expression." *Front Physiol* 10:1595. https://doi.org/10.3389/fphys.2019.01595.

Zhu, L., S. Ren, M. J. Daniels, W. Qiu, L. Song, T. You, D. Wang, and Z. Wang. 2021. "Exogenous HMGB1 Promotes the Proliferation and Metastasis of Pancreatic Cancer Cells." *Front Med (Lausanne)* 8:756988. https://doi.org/10.3389/fmed.2021.756988.

Section II. HMGB1 and Metabolic Interactions

Chapter 3

The Role of HMGB1 in Adipocyte Dysfunction Linking Obesity and Type 2 Diabetes

Sagar Barge[1] and Narayan Chandra Talukdar[1,2,*]

[1]Chemical Biology Lab I,
Institute of Advanced Study in Science and Technology,
Paschim Boragaon, Guwahati-35, Assam, India
[2]Assam Down Town University,
Panikhaiti, Guwahati, Assam 781006, India

Abstract

Overconsumption of calories leads to excess fat deposition in adipose tissue, which induces consequent metabolic disturbances like inflammation in cells and generates a state of obesity. Adipose tissue plays various physiological roles associated with regulations of the homeostatic process, such as maintenance of blood glucose and lipid metabolism. Adipose tissue also produces several adipokines and cytokines, further regulating physiological functions such as insulin secretion and insulin sensitivity, immunity, and appetite. The dysfunction of adipocytes leads to dysregulation in metabolic processes and the promotion of insulin resistance and type 2 diabetes. HMGB1, a high mobility group box one protein, called amphoterin, is a non-histone chromatin-associated protein regulating gene expression. Infiltrated macrophages mainly release HMGB1 in adipocytes upon the stressful situation exerted to the extracellular milieu, and the released HMGB1 triggers an inflammatory response in obesity. Extracellular HMGB1 is primarily studied to activate immune cells in adipose tissue, with the aim of making it an innate alarmin protein. The activation of those immune

* Corresponding Author's Email: nctalukdar@yahoo.com.

In: HMGB1
Editors: Eylem Taskin, Celal Guven and Salih Tunc Kaya
ISBN: 979-8-88697-408-9

cells releases more HMGB1, which further recruits additional immune cells and triggers the sustained inflammatory response, creating inflammatory feedback in the tissue. This signaling further triggers the insulin resistance-like condition. This chapter focuses on the HMGB1 induced inflammation and its associated changes in the cells and impaired molecular mechanisms in adipose tissue related to obesity and type 2 diabetes. Several studies demonstrate that (reversal of) HMGB1 secretion can be targeted by directly binding with many antagonist drugs.

Keywords: HMGB1, obesity, adipose tissue, inflammation, insulin resistance, type 2 diabetes

Introduction

Obesity is the most significant health concern in today's modern society. Lifestyle changes with increased caloric food intake and low physical activity result in metabolic alterations in the body and worsen the ongoing obesity epidemic. Obesity is associated with many health conditions, but the most affected health condition is insulin resistance and type 2 diabetes (Schwartz and Kahn 1999; Kahn and Flier 2000; Kahn, Hull, and Utzschneider 2006). However, the relationship between obesity and insulin resistance is confusing, as all obese people do not develop insulin resistance (Ferrannini et al. 1997).

The alarm system in the cells can sense stressful situations and activate performed endogenous molecules called alarmins or damage-associated molecular pattern molecules (DAMPs) against stress stimulus (Matzinger 1994). Several HMG proteins, including HMGB1 (Scaffidi, Misteli, and Bianchi 2002; Yang, Wang, and Andersson 2020a), HMGB2 (Pusterla et al. 2009), and HMGN1 (Yang et al. 2012), exhibit DAMPs properties. These are the second most abundant proteins in the human body and are highly conserved to the DNA histone throughout the evolution (Einck and Bustin 1985). HMGB1 proteins play an essential role in different diseases such as infectious and autoimmune diseases and cancer (Kang et al. 2014). They also perform a specific intracellular function in homeostasis and promote inflammation in response to danger signals, leading to tissue injury or organ dysfunction (Andersson, Yang, and Harris 2018; Xue et al. 2021; Zhang et al. 2017). Primarily HMGB1 are considered nuclear proteins while performing a nuclear function like transcriptional regulation of various genes (Bianchi, Beltrame, and Paonessa 1989).

In recent years, it has been observed that HMGB1 causes chronic inflammation and insulin resistance in the adipocyte (Zatterale et al. 2020; Guzmán-Ruiz et al. 2014), suggesting HMGB1 plays a role in adipocyte dysfunction in obesity and the development of type 2 diabetes (Giacobbe et al. 2015; Wang, Qu, and Deng 2015). This chapter focuses on the implication of the role of HMGB1 in the initiation of adipose tissue dysfunction and the pathogenesis of obesity and type 2 diabetes.

Dysfunction of Adipocyte in Obesity and Type 2 Diabetes

The adipose tissue is dedicated to storing large amounts of triglyceride during feeding and releases triglyceride as free fatty acids (FFAs) and glycerol during fasting (Rosen and Spiegelman 2006; Kalderon et al. 2000). According to one hypothesis, FFAs mediate insulin resistance in obesity due to its (FFA) high circulating levels (Figure 2) (Savage, Petersen, and Shulman 2007). It is also found that FFAs can circulate in peripheral organs such as skeletal muscles and can cause an insulin resistance-like state in animals and humans, which further supports this idea (Kusminski et al. 2009; Spalding et al. 2008). Adipokines such as adiponectin, primarily secreted by adipocytes, have an ability to sense insulin sensitivity and mainly depend on the size of adipocytes and the presence of a number of adipocytes. Excess adiposity and adipocyte dysfunction further cause downregulation of the anti-inflammatory adiponectin, increasing inflammatory responses and metabolic dysfunction (Ouchi et al. 2011).

According to one study, lipoatrophy, a disease in which humans and mice have too little fat, promotes decreased adipokine release in lipodystrophies (25, 26). Excess fat consumption increases adipocyte mass with consequent macrophage recruitment in the tissue and alternation in the secretion of pro-inflammatory cytokines (Kohlgruber and Lynch 2015; Burhans et al. 2018; Zatterale et al. 2020). An increase in adiposity may affect the ability of adipocyte function and secrete other adipokines such as monocyte chemoattractant protein-1 (MCP-1) and tumor-necrosis factor-α (TNFα); in addition, the distribution of fat is correlated with the disturbance in the metabolic pathways that leads to metabolic abnormalities. During this process, adipocyte gets overloaded with lipid and reaches its critical level causing insulin resistance (Blüher 2016). The study carried out by Kim et al. demonstrates that adipocyte hypertrophy is responsible for adult fat mass expansion and loss of adipocyte plasticity resulting in insulin resistance (Kim et al. 2014). Furthermore, the study by Kim Ji et al. also states that fat expansion also results in insulin resistance in adipocytes without causing inflammation (Kim et al. 2015). However, other studies mainly focus on adipocyte hypertrophy causing inflammation and its associated pathways connected with insulin resistance (Huh et al. 2014). Pancratic β cells tend to secrete insulin during its need. However, in obese conditions, they secrete insufficient insulin, which is one of the main reasons for obesity-driven insulin resistance in peripheral organs such as muscle cells, liver, and adipose tissues. (Kusminski et al. 2009). These studies are evident that an increase in fat mass and hypertrophy is mainly responsible for metabolism-related disorders.

HMGB1 Protein Structure and Function

HMGB1 is a nuclear, non-histone DNA-binding protein that was discovered in 1973 and named for its high electrophoretic mobility on polyacrylamide gels (Kang et al. 2014). It is a member of the histone family of DNA-binding proteins. DNA replication, transcription, and repair are all controlled by HMGB1, a highly conserved protein that performs various functions in the nucleus (Travers 2003). HMGBs comprise four different types (i.e. HMGB1-4) but HMGB1 is the most highly abundant protein expressed in all nucleated cells and platelets (Stros 2010). The release of HMGB1 from lipopoly-saccharide-induced macrophages was first reported in 1919 by Wang et al. They also observed that the HMGB1 secretion causes an increase in pro-inflammatory cytokines in sepsis (Wang et al. 1999). This discovery led to the exploration of the role of HMGB1 in inflammation and inflammatory diseases. Furthermore, the therapeutic regulation of HMGB1 was explored by various research groups in preclinical models (Yang, Wang, and Andersson 2020b).

The structure of the HMGB1 is highly conserved and originated around 525 million years ago during the divergence of protostomes and deuterostomes (Sharman, Hay-Schmidt, and Holland 1997). HMGB1 has a 99% sequence identity with mammals, and its structure is made up of 215 amino acid residues.

Furthermore, HMGB1 protein comprises three domains HMG-box A and box B, and it has a C-terminal tail that is highly negatively charged with 30 repeated glutamic aspartic acid residues (Figure 1) (Kang et al. 2014; Bianchi et al. 1992). There are two amino acids present at C terminus, which differs between the mice and human. Furthermore, HMGB1 can travel between the nucleus and cytoplasm but binds to the chromatin in normal conditions

(Isackson et al. 1980). HMGB1 has been found in the different organelles of cell cytosol, e.g., mitochondria (Stumbo et al. 2008), lysosome (Gardella et al. 2002), and cellular membrane during nuclear localization signal (NLS) (Kuehl, Rechsteiner, and Wu 1985). The HMGB1 protein is widely spread in the different tissue spleen, and the thymus comprises high levels of HMGB1 protein (Prasad and Thakur 1990). HMGB1 is mainly secreted by dead cells in the extracellular space upon signal. These extracellular activities are primarily regulated by its receptors as well as redox state (Tang, Billiar, and Lotze 2012). According to several studies, HMGB1 has three different forms and based on its redox state condition, the first is a fully reduced HMGB1 state, the second is a disulfide HMGB1, and the third is sulfonyl HMGB1 state (Antoine et al. 2014). The immune function of HMGB1 is regulated when cysteine residues are at a position of 23, 45, and 106 during its redox state (Magna and Pisetsky 2016). The oxidized condition of Cys106 leads to an inactive HMGB1, which increases immunological tolerance by releasing proinflammatory mediators and aids in DNA damage repair (Kwak et al. 2021). The stress redox signals are significantly less in healthy obesity than unhealthy obesity; also, the formation of new adipose tissue and lipids via adipogenesis and lipogenesis are sensitive to the redox signals (Bañuls et al. 2017; Böhm et al. 2020). The study on HMGB1, its redox forms, and their targeted pathways in mediating obesity and insulin resistance are still progressing.

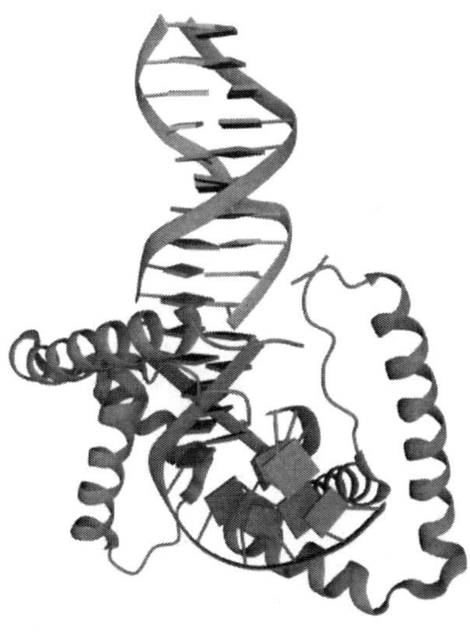

Figure 1. Structure of HMGB1 protein containing two HMG boxes.

HMGB1 Receptor and Signaling Pathways in Adipose Tissue

HMGB1 acts through different membrane receptors and generates pro-inflammatory signaling pathways. There are two primary HMGB1 receptors, advanced glycation end-products (RAGE) that bind at residues 150–183 and toll-like receptor 4 (TLR4) that binds at residues 89–108 within the B box of the HMGB1 molecule. (Huttunen et al. 2002; Yang et al. 2010). Among the other receptors of HMGB1, RAGE and TLRs are primarily involved in obesity and insulin resistance; therefore, here we are discussing the role of RAGE and TLRs receptors on signaling pathways of adipose tissue.

RAGE

RAGE is the primary receptor for the HMGB1 protein, and it is the first discovered receptor for the RAGEs (Hori et al. 1995). Due to its ability to bind to RAGEs, it accumulates in the states of hyperglycemia, aging, inflammation, renal failure, and oxidative stress (López-Díez et al. 2016; Shekhtman, Ramasamy, and Schmidt 2017). In adipose tissue, HMGB1 via binding through a RAGE receptor induces the expression of transcription factor nuclear factor κB (NFκB), which is a classical inflammatory molecule (van Beijnum, Buurman, and Griffioen 2008). Besides activation of NFκB, it also increases the production of cytokines (Figure 2). NFκB activation further induces HMGB1 expression and receptors, compounding a more inflammatory response to cells and creating a positive feedback loop (Figure 2) (Yan et al. 1994; van Beijnum, Buurman, and Griffioen 2008). HMGB1 also secrets the IL6 inflammatory molecule through the RAGE receptor. IL6 is a mediator of obesity which increases the inflammation in the obese state and ultimately causes insulin resistance (Nativel et al. 2013). Around 35 percent of the Interleukin-6 is found in adipose tissue, leading to low-grade inflammation, disrupting insulin signaling, and decreasing sensitivity to insulin in visceral fat (Nativel et al. 2013; Wang, Qu, and Deng 2015).

RAGE signaling is also involved in Epicardial adipose tissue (EAT) in the patient having progression of coronary artery disease and found to be promoting EAT and impaired insulin signaling (Dozio et al. 2016). RAGE receptors are also found to regulate adiposity. An insulin sensitivity study suggests that an increase in the expression of RAGE using adenoviral gene transfer in 3T3-L1 pre-adipocyte stimulates adipocyte hypertrophy, whereas blocking the gene expression by small interfering RNA significantly decreases adipocyte hypertrophy (Monden et al. 2013). During diet-induced obesity in mice, the expression of HMGB1 was found to be upregulated in the liver and the adipose tissues, but in the absence of RAGE receptor, it did not cause any weight gain, insulin resistance, or inflammation in adipose tissue (Song et al. 2014). These studies suggest that RAGE can be a therapeutic target for inflammation and diabetes (Nativel et al. 2013).

TLRs

TLR2, TLR4, and TLR 9 are among the thirteen TLRs reported with HMGB1 receptors in TLR2, TLR4, and TLR 9. (64, 65). TLRs are a type of pattern recognition receptor that may bind to external ligands and recognize danger signals such as pathogen-associated molecular patterns (PAMPs) and damage-associated molecular patterns (DAMPs) and are involved in the generation of immunological response to illnesses (66, 67). TLR4 receptors mainly increase the inflammation in adipose tissue, which results in insulin resistance (Orr et al. 2012; Poggi et al. 2007; Tsukumo et al. 2016; Suganami et al. 2007; Saberi et al. 2009). Similarly, TLR2 has been reported to have an inflammatory response in FFA-induced cultured adipocytes and skeletal muscles (Figure 2) (Senn 2006; Nguyen et al. 2007). Also, downregulation of TLR2 signaling in the liver and adipose tissue was found to be effective against the HFD-induced lipid accumulation and inflammation and in the improvement of peripheral insulin sensitivity (Jin, Henao-Mejia, and Flavell 2013). HMGB1 binding to receptors of TLR2 and TLR4 also activates the NFκB to induce the inflammatory response in adipose tissue via TNF-α, IL6, and IL-1β cytokines (Figure 2) (Park et al. 2003; Gunasekaran et al. 2016). Other than adipose tissue, the loss of TLR4 improved HFD- induced hepatic insulin sensitivity and reduced hepatic steatosis in liver (Poggi et al. 2007; Tsukumo et al. 2016; Saberi et al. 2009). Reduction in TLR4 receptor was also found effective against lipid-induced insulin resistance in skeletal muscles (Tsukumo et al. 2016; Holland et al. 2011) (Figure 2). Research suggests that level of HMGB1 was elevated in type 2 diabetic subjects upon an increase in the TLR2 and TLR4 receptor binding, BMI and HOMA-IR; furthermore, the expression of the MyD88 and NF-κB p65 were also increased (Dasu et al. 2010). TLR4 signaling is responsible for increasing the very-long-chain fatty acids (VLCFA) and GM3 ganglioside in adipose tissue of obese mice, whereas mice lacking TLR4 devoid of VLCFA-GM3 (Kanoh et al. 2020).

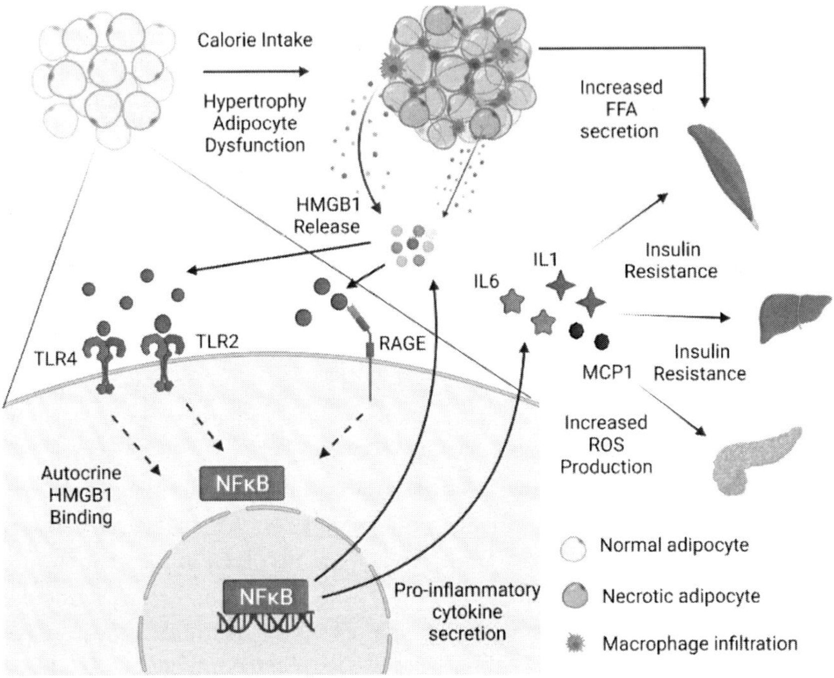

Figure 2. Pathway linking of the HMGB1 signaling and release from adipose tissue during obesity and insulin resistance. Excess calorie intake in obesity leads to hypertrophic adipocytes, which causes infiltration of macrophages and necrosis of adipocytes. Necrotic adipocyte and macrophage secrete HMGB1 protein, which in turn binds to activated HMGB1receptors (TLR4, TLR2, and RAGE) present on adipose tissue via autocrine signaling. Activation of receptors initiates NFkB signaling and transcription of inflammatory cytokines. These receptors also further activate HMGB1 release in the extracellular milieu leading to sustained activation of inflammatory cytokines. Secretion of more cytokines causes oxidative stress and ROS production to nearby cells, leading to insulin resistance via activation of the JNK pathway and inhibition of insulin signaling in muscle, liver, and pancreatic cells. Dysfunction in adipocytes also causes the release of FFA into the blood. It accumulates in liver and muscle cells, which further disturbs the insulin pathway, leading to type 2 diabetes.

Adipocyte Cell Death and HMGB1 Release

Several reasons can trigger cell death, but physical trauma is common. Apoptotic cell death can occur as a physiological process that is characterized by the shrinkage of the cell or maintenance of the membrane integrity of the plasma membrane (Hotchkiss et al. 2009). On the other hand, during the pathological condition, necrotic cell death occurs with the resultant loss of membrane integrity and swelling (Hotchkiss et al. 2009). These events produce a danger signal by intracellular components, which triggers the inflammatory response. Increased adiposity and hypertrophy can cause membrane rupture of the fragile adipocytes in the abdominal cavity (Monteiro et al. 2006). Furthermore, Intraabdominal pressure can cause adipose tissue cell death due to the formation of varying pressure of physical activity and cough in obese subjects (Wilson et al. 2010; Cobb et al. 2005). The necrotic adipose tissues are identified by microscopic observation for accumulation of extracellular debris, ruptured membranes, dilated endoplasmic reticulum, and small cytoplasmic lipid droplets in them (Cinti et al. 2005; Ziegler and Groscurth 2004). Necrotic cells can release HMGB1 protein, activating the inflammatory response by recruiting immune cells in the extracellular milieu (Wagner 2014; Rockenfeller et al. 2010). Visceral adipose tissues are more susceptible to adipose cell death than subcutaneous tissue (Kammoun, Kraakman, and Febbraio 2014).

Adipocyte cell death can also happen due to decreased oxygen supply to the surrounding tissue creating the hypoxic condition (Ye 2009; Trayhurn 2013). Excess adiposity leads to an increase in the expression of hypoxia-inducible genes like leptin, plasminogen activator inhibitor type-1, and matrix metalloproteinase-2 (Hosogai et al. 2007). Hypoxic condition directly affects intracellular organelles causing endoplasmic reticulum stress and oxidative stress (Koumenis et al. 2002). Increasing adiposity increases adipose tissue size in obese subjects, which reach up to 180 μm, while for oxygen diffusion, most adipose tissues are at 100 μm in size (Goossens 2008; Hosogai et al. 2007). Additionally, an increase in the FFA concentration in adipose tissue activates the stress pathways JNK and TNFα, which may lead to adipocyte death (Nguyen et al. 2005).

HMGB1 Linked Macrophage Associated Inflammation in Adipose Tissue

Macrophages are the professional phagocytic immune cells and perform the important functions executed by adipose tissues to maintain tissue homeostasis. In lean adipose tissue, macrophages serve different roles, including lipid buffering, efferocytosis, and angiogenesis (Kosteli et al. 2010; Schipper et al. 2012). Macrophages make up to 5% of the cells in adipose tissue of lean mice and humans; during obesity, the percentage of macrophages increases up to 50% in adipose tissue (Weisberg et al. 2003). Adipose tissue consists of two distinct types of macrophages and is categorized based on its two activation states, classical (M1) and alternative (M2). M1 macrophages are highly inflammatory and are mainly activated by lipopolysaccharide (LPS) and interferon-γ (IFN-γ) (Sica and Mantovani 2012). While alternatively, M2 macrophages are activated by IL-4 or IL-13, promoting tissue repair, inhibiting M1 macrophages, and showing anti-inflammatory properties. (Hill, Reid Bolus, and Hasty 2014; Rhee 2016). The state of imbalance between the M1 and M2 macrophages in adipose tissue can cause increased production of pro-inflammatory factors such as TLR4 (Maury et al. 2009). The M2 macrophages are predominant in lean individuals and are insulin-sensitive, whereas M1 macrophages are high in an obese state with an increasing adipose tissue mass (Weisberg et al. 2003; Kosteli et al. 2010)

Many scientific reports provide evidence that in adipose tissue, HMGB1 causes increased pro-inflammation markers. A recent study observed that DNA repair genes HMGB1 helps in fat distribution in humans. This further suggests that DNA damage can cause increased inflammation in adipose tissue; however, the functional role of HMGB1 on DNA repair is not evaluated in obesity (Rohde et al. 2020). The serum level of HMGB1 is also found to be upregulated in obese children, which further affects the production and release of pro-inflammatory cytokines and lowers the release of adipokines (Arrigo et al. 2013). Yang et al. showed the effect of the HMGB1 in the stimulation of macrophages with an increase in the pro-inflammatory cytokines (Yang et al. 2005). This evidence suggests that HMGB1 has a role in the macrophage infiltration in adipose tissue and the development of inflammation.

HMGB1 Secretion in Adipocyte

Obesity, insulin resistance, and type 2 diabetes are the silent killers and possess a greater metabolic health risk. To early diagnose, manage, and prevent these comorbidities, many biomarkers have been therapeutically targeted. HMGB1 and its receptors are also important biomarkers in the induction of adipose dysfunction in obesity and type 2 diabetes. These biomarkers are mainly found in blood sources and serum levels; therefore, they have emerged as a therapeutic target in obesity. Clinical studies have shown the role of HMGB1 as a biomarker in obesity and diabetes of pregnant women and plasma of the Chinese population (Wang, Qu, and Deng 2015; Giacobbe et al. 2015).

In this section, we have discussed the studies on HMGB1 inhibition and its effect. The treatment with Anti-HMGB1 antibody reduced weight gain and liver inflammation in high-fat-fed mice (Montes et al. 2015). A recent study shows that the salidroside significantly reduced HMGB1 induced hepatic inflammation in db/db mice via downregulating the HMGB1/ RAGE/NF-κB and HMGB1/TLR4/NLRP3 signaling pathways and also controlled hyperglycemia (Yang et al. 2020). Riboflavin, a vitamin B2, has anti-inflammatory activity and research suggests that following stimulation of lipopolysaccharide-induced inflammation in adipocyte-macrophage co-culture (3T3 L1-RAW 264.7), effectively reduces the inflammatory factors such as HMGB1 and TNF α upon riboflavin supplementation (Mazur-Bialy and Pocheć 2016). Kallistatin is an adequate anti-inflammatory endogenous protein found to be effective against cytokine signaling by binding to heparin. The level of kallistatin was found to be reduced in many diseases, including obesity and diabetes, causing increased inflammation in different tissues (Chao, Bledsoe, and Chao 2016).

Natural product compounds are also effective against HMGB1 protein; among them, curcumin has a beneficial role in reducing the HMGB1 signaling pathway in db/db mice and inhibits the production of inflammatory cytokines in diabetic cardiomyopathy (Yan et al. 2020). Another natural compound, metformin, directly binds to the HMGB1 protein and reduces cytokine production *in vitro* and *in vivo* (Horiuchi et al. 2017). Metformin also acts through receptors for RAGE of HMGB1 to reduce hyperglycemia in cardiocytes (Zhang et al. 2014). RAGE receptors can also be found in adipose-specific tissue, and they encode for the Ager gene. The ligand of RAGE receptors is bound explicitly to HMGB1, and AGEs exhibit properties of diabetes and inflammation. However, the report suggests that deletion of Ager RAGE encoding gene in mice adipose tissue protected against HFD-induced obesity and insulin resistance (Hurtado Del Pozo et al. 2019).

Conclusion

Obesity and type 2 diabetes are growing concerns in public health. These diseases arise from the metabolic complications and dysfunction of adipose tissues. Many factors such as low-grade inflammation, cytokines, and adipokines make these diseases more severe. The alarmin protein HMGB1 secreted from dead cells or adipokines in adipose tissues activates the immune response, which further stimulates inflammation and causes dysfunction of tissues. Therefore, it acts as an inflammatory marker in obesity and diabetic subjects and can be therapeutically targeted by many drugs to reduce its secretion. In this direction, more clinical studies and evidence need to be generated using potential drugs along with further studies to understand its signaling pathways.

References

Andersson, U., Yang, H., and Harris, H. (2018). "Extracellular HMGB1 as a therapeutic target in inflammatory diseases." *Expert Opin Ther Targets*, 22 (3), 263-277. https://doi.org/10.1080/14728222.2018.1439924.

Antoine, D. J., Harris, H. E., Andersson, U., Tracey, K. J., and Bianchi, M. E. (2014). "A systematic nomenclature for the redox states of high mobility group box (HMGB) proteins." *Mol Med*, 20 (1), 135-7. https://doi.org/10.2119/molmed.2014.00022.

Arrigo, T., Chirico, V., Salpietro, V., Munafò, C., Ferraù, V., Gitto, E., Lacquaniti, A., and Salpietro, C. (2013). "High-mobility group protein B1: a new biomarker of metabolic syndrome in obese children." *Eur J Endocrinol*, 168 (4), 631-8. https://doi.org/10.1530/eje-13-0037.

Bañuls, C., Rovira-Llopis, S., Lopez-Domenech, S., Diaz-Morales, N., Blas-Garcia, A., Veses, S., Morillas, C., Victor, V. M., Rocha, M., and Hernandez-Mijares, A. (2017). "Oxidative and endoplasmic reticulum stress is impaired in leukocytes from metabolically unhealthy vs healthy obese individuals." *Int J Obes (Lond)*, 41 (10), 1556-1563. https://doi.org/10.1038/ijo.2017.147.

Bianchi, M. E., Beltrame, M., and Paonessa, G. (1989). "Specific recognition of cruciform DNA by nuclear protein HMG1." *Science*, 243 (4894), 1056-1059. https://doi.org/10.1126/science.2922595.

Bianchi, M. E., Falciola, L., Ferrari, S., and Lilley, D. M. (1992). "The DNA binding site of HMG1 protein is composed of two similar segments (HMG boxes), both of which have counterparts in other eukaryotic regulatory proteins." *Embo J.*, 11 (3), 1055-63.

Blüher, M. (2016). "Adipose tissue inflammation: a cause or consequence of obesity-related insulin resistance?" *Clin Sci (Lond)* 130 (18), 1603-14. https://doi.org/10.1042/cs20160005.

Böhm, A., Keuper, M., Meile, T., Zdichavsky, M., Fritsche, A., Häring, H. U., de Angelis, M. H., Staiger, H., and Franko, A. (2020). "Increased mitochondrial respiration of adipocytes from metabolically unhealthy obese compared to healthy obese individuals." *Sci Rep*, 10 (1), 12407. https://doi.org/10.1038/s41598-020-69016-9.

Burhans, M. S., Hagman, D. K., Kuzma, J. N., Schmidt, K. A., and Kratz, M. (2018). "Contribution of Adipose Tissue Inflammation to the Development of Type 2 Diabetes Mellitus." *Compr Physiol*, 9 (1), 1-58. https://doi.org/10.1002/cphy.c170040.

Chao, J., Bledsoe, G., and Chao, L. (2016). "Protective Role of Kallistatin in Vascular and Organ Injury." *Hypertension*, 68 (3), 533-41. https://doi.org/10.1161/hypertensionaha.116.07861.

Cinti, S., Mitchell, G., Barbatelli, G., Murano, I., Ceresi, E., Faloia, E., Wang, S., Fortier, M., Greenberg, A. S., and Obin, M. S. (2005). "Adipocyte death defines macrophage localization and function in adipose tissue of obese mice and humans." *J Lipid Res*, 46 (11), 2347-55. https://doi.org/10.1194/jlr.M500294-JLR200.

Cobb, W. S., Burns, J. M., Kercher, K. W., Matthews, B. D., James Norton, H., and Todd Heniford, B.. (2005). "Normal intraabdominal pressure in healthy adults." *J Surg Res*, 129 (2), 231-5. https://doi.org/10.1016/j.jss.2005.06.015.

Dasu, M. R., Devaraj, S., Park, S., and Jialal, I. (2010). "Increased toll-like receptor (TLR) activation and TLR ligands in recently diagnosed type 2 diabetic subjects." *Diabetes Care*, 33 (4), 861-8. https://doi.org/10.2337/dc09-1799.

Dozio, E., Vianello, E., Briganti, S., Lamont, J., Tacchini, L., Schmitz, G., and Corsi Romanelli, M. M. (2016). "Expression of the Receptor for Advanced Glycation End Products in Epicardial Fat: Link with Tissue Thickness and Local Insulin Resistance in Coronary Artery Disease." *J Diabetes Res*, 2016, 2327341. https://doi.org/10.1155/2016/2327341.

Einck, L., and Bustin, M. (1985). "The intracellular distribution and function of the high mobility group chromosomal proteins." *Exp Cell Res*, 156 (2), 295-310. https://doi.org/10.1016/0014-4827(85)90539-7.

Ferrannini, E., Natali, A., Bell, P., Cavallo-Perin, P., Lalic, N., and Mingrone, G. (1997). "Insulin resistance and hypersecretion in obesity. European Group for the Study of Insulin Resistance (EGIR)." *J Clin Invest*, 100 (5), 1166-73. https://doi.org/10.1172/jci119628.

Gardella, S., Andrei, C., Ferrera, D., Lotti, L. V., Torrisi, M. R., Bianchi, M. E., and Rubartelli, A. (2002). "The nuclear protein HMGB1 is secreted by monocytes via a non-classical, vesicle-mediated secretory pathway." *EMBO Rep*, 3 (10), 995-1001. https://doi.org/10.1093/embo-reports/kvf198.

Giacobbe, A., Grasso, R., Imbesi, G., Salpietro, C. D., Grasso, L., Laganà, A. S., Triolo, O., and Di Benedetto, A. (2015). "High mobility group protein B1: a new biomarker of obesity in pregnant women?" *Gynecol Endocrinol*, 31 (2), 113-5. https://doi.org/10.3109/09513590.2014.964637.

Goossens, G. H. (2008). "The role of adipose tissue dysfunction in the pathogenesis of obesity-related insulin resistance." *Physiol Behav*, 94 (2), 206-18. https://doi.org/10.1016/j.physbeh.2007.10.010.

Gunasekaran, M. K., Virama-Latchoumy, A. L., Girard, A. C., Planesse, C., Guérin-Dubourg, A., Ottosson, L., Andersson, U., Césari, M., Roche, R., and Hoareau, L.. (2016). "TLR4-dependant pro-inflammatory effects of HMGB1 on human adipocyte." *Adipocyte*, 5 (4), 384-388. https://doi.org/10.1080/21623945.2016.1245818.

Guzmán-Ruiz, R., Ortega, F., Rodríguez, A., Vázquez-Martínez, R., Díaz-Ruiz, A., Garcia-Navarro, S., Giralt, M., Garcia-Rios, A., Cobo-Padilla, D., Tinahones, F. J., López-Miranda, J., Villarroya, F., Frühbeck, G., Fernández-Real, J. M., and Malagón, M. M. (2014). "Alarmin high-mobility group B1 (HMGB1) is regulated in human adipocytes in insulin resistance and influences insulin secretion in β-cells." *International Journal of Obesity*, 38 (12), 1545-1554. https://doi.org/10.1038/ijo.2014.36.

Hill, A. A., Reid Bolus, W., and Hasty, A. H. (2014). "A decade of progress in adipose tissue macrophage biology." *Immunol Rev*, 262 (1), 134-52. https://doi.org/10.1111/imr.12216.

Holland, W. L., Bikman, B. T., Wang, L. P., Yuguang, G., Sargent, K. M., Bulchand, S., Knotts, T. A., Shui, G., Clegg, D. J., Wenk, M. R., Pagliassotti, M. J., Scherer, P. E., and Summers, S. A. (2011). "Lipid-induced insulin resistance mediated by the proinflammatory receptor TLR4 requires saturated fatty acid-induced ceramide biosynthesis in mice." *J Clin Invest*, 121 (5), 1858-70. https://doi.org/10.1172/jci43378.

Hori, O., Brett, J., Slattery, T., Cao, R., Zhang, J., Chen, J. X., Nagashima, M., Lundh, E. R., Vijay, S., Nitecki, D. et al. (1995). "The receptor for advanced glycation end products (RAGE) is a cellular binding site for amphoterin. Mediation of neurite outgrowth and co-expression of rage and amphoterin in the developing nervous system." *J Biol Chem*, 270 (43), 25752-61. https://doi.org/10.1074/jbc.270.43.25752.

Horiuchi, T., Sakata N., Narumi Y., Kimura T., Hayashi T., Nagano K., Liu K., Nishibori M., Tsukita S., Yamada T., Katagiri H., Shirakawa R., and Horiuchi H. (2017). "Metformin directly binds the alarmin HMGB1 and inhibits its proinflammatory activity." *J Biol Chem*, 292 (20), 8436-8446. https://doi.org/10.1074/jbc.M116.769380.

Hosogai, N., Fukuhara, A., Oshima, K., Miyata, Y., Tanaka, S., Segawa, K., Furukawa, S., Tochino, Y., Komuro, R., Matsuda, M., and Shimomura, I. (2007). "Adipose tissue hypoxia in obesity and its impact on adipocytokine dysregulation." *Diabetes*, 56 (4), 901-11. https://doi.org/10.2337/db06-0911.

Hotchkiss, R. S., Strasser, A., McDunn, J. E., and Swanson, P. E. (2009). "Cell death." *N Engl J Med*, 361 (16), 1570-83. https://doi.org/10.1056/NEJMra0901217.

Huh, J. Y., Park, Y. J., Ham, M., and Kim, J. B. (2014). "Crosstalk between adipocytes and immune cells in adipose tissue inflammation and metabolic dysregulation in obesity." *Mol Cells*, 37 (5), 365-71. https://doi.org/10.14348/molcells.2014.0074.

Hurtado Del Pozo, C., Ruiz, H. H., Arivazhagan, L., Aranda, J. F., Shim, C., Daya, P., Derk, J., MacLean, M., He, M., Frye, L., Friedline, R. H., Noh, H. L., Kim, J. K., Friedman, R. A., Ramasamy, R., and Schmidt, A. M. (2019). "A Receptor of the Immunoglobulin Superfamily Regulates Adaptive Thermogenesis." *Cell Rep*, 28 (3), 773-791.e7. https://doi.org/10.1016/j.celrep.2019.06.061.

Huttunen, H. J., Fages, C., Kuja-Panula, J., Ridley, A. J., and Rauvala, H. (2002). "Receptor for advanced glycation end products-binding COOH-terminal motif of amphoterin inhibits invasive migration and metastasis." *Cancer Res*, 62 (16), 4805-11.

Isackson, P. J., Bidney, D. L., Reeck, G. R., Neihart, N. K., and Bustin, M. (1980). "High mobility group chromosomal proteins isolated from muclei and cytosol of cultured hepatoma cells are similar." *Biochemistry*, 19 (19), 4466-71. https://doi.org/10.1021/bi00560a013.

Jin, C., Henao-Mejia, J., and Flavell, R. A. (2013). "Innate immune receptors: key regulators of metabolic disease progression." *Cell Metab*, 17 (6), 873-882. https://doi.org/10.1016/j.cmet.2013.05.011.

Kahn, B. B., and Flier, J. S. (2000). "Obesity and insulin resistance." *J Clin Invest*, 106 (4), 473-81. https://doi.org/10.1172/jci10842.

Kahn, S. E., Hull, R. L., and Utzschneider, K. M. (2006). "Mechanisms linking obesity to insulin resistance and type 2 diabetes." *Nature*, 444 (7121), 840-6. https://doi.org/10.1038/nature05482.

Kalderon, B., Mayorek, N., Berry, E., Zevit, N., and Bar-Tana, J. (2000). "Fatty acid cycling in the fasting rat." *Am J Physiol Endocrinol Metab*, 279 (1), E221-7. https://doi.org/10.1152/ajpendo.2000.279.1.E221.

Kammoun, H. L., Kraakman, M. J., and Febbraio, M. A. (2014). "Adipose tissue inflammation in glucose metabolism." *Rev Endocr Metab Disord*, 15 (1), 31-44. https://doi.org/10.1007/s11154-013-9274-4.

Kang, R., Chen, R., Zhang, Q., Hou, W., Wu, S., Cao, L., Huang, J., Yu, Y., Fan, X. G., Yan, Z., Sun, X., Wang, H., Wang, Q., Tsung, A., Billiar, T. R., Zeh, H. J., 3rd, Lotze, M. T., and Tang, D. (2014). "HMGB1 in health and disease." *Mol Aspects Med*, 40, 1-116. https://doi.org/10.1016/j.mam.2014.05.001.

Kanoh, H., Nitta, T., Go, S., Inamori, K. I., Veillon, L., Nihei, W., Fujii, M., Kabayama, K., Shimoyama, A., Fukase, K., Ohto, U., Shimizu, T., Watanabe, T., Shindo, H., Aoki, S., Sato, K., Nagasaki, M., Yatomi, Y., Komura, N., Ando, H., Ishida, H., Kiso, M., Natori, Y., Yoshimura, Y., Zonca, A., Cattaneo, A., Letizia, M., Ciampa, M., Mauri, L., Prinetti, A., Sonnino, S., Suzuki, A., and Inokuchi, J. I. (2020). "Homeostatic and pathogenic roles of GM3 ganglioside molecular species in TLR4 signaling in obesity." *Embo j*, 39 (12), e101732. https://doi.org/10.15252/embj.2019101732.

Kim, J. I., Huh, J. Y., Sohn, J. H., Choe, S. S., Lee, Y. S., Lim, C. Y., Jo, A., Park, S. B., Han, W., and Kim, J. B. (2015). "Lipid-overloaded enlarged adipocytes provoke insulin resistance independent of inflammation." *Mol Cell Biol*, 35 (10), 1686-99. https://doi.org/10.1128/mcb.01321-14.

Kim, S. M., Lun, M., Wang, M., Senyo, S. E., Guillermier, C., Patwari, P., and Steinhauser, M. L. (2014). "Loss of white adipose hyperplastic potential is associated with enhanced susceptibility to insulin resistance." *Cell Metab*, 20 (6), 1049-58. https://doi.org/10.1016/j.cmet.2014.10.010.

Kohlgruber, Ayano, and Lydia Lynch. (2015). "Adipose Tissue Inflammation in the Pathogenesis of Type 2 Diabetes." *Current Diabetes Reports*, 15 (11), 92. https://doi.org/10.1007/s11892-015-0670-x.

Kosteli, A., Sugaru, E., Haemmerle, G., Martin, J. F., Lei, J., Zechner, R., and Ferrante, A. W., Jr. (2010). "Weight loss and lipolysis promote a dynamic immune response in murine adipose tissue." *J Clin Invest*, 120 (10), 3466-79. https://doi.org/10.1172/jci42845.

Koumenis, C., Naczki, C., Koritzinsky, M., Rastani, S., Diehl, A., Sonenberg, N., Koromilas, A., and Wouters, B. G. (2002). "Regulation of protein synthesis by hypoxia via activation of the endoplasmic reticulum kinase PERK and phosphorylation of the translation initiation factor eIF2alpha." *Mol Cell Biol*, 22 (21), 7405-16. https://doi.org/10.1128/mcb.22.21.7405-7416.2002.

Kuehl, L., Rechsteiner, M., and Wu, L. (1985). "Relationship between the structure of chromosomal protein HMG1 and its accumulation in the cell nucleus." *J Biol Chem*, 260 (18), 10361-8.

Kusminski, C. M., Shetty, S., Orci, L., Unger, R. H, and Scherer, P. E. (2009). "Diabetes and apoptosis: lipotoxicity." *Apoptosis*, 14 (12), 1484-95. https://doi.org/10.1007/s10495-009-0352-8.

Kwak, M. S., Rhee, W. J., Lee, Y. J., Kim, H. S., Kim, Y. H., Kwon, M. K., and Shin, J. S.. (2021). "Reactive oxygen species induce Cys106-mediated anti-parallel HMGB1 dimerization that protects against DNA damage." *Redox Biol*, 40, 101858. https://doi.org/10.1016/j.redox.2021.101858.

López-Díez, R., Shekhtman, A., Ramasamy, R., and Schmidt, A. M. (2016). "Cellular mechanisms and consequences of glycation in atherosclerosis and obesity." *Biochim Biophys Acta*, 1862 (12), 2244-2252. https://doi.org/10.1016/j.bbadis.2016.05.005.

Magna, M., and Pisetsky, D. S. (2016). "The Alarmin Properties of DNA and DNA-associated Nuclear Proteins." *Clin Ther*, 38 (5), 1029-41. https://doi.org/10.1016/j.clinthera.2016.02.029.

Matzinger, P. (1994). "Tolerance, Danger, and the Extended Family." *Annual Review of Immunology*, 12 (1), 991-1045. https://doi.org/10.1146/annurev.iy.12.040194.005015.

Maury, E., Noël, L., Detry, R., and Brichard, S. M. (2009). "*In vitro* hyper responsiveness to tumor necrosis factor-alpha contributes to adipokine dysregulation in omental adipocytes of obese subjects." *J Clin Endocrinol Metab*, 94 (4), 1393-400. https://doi.org/10.1210/jc.2008-2196.

Mazur-Bialy, A. I., and Pocheć, E. (2016). "Riboflavin Reduces Pro-Inflammatory Activation of Adipocyte-Macrophage Co-culture. Potential Application of Vitamin B2 Enrichment for Attenuation of Insulin Resistance and Metabolic Syndrome Development." *Molecules*, 21 (12). https://doi.org/10.3390/molecules21121724.

Monden, M., Koyama, H., Otsuka, Y., Morioka, T., Mori, K., Shoji, T., Mima, Y., Motoyama, K., Fukumoto, S., Shioi, A., Emoto, M., Yamamoto, Y., Yamamoto, H., Nishizawa, Y., Kurajoh, M., Yamamoto, T., and Inaba, M. (2013). "Receptor for advanced glycation end products regulates adipocyte hypertrophy and insulin sensitivity in mice: involvement of Toll-like receptor 2." *Diabetes*, 62 (2), 478-89. https://doi.org/10.2337/db11-1116.

Monteiro, R., de Castro, P. M., Calhau, C., and Azevedo, I. (2006). "Adipocyte size and liability to cell death." *Obes Surg*, 16 (6), 804-6. https://doi.org/10.1381/096089206777346600.

Montes, V. N., Subramanian, S., Goodspeed, L., Wang, S. A., Omer, M., Bobik, A., Teshigawara, K., Nishibori, M., and Chait, A. 2015. "Anti-HMGB1 antibody reduces weight gain in mice fed a high-fat diet." *Nutr Diabetes*, 5 (6), e161. https://doi.org/10.1038/nutd.2015.11.

Nativel, B., Marimoutou, M., Thon-Hon, V. G., Gunasekaran, M. K., Andries, J., Stanislas, G., Planesse, C., Da Silva, C. R., Césari, M., Iwema, T., Gasque, P., and Viranaicken, W. (2013). "Soluble HMGB1 is a novel adipokine stimulating IL-6 secretion through RAGE receptor in SW872 preadipocyte cell line: contribution to chronic inflammation in fat tissue." *PLoS One*, 8 (9), e76039. https://doi.org/10.1371/journal.pone.0076039.

Nguyen, M. T., Favelyukis, S., Nguyen, A. K., Reichart, D., Scott, P. A., Jenn, A., Liu-Bryan, R., Glass, C. K., Neels, J. G., and Olefsky, J. M. (2007). "A subpopulation of macrophages infiltrates hypertrophic adipose tissue and is activated by free fatty acids via Toll-like receptors 2 and 4 and JNK-dependent pathways." *J Biol Chem*, 282 (48), 35279-92. https://doi.org/10.1074/jbc.M706762200.

Nguyen, M. T., Satoh, H., Favelyukis, S., Babendure, J. L., Imamura, T., Sbodio, J. I., Zalevsky, J., Dahiyat, B. I., Chi, N. W., and Olefsky, J. M. (2005). "JNK and tumor necrosis factor-alpha mediate free fatty acid-induced insulin resistance in 3T3-L1 adipocytes." *J Biol Chem*, 280 (42), 35361-71. https://doi.org/10.1074/jbc.M504611200.

Orr, J. S., Puglisi, M. J., Ellacott, K. L., Lumeng, C. N., Wasserman, D. H., and Hasty, A. H. (2012). "Toll-like receptor 4 deficiency promotes the alternative activation of adipose tissue macrophages." *Diabetes*, 61 (11), 2718-27. https://doi.org/10.2337/db11-1595.

Ouchi, N., Parker, J. L., Lugus, J. J., and Walsh, K. (2011). "Adipokines in inflammation and metabolic disease." *Nat Rev Immunol*, 11 (2), 85-97. https://doi.org/10.1038/nri2921.

Park, J. S., Arcaroli, J., Yum, H. K., Yang, H., Wang, H., Yang, K. Y., Choe, K. H., Strassheim, D., Pitts, T. M., Tracey, K. J., and Abraham, E. (2003). "Activation of gene expression in human neutrophils by high mobility group box 1 protein." *Am J Physiol Cell Physiol*, 284 (4), C870-9. https://doi.org/10.1152/ajpcell.00322.2002.

Poggi, M., Bastelica, D., Gual, P., Iglesias, M. A., Gremeaux, T., Knauf, C., Peiretti, F., Verdier, M., Juhan-Vague, I., Tanti, J. F., Burcelin, R., and Alessi, M. C. (2007). "C3H/HeJ mice carrying a toll-like receptor 4 mutation are protected against the development of insulin resistance in white adipose tissue in response to a high-fat diet." *Diabetologia*, 50 (6), 1267-76. https://doi.org/10.1007/s00125-007-0654-8.

Prasad, S., and Thakur, M. K. (1990). "Distribution of high mobility group proteins in different tissues of rats during aging." *Biochem Int*, 20 (4), 687-95.

Pusterla, T., de Marchis, F., Palumbo, R., and Bianchi, M. E. (2009). "High mobility group B2 is secreted by myeloid cells and has mitogenic and chemoattractant activities similar to high mobility group B1." *Autoimmunity*, 42 (4), 308-10. https://doi.org/10.1080/08916930902831845.

Rhee, I. (2016). "Diverse macrophages polarization in tumor microenvironment." *Arch Pharm Res* 39 (11), 1588-1596. https://doi.org/10.1007/s12272-016-0820-y.

Rockenfeller, P., Ring, J., Muschett, V., Beranek, A., Buettner, S., Carmona-Gutierrez, D., Eisenberg, T., Khoury, C., Rechberger, G., Kohlwein, S. D., Kroemer, G., and Madeo, F. (2010). "Fatty acids trigger mitochondrion-dependent necrosis." *Cell Cycle*, 9 (14), 2836-42. https://doi.org/10.4161/cc.9.14.12267.

Rohde, K., Rønningen, T., la Cour Poulsen, L., Keller, M., Blüher, M., and Böttcher, Y. (2020). "Role of the DNA repair genes H2AX and HMGB1 in human fat distribution and lipid profiles." *BMJ Open Diabetes Res Care*, 8 (1). https://doi.org/10.1136/bmjdrc-2019-000831.

Rosen, E. D., and Spiegelman, B. M. (2006). "Adipocytes as regulators of energy balance and glucose homeostasis." *Nature*, 444 (7121), 847-53. https://doi.org/10.1038/nature05483.

Saberi, M., Woods, N. B., de Luca, C., Schenk, S., Lu, J. C., Bandyopadhyay, G., Verma, I. M., and Olefsky, J. M. (2009). "Hematopoietic cell-specific deletion of toll-like receptor 4 ameliorates hepatic and adipose tissue insulin resistance in high-fat-fed mice." *Cell Metab*, 10 (5), 419-29. https://doi.org/10.1016/j.cmet.2009.09.006.

Savage, D. B., Petersen, K. F., and Shulman, G. I. (2007). "Disordered lipid metabolism and the pathogenesis of insulin resistance." *Physiol Rev*, 87 (2), 507-20. https://doi.org/10.1152/physrev.00024.2006.

Scaffidi, P., Misteli, T., and Bianchi, M. E. (2002). "Release of chromatin protein HMGB1 by necrotic cells triggers inflammation." *Nature*, 418 (6894), 191-5. https://doi.org/10.1038/nature00858.

Schipper, H. S., Prakken, B., Kalkhoven, E., and Boes, M. (2012). "Adipose tissue-resident immune cells: key players in immunometabolism." *Trends Endocrinol Metab*, 23 (8), 407-15. https://doi.org/10.1016/j.tem.2012.05.011.

Schwartz, M. W., and Kahn, S. E. (1999). "Insulin resistance and obesity." *Nature*, 402 (6764), 860-1. https://doi.org/10.1038/47209.

Senn, J. J. (2006). "Toll-like receptor-2 is essential for the development of palmitate-induced insulin resistance in myotubes." *J Biol Chem*, 281 (37), 26865-75. https://doi.org/10.1074/jbc.M513304200.

Sharman, A. C., Hay-Schmidt, A., and Holland, P. W. (1997). "Cloning and analysis of an HMG gene from the lamprey Lampetra fluviatilis: gene duplication in vertebrate evolution." *Gene*, 184 (1), 99-105. https://doi.org/10.1016/s0378-1119(96)00580-x.

Shekhtman, A., Ramasamy, R., and Schmidt, A. M. (2017). "Glycation & the RAGE axis: targeting signal transduction through DIAPH1." *Expert Rev Proteomics*, 14 (2), 147-156. https://doi.org/10.1080/14789450.2017.1271719.

Sica, Antonio, and Alberto Mantovani. (2012). "Macrophage plasticity and polarization: *in vivo* veritas." *The Journal of Clinical Investigation*, 122 (3), 787-795.

Song, F., Hurtado del Pozo, C., Rosario, R., Zou, Y. S., Ananthakrishnan, R., Xu, X., Patel, P. R., Benoit, V. M., Yan, S. F., Li, H., Friedman, R. A., Kim, J. K., Ramasamy, R., Ferrante, A. W., Jr., and Schmidt, A. M. (2014). "RAGE regulates the metabolic and inflammatory response to high-fat feeding in mice." *Diabetes*, 63 (6), 1948-65. https://doi.org/10.2337/db13-1636.

Spalding, K. L., Arner, E., Westermark, P. O., Bernard, S., Buchholz, B. A., Bergmann, O., Blomqvist, L., Hoffstedt, J., Näslund, E., Britton, T., Concha, H., Hassan, M., Rydén, M., Frisén, J., and Arner, P. (2008). "Dynamics of fat cell turnover in humans." *Nature*, 453 (7196), 783-7. https://doi.org/10.1038/nature06902.

Stros, M. (2010). "HMGB proteins: interactions with DNA and chromatin." *Biochim Biophys Acta*, 1799 (1-2), 101-13. https://doi.org/10.1016/j.bbagrm.2009.09.008.

Stumbo, A. C., Cortez, E., Rodrigues, C. A., Md Henriques, Porto L. C., Barbosa, H. S., and Carvalho, L. (2008). "Mitochondrial localization of non-histone protein HMGB1 during human endothelial cell-Toxoplasma gondii infection." *Cell Biol Int*, 32 (2), 235-8. https://doi.org/10.1016/j.cellbi.2007.08.031.

Suganami, T., Mieda, T., Itoh, M., Shimoda, Y., Kamei, Y., and Ogawa, Y. (2007). "Attenuation of obesity-induced adipose tissue inflammation in C3H/HeJ mice carrying a Toll-like receptor 4 mutation." *Biochem Biophys Res Commun*, 354 (1), 45-9. https://doi.org/10.1016/j.bbrc.2006.12.190.

Tang, D., Billiar, T. R., and Lotze, M. T. (2012). "A Janus tale of two active high mobility group box 1 (HMGB1) redox states." *Mol Med*, 18 (1), 1360-2. https://doi.org/10.2119/molmed.2012.00314.

Travers, A. A. (2003). "Priming the nucleosome: a role for HMGB proteins?" *EMBO Rep* 4 (2), 131-6. https://doi.org/10.1038/sj.embor.embor741.

Trayhurn, P. (2013). "Hypoxia and adipose tissue function and dysfunction in obesity." *Physiol Rev*, 93 (1), 1-21. https://doi.org/10.1152/physrev.00017.2012.

Tsukumo, D. M., Carvalho-Filho, M. A., Carvalheira, J. B., Prada, P. O., Hirabara, S. M., Schenka, A. A., Araújo, E. P., Vassallo, J., Curi, R., Velloso, L. A., and Saad, M. J. (2016). "Statement of Retraction. Loss-of-Function Mutation in Toll-Like Receptor 4 Prevents Diet-Induced Obesity and Insulin Resistance. Diabetes 2007, 56, 1986-1998. DOI: 10.2337/db06-1595." *Diabetes*, 65 (4), 1126-7. https://doi.org/10.2337/db16-rt04a.

van Beijnum, J. R., Buurman, W. A., and Griffioen, A. W. (2008). "Convergence and amplification of toll-like receptor (TLR) and receptor for advanced glycation end products (RAGE) signaling pathways via high mobility group B1 (HMGB1)." *Angiogenesis*, 11 (1), 91-9. https://doi.org/10.1007/s10456-008-9093-5.

Wagner, Marek. (2014). "A dangerous duo in adipose tissue: high-mobility group box 1 protein and macrophages." *The Yale Journal of Biology and Medicine*, 87 (2), 127-133.

Wang, H., Bloom, O., Zhang, M., Vishnubhakat, J. M., Ombrellino, M., Che, J., Frazier, A., Yang, H., Ivanova, S., Borovikova, L., Manogue, K. R., Faist, E., Abraham, E., Andersson, J., Andersson, U., Molina, P. E., Abumrad, N. N., Sama, A., and Tracey, K. J.. (1999). "HMG-1 as a late mediator of endotoxin lethality in mice." *Science*, 285 (5425), 248-51. https://doi.org/10.1126/science.285.5425.248.

Wang, H., Qu, H., and Deng, H. (2015). "Plasma HMGB-1 Levels in Subjects with Obesity and Type 2 Diabetes: A Cross-Sectional Study in China." *PLoS One*, 10 (8), e0136564. https://doi.org/10.1371/journal.pone.0136564.

Weisberg, Stuart, P., Daniel McCann, Manisha Desai, Michael Rosenbaum, Rudolph L. Leibel, and Anthony W. Ferrante, Jr. (2003). "Obesity is associated with macrophage accumulation in adipose tissue." *The Journal of Clinical Investigation*, 112 (12), 1796-1808. https://doi.org/10.1172/JCI19246.

Wilson, A., Longhi, J., Goldman, C., and McNatt, S. (2010). "Intra-abdominal pressure and the morbidly obese patients: the effect of body mass index." *J Trauma* 69 (1), 78-83. https://doi.org/10.1097/TA.0b013e3181e05a79.

Xue, J., Suarez, J. S., Minaai, M., Li, S., Gaudino, G., Pass, H. I., Carbone, M., and Yang, H. (2021). "HMGB1 as a therapeutic target in disease." *J Cell Physiol*, 236 (5), 3406-3419. https://doi.org/10.1002/jcp.30125.

Yan, S. D., Schmidt, A. M., Anderson, G. M., Zhang, J., Brett, J., Zou, Y. S., Pinsky, D., and Stern, D. (1994). "Enhanced cellular oxidant stress by the interaction of advanced glycation end products with their receptors/binding proteins." *J Biol Chem*, 269 (13), 9889-97.

Yan, X., Xu, P., Zhou, L., Lu, J., Tang, H., Zheng, Y., and Cao, H. (2020). "Blockade of high mobility group box 1 involved in the protective of curcumin on myocardial injury in diabetes *in vivo* and *in vitro*." *IUBMB Life*, 72 (5), 931-941. https://doi.org/10.1002/iub.2226.

Yang, D., Postnikov, Y. V., Li, Y., Tewary, P., de la Rosa, G., Wei, F., Klinman, D., Gioannini, T., Weiss, J. P., Furusawa, T., Bustin, M., and Oppenheim, J. J. (2012). "High-mobility group nucleosome-binding protein 1 acts as an alarmin and is critical for lipopolysaccharide-induced immune responses." *J Exp Med*, 209 (1), 157-71. https://doi.org/10.1084/jem.20101354.

Yang, H., Hreggvidsdottir, H. S., Palmblad, K., Wang, H., Ochani, M., Li, J., Lu, B., Chavan, S., Rosas-Ballina, M., Al-Abed, Y., Akira, S., Bierhaus, A., Erlandsson-Harris, H., Andersson, U., and Tracey, K. J. (2010). "A critical cysteine is required for HMGB1 binding to Toll-like receptor 4 and activation of macrophage cytokine release." *Proc Natl Acad Sci U S A*, 107 (26), 11942-7. https://doi.org/10.1073/pnas.1003893107.

Yang, H., Wang, H., and Andersson, U. (2020a). "Targeting Inflammation Driven by HMGB1." *Front Immunol*, 11, 484. https://doi.org/10.3389/fimmu.2020.00484.

Yang, H., Wang, H., Czura, C. J., and Tracey, K. J. (2005). "The cytokine activity of HMGB1." *J Leukoc Biol*, 78 (1), 1-8. https://doi.org/10.1189/jlb.1104648.

Yang, Huan, Haichao Wang, and Ulf Andersson. (2020b). "Targeting Inflammation Driven by HMGB1." *Frontiers in Immunology*, 11 (484). https://doi.org/10.3389/fimmu.2020.00484.

Yang, L., Zhou, L., Wang, X., Wang, W., and Wang, J. (2020). "Inhibition of HMGB1 involved in the protective of salidroside on liver injury in diabetes mice." *Int Immunopharmacol*, 89 (Pt A), 106987. https://doi.org/10.1016/j.intimp.2020.106987.

Ye, J. (2009). "Emerging role of adipose tissue hypoxia in obesity and insulin resistance." *International Journal of Obesity*, (2005) 33 (1), 54-66. https://doi.org/10.1038/ijo.2008.229.

Zatterale, Federica, Michele Longo, Jamal Naderi, Gregory Alexander Raciti, Antonella Desiderio, Claudia Miele, and Francesco Beguinot. (2020). "Chronic Adipose Tissue Inflammation Linking Obesity to Insulin Resistance and Type 2 Diabetes." *Frontiers in Physiology*, 10 (1607). https://doi.org/10.3389/fphys.2019.01607.

Zhang, J., Zhang, L., Zhang, S., Yu, Q., Xiong, F., Huang, K., Wang, C. Y., and Yang, P. (2017). "HMGB1, an innate alarmin, plays a critical role in chronic inflammation of adipose tissue in obesity." *Mol Cell Endocrinol*, 454, 103-111. https://doi.org/10.1016/j.mce.2017.06.012.

Zhang, T., Hu, X., Cai, Y., Yi, B., and Wen, Z. (2014). "Metformin protects against hyperglycemia-induced cardiomyocytes injury by inhibiting the expressions of receptor for advanced glycation end products and high mobility group box 1 protein." *Mol Biol Rep*, 41 (3), 1335-40. https://doi.org/10.1007/s11033-013-2979-3.

Ziegler, U., and Groscurth, P. (2004). "Morphological features of cell death." *News Physiol Sci*, 19, 124-8. https://doi.org/10.1152/nips.01519.2004.

Chapter 4

The Roles of HMGB1 in Liver Injury

Gülsün Memi*

Department of Physiology,
Adıyaman University School of Medicine,
Adıyaman, Turkey

Abstract

The liver, an organ with a vital role in the organism, can become inflamed for various reasons and activate many intracellular and extracellular signalling pathways. To maintain homeostasis in the body, especially the regulation and treatment of liver damage are important. Regardless of the aetiology of liver damage, it causes cell death, causing the cellular contents to be transferred out of the cell and triggering the immune response. These cellular compounds and immune responses trigger the danger or death associated molecular patterns (DAMPs) molecules, especially high mobility group box 1 (HMGB1). HMGB1 has a significant modulatory role as a DAMP molecule, especially in the highly complex pathogenesis of liver injury, by regulating oxidative stress, inflammatory signalling, and autophagy in the hepatocytes, in the highly complex pathogenesis of liver injury by regulating oxidative stress, inflammatory signalling, and autophagy in hepatocytes. HMGB1 has contradictory effects on hepatocytes. While HMGB1 blockade reduced oxidative stress and inflammation in carbon tetrachloride-induced hepatic damage, it did not significantly affect chronic inflammatory responses (fibrosis, regeneration, and inflammation) in chronic liver failure. The effects can carry HMGB1 on hepatocytes. While it cannot significantly impact chronic liver prognosis and fibrosis in hepatocytes, its impact on cellular oxidative stress and inflammation is shown in acute periods. The chapter discussed the cellular and molecular mechanism of

---

* Corresponding Author's Email: glsnmemi@gmail.comю.

In: HMGB1
Editors: Eylem Taskin, Celal Guven and Salih Tunc Kaya
ISBN: 979-8-88697-408-9

hepatocellular death and injury that accounts for the inflammatory and pro-inflammatory role of HMGB1.

Keywords: HMGB1, DAMP, TLR, liver, inflammation

Introduction

The liver has vital roles in homeostatic control and metabolically major roles in the body. The liver has roles in nutrient metabolism and also purification of toxic chemicals. Due to routine metabolic activities or pathological conditions, liver can be overloaded, liver functions could be dysregulated, and liver damage could occur. In addition, chemical pollutants, drugs, and alcohol induce injuries in the liver (Kullak-Ublick et al., 2017; Peralta et al., 2013; Wang et al., 2016). Various factors, such as those mentioned above, can trigger an excessive apoptotic response in hepatic cells. Also, inflammatory and immune responses in hepatic and Kupffer cells against oxidative stress and dysregulation play an important role in liver injury (Catherine Brenner et al., 2013; Heymann et al., 2015; Li et al., 2015).

The liver is essential in combining immune reactions after severe systemic injury. Expressions of Toll-like receptors (TLRs) and advanced improved glycation end products (RAGE) receptors by parenchymal and non-parenchymal cell types in the liver can trigger innate immune cascades through HMGB1 (Chen et al., 2013). This article will explain the inflammatory and pro-inflammatory roles of HMGB1 in diverse liver diseases.

Liver Injury

Xenobiotic intoxications, cholestatic reasons, ischemia-reperfusion injury (I/R), and alcoholic or non-alcoholic hepatic steatosis for reasons other than inflammatory reasons can cause liver injury (Figure 1). The liver injury occurs by uncontrolled cell death (necrosis) or programmed cell death, apoptosis (Eguchi et al., 2014). Apart from etiological factors, acute or chronic liver injury can occur with various mechanisms kind as necroptosis, pyroptosis, and autophagic cell death. Beyond all that, liver injury remains a threatening condition.

Noxae either actively or passively initiates the release of intracellular compounds from hepatocytes. In cell death, intracellular compounds release into the extracellular area by plasma membrane disruptions, and these self-compounds are sensed by the host and trigger immunological responses (Eguchi et al., 2014). Regardless of which one of these noxae causes cell death, all the above conditions are triggered by danger or death-associated molecular patterns (DAMPs), nucleic acids, nuclear proteins such as HMGB1, cytosolic proteins, purine nucleotides, and mitochondrial compounds (Mihm, 2018).

DAMPs alert innate immunity responses either with the cytosolic receptors such as pattern recognition receptors, nucleotide-binding oligomerization domain-like receptors or retinoic acid-inducible gene-like receptors or by membranous TLRs and RAGE receptors (Dolasia et al., 2018). DAMPs stimulate immune responses by activating liver cells such as Kupffer cells, dendritic cells or other immune cells like monocytes, macrophages, neutrophils, and interferons. This is the first response of DAMPs, which is mostly seen in sterile liver inflammation. After the initial DAMPs first response, liver cell death may cause a secondary DAMPs response, resulting in a more significant inflammatory response (Mihm, 2018; Venereau et al., 2015).

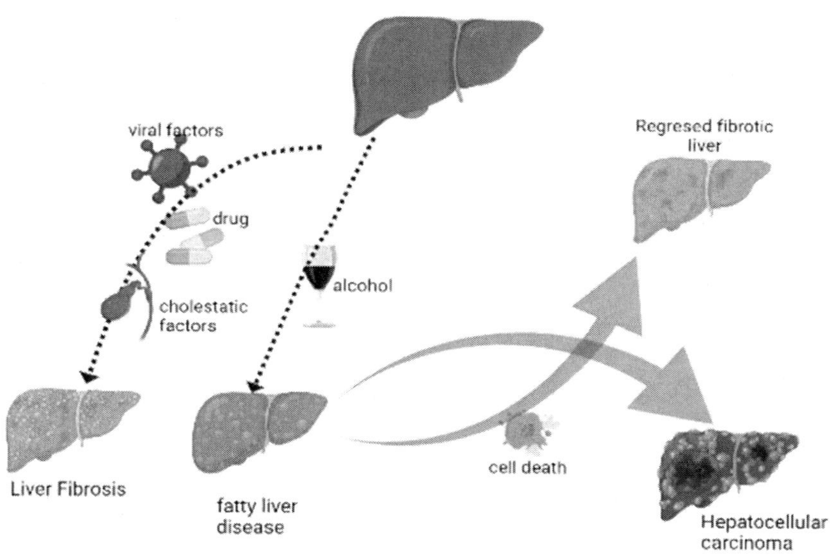

Figure 1. Ethological factors of liver injury. Viral factors, drug toxicity, cholestatic factors and alcohol consumption cause fibrosis, fatty liver, and eventually hepatocellular carcinoma. Appropriate treatment may lead to regression of the fibrosis. Figure modified from Nardo et al. study (Nardo et al., 2021).

HMGB1 and the Liver

HMGB1 is a DAMP molecule expressed in almost all nucleated eukaryotic cells (Lotze & Tracey, 2005; Vénéreau et al., 2015). HMGB1 does predominantly located in the nucleus and functions in DNA-binding and transcription regulation. Also, its translocation occurs in the cytosol and extracellular area. HMGB1 residues in the nucleus bind to DNA in a reduced form, interacting with DNA to involve transcription, nuclear replication, and repair. It plays a vital role in sepsis, tissue damage, inflammation and fibrosis by acting as a DAMP molecule outside the cell (Lotze & Tracey, 2005). HMGB1 is released in two forms, active and passive; it is expressed passively from dead cells and actively from macrophages (S. Gardella et al., 2002; B. Lu et al., 2012). Extracellular HMGB1 can act as both an inflammatory and a pro-inflammatory molecule; it can regulate inflammation by inducing cytokines and chemokine release and stimulating immune cells and promoting tissue repair and angiogenesis (Lotze & Tracey, 2005; Schiraldi et al., 2012; Vénéreau et al., 2015).

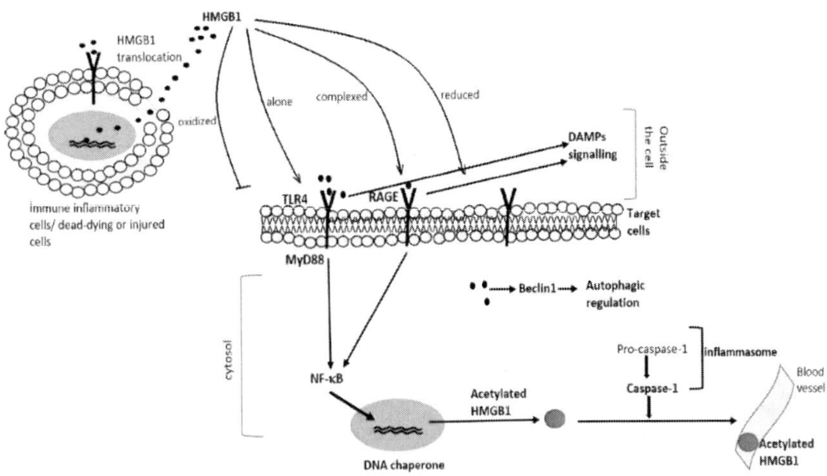

Figure 2. Translocation and functions of HMGB1. HMGB1 can migrate from the core to the external environment by active or passive means. HMGB1 acts as a DNA chaperone as a nuclear factor due to its influence on DNA functions such as replication, transcription, and repair. When HMGB1 release binds with TLRs and RAGE receptors and activates signalling DAMPs outside the target cell, cytosolic HMGB1 induces autophagy by binding to beclin-1 in fibrosis. Figure modified from Chen et al. and Fang et al. studies (Chen et al., 2013; Fang et al., 2021).

Previous findings suggest that, like HMGB1 itself, its specific residues can act as a DAMP molecule and initiate the immune response to mediate the redox state (P. Huebener et al., 2015). Wholly or partially reduced HMGB1 residues play an important role during necrosis by providing chemotaxis and cytokine release from innate immune cells. In addition, HMGB1 released from apoptotic cells in these necrotic areas is wholly oxidized and prevents the emergence of the innate immune response. It has also been shown that HMGB1 from innate immune cells have hyperacetylation activating DAMPs in drug-induced hepatic injury (Bonaldi et al., 2003; X. Ge et al., 2014). Serum alanine aminotransferase (ALT) activity is an important marker used to calculate the response of hepatocytes to the disease and determine the degree of injury (Nathwani et al., 2005). Some experimentations show that HMGB1 release precedes an increase in ALT activity (Rutherford et al., 2012). In another experimental study, total serum HMGB1 increased significantly earlier than ALT levels in acetaminophen-induced liver damage (Antoine et al., 2009). In the same study, an increase in serum hyperacetylated HMGB1 levels were observed simultaneously with the appearance of inflammatory cells in the liver. These studies show that necrosis in immune cells causes HMGB1 leakage and inflammation (Church & Watkins, 2017). The release and functions of HMGB1 are summarized in Figure 2.

Liver diseases, especially acute liver injury, are an important health problem targeting a large population, causing the death of more than one million people each year (Mokdad et al., 2014). Pathogenesis of liver injury is complicated and involves many mechanisms such as oxidative stress responses and cellular death mechanisms like autophagy, necrosis and apoptosis (Jaeschke et al., 2012; Shi et al., 2017). As a nuclear factor, HMGB1 appears to be involved in the pathophysiological process by regulating oxidative stress, inflammatory signalling, and autophagy in hepatocytes (Andersson & Tracey, 2011; Tang et al., 2011). It is thought that HMGB1 acts on NF-κB and MAPK (mitogen-activated protein kinase) through TLR-2, 4 and 9 modulation (Ding et al., 2017). Previous studies showed that HMGB1 has vital functions in diverse liver disorders by the aforementioned pathways or mechanisms yet to be elucidated (X. Ge et al., 2014; Li et al., 2016; Lundbäck et al., 2016; Zhang et al., 2008). Previous studies have shown that HMGB1 has a crucial part in the carbon tetrachloride (CCl4) induced hepatic injury in rodents. That HMGB1 blockade can reduce oxidative stress and inflammation (Chen et al., 2014). Studies show that some chemicals that inhibit the effects of HMGB1 also minimise liver damage. For example, Li et al. showed that HMGB1 has a role in fibrosis that develops with the effect of

CCl4. In this study, HMGB1 was inhibited by giving quercetin and protection from fibrosis was provided (Li et al., 2016). Chloroquine, a drug used in malaria for a long time, has also been shown to improve lipopolysaccharide-induced mortality by inhibiting the NF-KB-mediated inflammatory effects of HMGB1 (M. Yang et al., 2013). It has demonstrated that expression of HMGB1 from liver tissue and plasma HMGB1 concentration increased 2.6-fold in the first 24 hours with CCl4 administration and decreased in both areas with chloroquine therapy. Likewise, Cl4-induced translocation of the nuclear factor HMGB1, which is occurred from the nucleus to the cytoplasmic area, is also demonstrated by immunohistochemical staining (Dai et al., 2018).

HMBG1 and Inflammation

Inflammation triggered by various diseases, like many diseases mentioned above, can transform into liver fibrosis and, in later stages, such as cirrhosis and hepatic cellular carcinoma (HCC). HMGB1 is involved in these inflammatory processes by passive or active pathways mentioned before (Chen et al., 2013).

HMGB1 responses differ in various disease states from other pro-inflammatory cytokines. For example, while the response to inflammation takes 20 hours, it can take effect in 4 hours in cases such as ischemia-reperfusion injury (Xu et al., 2011). HMGB1 also acts as an inflammatory cytokine, generating and enhancing the inflammatory response. It also regulates cytokine secretion by interacting with pro-inflammatory also inflammatory pathways in signal transduction (Fiuza et al., 2003; Park et al., 2004).

Tsung et al. found that hypoxia induces HMGB1 release from hepatocytes via an active function which has promoted by TLR4-dependent reactive oxygen species. These reactive oxygen species induce HMGB1 secretion via a calcium/calmodulin-dependent kinase-dependent process (CaMK) (Tsung et al., 2007). These results showed that CaMK blockade or antioxidants could reduce HMGB1 secretion release and, of course, hepatocellular injury. In another mechanism, tumour necros factor-α and lipopolysaccharides induce an enormous quantity of HMGB1 secretion into the supernatants of liver cell cultures (Z. B. Huang et al., 2013). And this is one of the new physiological effects of HMGB1 on the liver. However, its mechanism of release is unknown.

Anderson et al. asked an essential question about sterile inflammation; how is innate immunity activated for an inflammatory response without a foreign invader? Latest discoveries have shown that the innate immune response can be activated not necessarily by a microbe or pathogen but also by a molecule. Regardless of the cause, released cellular damage products affect different organs and systems that give the appropriate metabolic, hemodynamic, or immunological responses to promote recovery and provoke the development of adaptive immunity (Andersson & Tracey, 2011). HMGB1 is secreted through immune cells as a protective reaction against pathogenic materials or injured or dying cells and regulates both sterile and infectious inflammation (Andersson & Tracey, 2011). Biologically active HMGB1 modulates cellular and systemic responses and other inflammatory cytokines. The HMGB1 response to pathogens or injury occurs in two different mechanisms. The first mechanism is passive release, which occurs rapidly in response to damage to cellular integrity (Tsung et al., 2007) (Scaffidi et al., 2002). And the second mechanism is the active release of HMGB1, which takes much more time than the first mechanism and activates by the interaction of the extracellular materials with plasma membrane receptors (Andersson & Tracey, 2011; S. Gardella et al., 2002; Wang et al., 1999).

Van Golen et al. showed increased plasma HMGB1, nucleosomes and mitochondrial DNA levels in the postoperative period regardless of liver intervention. However, in the same study, ALT levels positively correlated with HMBG1 levels and appear related to ischemic duration liver interventions (Van Golen et al., 2019). Following liver ischemia-reperfusion, HMGB1 levels were raised 1 hour after surgery (Evankovich et al., 2010) and returned to baseline in 6 hours (Van Golen et al., 2019) in mice subjected to I/R. HMGB1 has shown to diffuse from non-immune hepatocytes with ischemia and persist throughout the reperfusion phase (Evankovich et al., 2010).

In the studies of Andersson et al. in which they induced ischemia and inflammatory state in rats, it showed that the use of agents that specifically inhibit the activity of HMGB1 and other inflammatory cytokines (antibodies, antagonists, release inhibitors) interrupt the advancement of tissue damage and suppress inflammatory process (Andersson & Tracey, 2011).

HMGB1 has shown its effects through TLR4 (H. Yang et al., 2010) to moderate cytokine secretion and cellular damage with experimental studies which cause infection, ischemia, and cellular injury. And the use of anti-HMGB1 antibodies in these studies has been shown to decrease the injury in a variety of acute and chronic damages caused by pathogenic factors, immune activation or ischemia (Apetoh et al., 2007) (Andersson & Tracey, 2011). There is strong evidence that inhibiting of HMBG1 in sterile inflammations will reduce the inflammatory damage in the liver. We summarized the release mechanism and biological activities of HMGB1 with sterile and non-sterile factors in Figure 3.

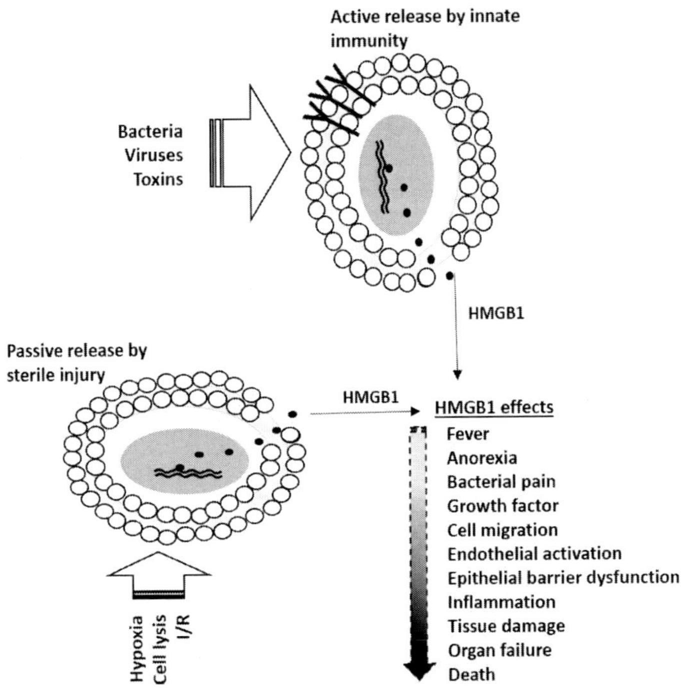

Figure 3. HMGB1 release actively by innate immunity or passively through sterile injury. Infection, necrosis, and apoptosis may elevate HMGB1 release. HMGB1 mediates inflammatory responses. A high level of HMGB1 release is related to epithelial wall dysfunction, organ failure and death. Figure modified from Andersson & Tracey study (Andersson & Tracey, 2011).

Drug Induced Liver Injury

Drug-induced liver injury, a sterile inflammatory liver disease, is one of the leading causes of acute liver failure (ALF) and is responsible for more than half of ALF cases in the United States alone (Lee, 2012). Most drug-induced liver injury cases that lead to ALF has caused by an overdose of acetaminophen (Iorga et al., 2017). Intrinsic hepatotoxic agents such as acetaminophen cause direct chemical damage to liver cells in a dose-dependent manner and with a short latency period (Corsini et al., 2012). The toxicity of acetaminophen is due to the development of the reactive metabolite N-acetyl- p-benzoquinone imine, and its causes the reduction of cellular glutathione (GSH). Mitochondrial damage, oxidative stress, activation of mitogen-activated protein kinases, and hepatocyte necrosis occur by GSH depletion (Iorga et al., 2017).

Activation of the immune system by necrotic cells causes the release of HMGB1 via TLR and RAGE signalling pathways (Antoine et al., 2013). In studies given HMGB1 monoclonal antibody treatment, drug-induced resistance development was observed. According to the results of various studies, the isoforms of HMGB1 may increase differently in diseases. These data may imply that HMGB1 is a more predictive prognostic marker than ALT across multiple liver diseases (Rockey et al., 2010). Based on these data, we may imply that HMGB1 is a possible prognostic biomarker to reveal the adaptive immune process in the progression of drug-induced liver injury (Kubes & Mehal, 2012).

ALD (Alcoholic Liver Disease) and HMGB1

Alcohol-induced liver disease is a preventable form of liver failure. The pathophysiology of ALD is complicated and mainly develops alcohol-induced fatty liver, but very few patients develop advanced stages such as cirrhosis or hepatocellular carcinoma (Rehm et al., 2010). It has been reported that genetic and epigenetic factors are associated with the transition to advanced stages. Especially PNPLA3, TM6SF2, MBOAT7 and HSD17B13 genes are associated with ALD progression (Habash et al., 2022; Seitz et al., 2018).

ALD is caused by the effects of alcohol or its metabolite acetaldehyde's direct effects on diverse hepatic cell types. Another consequence of ethanol is that it suppresses ALT enzyme activity endogenously, so a minimal increase in serum ALT activity is also observed (Woolbright & Jaeschke, 2018). TLR4 signal activation, especially seen in Kupffer cells, constitutes the basic pathogenesis of liver damage. As ethanol disrupts the intestinal barrier and allows the passage of macromolecules, there is an increase in the number of lipopolysaccharides in systemic and portal circulation (Parlesak et al., 2000). So, we can categorise ALD as a non-sterile inflammatory disease of the liver. TLR4, CD14, and lipopolysaccharides binding protein (LBP) knockout mice have found resistance to ALD development (Petrasek et al., 2010; Szabo & Bala, 2010). These data showed that these proteins were important in the pathogenesis of ALD. Lipopolysaccharides, which have circulated from the intestinal microflora, activate the hepatic TLR4 signal in alcoholic liver disease, which may enable the activation of HMBG1. TLR4 activates MyD88-dependent or TRIF-dependent intracellular pathways in Kupffer cells to operate various transcription factors with lipopolysaccharides (Szabo & Bala, 2010). It also shows that the HMGB1-TLR4-MyD88 signalling pathway is critical in the development and especially the progression of NAFLD (Li et al., 2011).

According to current research, sirtuins, particularly sirtuin1 (SIRT1), play a crucial role in the pathogenesis of ALD. SIRT1 is mainly located in the nucleus and enhances the acetic acid synthesis of fatty acids (Zhao et al., 2022) to regulate glucose metabolism and insulin sensitivity (F. Li et al., 2019). The overexpression of SIRT1 down-regulates TLR4, and it also demonstrated that it partially inhibits another pathway, the JNK/NF-KB pathway, through the HMGB1/TLR4/NF-KB pathway (F. Li et al., 2019; Zhao et al., 2022). That means HMGB1 activity is also affected by SIRT1.

Non-Alcoholic Fatty Liver Disease (NAFLD)

Non-alcoholic fatty liver disease (NAFLD) is a disease that is mostly associated with being overweight and increased fatty body weight and is common in developed countries. This disease usually causes inflammation or necrosis in the liver cells following the accumulation of lipids. In addition to cirrhosis, different degrees of fibrosis can also have seen. NAFLD causes lipid accumulation (steatosis) in liver cells, causing damage to the liver at various stages, primarily causing steatohepatitis or fibrosis. Chronic disease can cause cirrhosis, a stage that is difficult to reverse. In the pathogenesis of NAFLD, hepatic triglyceride accumulation, namely steatosis, is seen first. In advanced stages, local or systemic cytokines, adipokines, and endotoxins cause oxidative stress response and drive the mitochondrial dysfunction in hepatic cells. The development of steatosis leaves the liver vulnerable by causing the liver to develop sensitivity to multiple responses in this advanced stage. In the steatosis setting, a de-novo synthesis of lipogenesis has increased (Day & James, 1998). Serum ALT activity, an important parameter used in liver injury, often remains at normal levels in NAFLD and is not perfectly sensitive and specific for NAFLD as a biomarker (Mofrad et al., 2003). However,

HMGB1 translocation to the extracellular fluid in the initial phases of NAFLD increases local hepatocellular damage and systemic inflammation. In the study by Li L et al., TLR4 and cytosolic MyD88 protein-deficient mice given a fatty diet had decreased plasma HMGB1 levels and improved liver functions. And the fact that HMGB1-antibody treatment inhibits the release of TNF-α and IL-6 also shows its effect on proinflammatory cytokines (Li et al., 2011). The HMGB1-TLR4-MyD88 signalling pathway is important in the advancement and evolvement of NAFLD (Li et al., 2011). In the initial phase of NAFLD, the secretion of proinflammatory cytokines is generated by free fatty acids (FFA) by liver parenchymal cells (Rivera et al., 2010). Multiple TLRs mediate NAFLD (Li et al., 2011; Miura et al., 2010; Rivera et al., 2010). Inflammatory cytokine signalling through TLR4 supports NAFLD, such as TNF-α and IL-1β. Also, TLR4 activates caspase, which cooperates with TLR9 to induce IL-1b in Kupffer cells to promote lipid accumulation and cell death in hepatocytes. TLR4 and TLR9 synergistically induce the expression of TNF-α. TLR9-MyD88 signalling pathways mediates IL-1b production in Kupffer cells, advancing NAFLD by stimulating hepatocytes and stellate liver cells (HSCs) (Miura et al., 2010). TLR-2 deficiency causes an increase in TLR-4 signalling, suggesting that TLR-2 expression has a protective role against the progression of NAFLD (Rivera et al., 2010). Among the other TLRs, TLR4 acts as a functional receptor for HMGB1 and has an important role in the pathogenesis of NAFLD (Chen et al., 2013; Li et al., 2011; Rivera et al., 2010).

Viral Hepatitis and HBMG1

Five different types of viruses (hepatitis A to E) have been identified based on viral hepatitis in humans. Among these viruses, hepatitis B (HBV), C (HCV), and D (HDV) viruses may also drive constant infection and hepatitis (El-Serag & Rudolph, 2007). In addition, viral hepatitis has multiple progressive effects with a high chance of advancement to chronic disease that can cause progressive damage to the liver, cirrhosis and even hepatic cancer (Albayrak et al., 2010).

HMGB1 is important role in the inflammatory response and coordination of viral hepatitis. HCV and HBV infection causes translocation of HMGB1 from the nucleus into the cytoplasm and to the extracellular environment (Jung et al., 2011; Liu et al., 2007; Zhou et al., 2011). And extracellular HMGB1 may trigger the interferon antiviral response by binding to TLR-4 during HCV infection (Jung et al., 2011; Zhou et al., 2011). Serum levels of HMGB1 were found to be significantly high in chronic HBV patients and have further increased in acute, chronic liver failure (ACLF) conditions (Cao et al., 2015; Cheng et al., 2008). HMGB1 plays an important part in pathogenesis in liver failure by down-regulation the expression of the Foxp3 and also regulating the immune activity of regulatory T cells in chronic HBV patients (Wang et al., 2010). Most Hepatitis-B viral infections can remain silent for many years (Albayrak et al., 2010). No extranuclear translocation of HMGB1 in such quiet HBV patients and healthy individuals' hepatocytes. HMGB1 translocation is an important clue which is showing that it will be secreted to the extracellular fluid; this situation gives us important information about the hepatic pathophysiology (Zhou et al., 2011). As mentioned at the beginning of our chapter, HMGB1, which acts as a DAMP factor, increases significantly in extracellular areas, especially in severe liver injury, and plays a significant role in diverse hepatocellular damages (Khambu et al., 2019).

Sepsis

Sepsis is the body's inflammatory response to external pathogens such as bacteria, viruses, or parasites. The liver is one of the most frequently affected organs involved in the intervention of sepsis (Chen et al., 2020; Hotchkiss & Karl, 2003). HMGB1 has shown to be overregulated in sepsis (Karlsson et al., 2008).

MicroRNAs (miRNAs) such as miR-103a-3p are associated with inflammation and immunity, and changes in their expression can affect the expression of many proteins. It was observed that miR-103a-3p gene therapy significantly reduced HMGB1 expression in mice treated with a sepsis model. In this study, Chen et al. miR-103a-3p have been shown to protect against liver injury in sepsis by targeting HMGB1. In light of these data, HMGB1 appears to be a gene targeted by miR-103a-3p in regulating the sepsis response 3p in regulating the sepsis response (Chen et al., 2020).

Cell culture studies showed that liver cells secreted HMGB1 protein when exposed to lipopolysaccharides. According to Huang et al., lipopolysaccharides-induced HMGB1 secretion is not caused by cellular damage. Furthermore, damage-related factor lactate dehydrogenase (LDH) did not differ in cultured cells. This finding suggests that non-apoptotic liver cells could release HMBG1 to trigger an inflammatory cascade (Z.-B. Huang et al., 2013).

The activity of HMGB1 varies according to its presence in the nucleus or cytosol or extracellular fluid. HMGB1 in the nucleus contributes to gene expression. When HMGB1 moves to the cytosol, it acts as a late inflammatory mediator after proinflammatory cytokines such as interleukins and tumour necrosis factor (Erlandsson Harris & Andersson, 2004; Z.-B. Huang et al., 2013; Wang, 1999). With lipopolysaccharides induction cultured cells, HMBG1 has released into the extracellular space. This order may be a caution alert to the hepatocytes and affects endothelial cells taxis for the elevated aggregation for tissue repair. Also, external stimuli induced HMGB1 release into the cytosolic liver area, such as lipopolysaccharide induction, may be important for cellular protection like viral hepatitis (Z.-B. Huang et al., 2013).

Previous researchers thought that HMGB1 was a late-phase cytokine associated with sepsis. Most data from animal studies imply that HMGB1 elevation is also related to sepsis severity (Wang et al., 1999). However, few human studies have shown a lower association between sepsis and HMGB1. According to the Karlsson study, HMGB1 has moderately elevated in sepsis and septic shock patients compared to healthy volunteers (Karlsson et al., 2008). Therefore, current studies suggest plasma HMGB1 is not a predictive marker for sepsis severity.

Acute Chronic Liver Failure (ACLF)

In terms of the formation process, liver damage can be acute or chronic. In addition, exacerbated jaundice and diffuse coagulopathy were seen in acute periods due to chronic liver disease. The occurrence of severe acute episodes based on such chronic illness is called acute chronic liver failure (ACLF) (Sarin et al., 2019). ACLF has determined by diverse complications like ascites, hepatic encephalopathy, infection, and extrahepatic organ failure (Hernaez et al., 2017). ACLF can cause death in a short time by causing a bad prognosis (Hernaez et al., 2017; Levesque et al., 2017). While ACLF generally occurs due to alcoholic liver disease, inappropriate drug use and infection in western societies, hepatitis and HBV in eastern cultures (Huang et al., 2019). There are unknown processes in the pathogenesis of ACLF. Still, it is known that inappropriate immune systems or inflammatory responses and, of course, DAMP molecules increase the disease process and severity (Du et al., 2018). HMGB1 has significant intracellular and extracellular biological activities (Wang et al., 1999; H. Yang et al., 2013), and it affects the pathological process of ACLF by acting as DAMPs when it has released outside the cell membrane (Yamamoto & Tajima, 2017; Yang et al., 2015).

The studies showed that the total amount of HMBG1 and cytoplasmic HMBG1 ratio has increased by ACLF in rats (Fang et al., 2021). HMBG1 receptor expression levels increased in the cytoplasm of the damaged hepatocytes. Also, ACLF increases HMGB1-mediated signalling pathway proteins like NF-kB-p65, iNOS, and Cox-2 (Fang et al., 2021). HMGB1 has accelerating effects on ACLF by activating the caspase-1 pathway, which is causing pyroptosis. Previous studies have shown that HMGB1 inhibition downregulates the Caspase-1-dependent mechanism and reduces liver damage in ACLF by inducing pyroptosis in hepatic cells. With inhibition HMGB1 and Caspase-1 in ACLF has an improving downregulatory effect (W. Hou et al., 2021). Quercetin, which affects HMGB1, decreases ACLF-induced HMBG1 expression and translocation in rats (Fang et al., 2020). And quercetin inhibits the HMGB1-mediated signalling pathway by changing its protein expression (Fang et al., 2021). As a result, HMGB1 has critical functions in the pathophysiology of ACLF, and its inhibition reduces the degree of ACL (W. Hou et al., 2021).

Hepatocellular Death and HMGB1

Autophagy, a process in which the cell digests itself, is seen in the pathogenesis of liver cells like other mammalian cells. Under stressful intracellular conditions, autophagy allows organelles to expand and decompose, and residues of proteins accumulate, but the cell survives (Kewei Wang, 2015). Autophagy renews the organelles in the liver cells and contributes to its more effective and quality functioning. Based on the autophagic activities and functions, it seems that a disruption or failure in autophagic activity will cause diseases, such as hepatitis or carcinoma, and vice versa. On the other hand, as a normal physiological process, autophagic activity decreases with the activation of various metabolic pathways with ageing. This decreased autophagic activity plays an important role in the pathogenesis of many liver diseases (Ueno & Komatsu, 2017). And there is some crosstalk between the different programmed cellular death mechanisms like apoptosis, pyroptosis, necroptosis and necrosis. Autophagy is the most common cell death mechanism and can be induced in stressful conditions. Still, also it can act together or independently with apoptosis to lead to cell death. Autophagy response is observed at an earlier than apoptotic cell death

(Kewei Wang, 2015). And the mechanism that regulates hepatic apoptosis is autophagic activity. Although the mechanisms moderating the relations between apoptosis and autophagy have not yet been determined. As mentioned above, basal autophagy provides the transformation of the damaged and aged organelles and cellular homeostasis. In addition, the proportion between the autophagic answer and apoptotic response determines the severity and also the progression of the diseases (K. Wang, 2015).

Apoptosis and autophagy processes are partners; cytoprotective apoptosis may suppress autophagy and vice versa. Autophagy causes mass degradation of the cytoplasm and mitochondria (Rabinowitz & White, 2010; Wu et al., 2010). The influences on the mitochondrial process may modulate hepatic apoptosis (Kewei Wang, 2015). Pathological changes could lead to deficiency in autophagic response in the liver, leading to hepatomegaly, cellular injury, inflammation, fibrosis, cellular expansion, and tumour development (Stauffer et al., 2012). As it is known, cell death can cause inflammation, and tissue repair and fibrosis formation, which repeats in a vicious cycle, causes tissue damage in chronic diseases (Kuraishy et al., 2011). Imbalances between autophagy and apoptosis in liver diseases cause liver cirrhosis and HCC development. For example, viral hepatitis, NAFLD and hepatic fibrosis cause autophagic dysfunction and adversely affect the prognosis of the disease (Ueno & Komatsu, 2017). In addition to diseases, nutritional status and liver metabolism influence cellular death response. Recent studies suggest that there is an emerging relationship between autophagy and hepatocellular metabolism. For example, high-fat diet feed mice have reduced autophagic protein expression and decreased autophagic activity in the liver (L. Yang et al., 2010).

Hepatic HMGB1 may involve inflammation, fibrosis, and damage in different circumstances such as viral infection (hepatitis C) (Arriazu et al., 2017), drug-induced injury (Xiaodong Ge et al., 2014; Peter Huebener et al., 2015), or ischemia-reperfusion (Tsung et al., 2005). Khambu et al. showed that in autophagy-deficient mice, HMBG1 mediates the hepatocellular response to injuries. HMGB1 activates molecular pathways through nuclear factor erythroid 2-related factor 2 (Nrf2) and inflammasomes to promote dendritic cells, hepatic progenitor cells, and tumour growth in the autophagy-deficient liver (Khambu et al., 2018). Various post-translational modifications in the cytosol, such as phosphorylation and methylation or acetylation, cause HMGB1 to translocation from the nucleus to the cytosol, resulting in its accumulation in the cytosolic area (Lu et al., 2014). During the translocation, HMGB1 is transported to the cytosol or extracellular fluids from the nucleus;

it must first undergo acetylation through an active process (Bonaldi et al., 2003). However, HMG1 is secreted from dead or dying cells by a passive approach and is not acetylated (Scaffidi et al., 2002). Constant activation of Nrf2 has been demonstrated in autophagy-deficient hepatocytes, which are thought to inhibit the nucleus-cytosol translocation of HMGB1 and the release of HMGB1 (Komatsu et al., 2010). The p62, the selective autophagic substrate, induces the Nrf2 proteins endogenously. Excessive accumulation of p62 in autophagy-defective livers was responsible for persistent activation of Nrf2 (Khambu et al., 2018; Komatsu et al., 2010). The p62 binds to Keap1, a Cullin-3 ubiquitin ligase component, to bind to Nrf2 (Komatsu et al., 2010). Overproduction of p62 or a deficiency in autophagy processes can create a competition against this binding of Nrf2 and Keap1 that we mentioned can significantly reduce its activity. This reduction in binding-related activity results in Nrf2 consolidation and activates Nrf2 target gene transcriptions (Komatsu et al., 2010). Khambu et al. observed the systemic treatments of potent synthetic triterpenoid derivatives triggering Nrf2 activity while decreasing HMBG1 levels in the liver for 4-6 days by binding to KEAP1 and disrupting its interaction with NRF2. On the contrary, serum HMBG1 elevated (Khambu et al., 2018). An interesting finding of this study is that the amount of HMGB1 release of hepatocytes is susceptible to the treating of triterpenoid derivatives. From this point of view, activation of Nrf2 by systemic treatment with a pharmacological agent is possible. Furthermore, the induction of this activation may result in NRF2-mediated HMGB1 release from hepatocytes without tissue and hepatocyte damage in the organism (Khambu et al., 2018).

HMBG1, an important DAMPs molecule, is important in inflammation, fibrosis, and tissue repair mechanisms. Despite the absence of HMGB1 in the apoptosis-deficient liver, the appearance of inflammation and fibrosis may indicate that multiple mechanisms are involved. It is thought that HMBG1 has a role in tissue repair effect on the enlargement of dendritic cells and hepatic progenitor cells (Lu et al., 2015; Raven et al., 2017; Venereau et al., 2015). Besides immune cells, monocytes and macrophages can also actively secrete HMGB1 through inflammasome-mediated mechanisms (Lamkanfi et al., 2010; Ben Lu et al., 2012), which regulates the release of caspase activation-dependent cytokines (Ben Lu et al., 2012). The liver requires the participation of inflammasomes to actively release HMBG1 in autophagy deficiency. Caspase-1 act as a common effector molecule for releasing HMBG1, IL-1β, and IL-18 (Stefania Gardella et al., 2002; Ben Lu et al., 2012). Caspase-1 encoding in the apoptosis-deficient liver suppressed nuclear HMBG1 loss without being affected by p62 accumulation or Nrf2 activation (Khambu et al., 2018).

Acetaminophen, one of the drugs that cause the most liver damage due to drug toxicity, induces mitochondrial oxidative stress during hepatotoxicity, causing mitochondrial dysfunction and, thus, cell necrosis observed. Oxidative stress causes mitochondrial peroxynitrite formation, which triggers cell necrosis by increasing the membrane permeability transition, which causes mitochondrial DNA damage and nitration of mitochondrial proteins. In addition, existing studies have shown that apoptosis is also associated with cell death by a caspase-independent pathway, leading to necrosis and necroptosis (K. Wang, 2015; Wen et al., 2020).

Pyroptosis is a cell death mechanism, an inflammation-related necrosis type, is caspase-1-dependent and triggers the secretion of IL-1b, IL-18, and IL-33 in immune cells like macrophages, monocytes, or phagocytes (Bergsbaken et al., 2009; Bortoluci & Medzhitov, 2010; Yao Wang et al., 2020). Apart from Caspase-1, pyroptosis has also triggered by Caspase-4/-5/-11- activity, and this activity is mediated by inflammation and Gasdermin D (GSDMD) (Fink & Cookson, 2006; Hu et al., 2020; H. Li et al., 2019). With GSDM activity, cells swell with endogenous or exogenous stimuli and eventually rupture, forming multiple pores (Fink & Cookson, 2006). Then lactate dehydrogenase (LDH) and electrolyte compounds of the intracellular environment flow outside the cellular membrane and induce the synthesis/release of cytokines, inflammatory components, and adhesion molecules. And these factors cause expansion locally and produce systemic inflammatory responses to worsening tissue damage (Deng et al., 2021). In addition, during gasdermin-D activity, proinflammatory cytokines, like IL-1β and IL-18, are released from the damaged cells (Lacey et al., 2018). Also, in pyroptosis, the cysteine protease pro-caspase-1 integrates with NLRP3 (nucleotide-binding oligomerization domain-like receptor protein-3), creating inflammasomes (Cookson & Brennan, 2001). Hepatic pyroptosis mainly occurs in Kupffer cells (Labbé & Saleh, 2008; Y. Wang et al., 2020). Previous studies showed that plasma HMGB1 levels increased in liver injury and ACLF, and HMGB1 can induce pyroptosis in these disease scenarios. The primary pathophysiological response in these diseases is an inflammatory response, which is very similar to the pyroptosis mechanism. Wang et al. showed the role of HMGB1 on acute liver injury-induced pyroptotic response in hepatocytes (Weixin Hou et al., 2021; Y. Wang et al., 2020). When HMGB1 has translocated out of the cell, it takes on a new identity as a DAMP molecule and activates immune cells to produce tumour necrosis factor-α (TNF-α) (Xu et al., 2014). M1 macrophages are the primary source of proinflammatory factors released with cell damage during pyroptosis and play an essential role in acute liver failure. The study by Wang et al. showed that TNF-α can induce pyroptosis and the release of HMGB1 in M1 macrophages (Y. Wang et al., 2020). The effects of HMGB1 can be suppressed by inhibition and mutation of TNF-a and inhibit the pyroptosis process. The primary mechanism of action of HMGB1 in pyroptosis is to initiate the process by binding to RAGE and TLR4 receptors (Sun et al., 2020) in ACLF. The inhibitions of HMGB1 in ACLF showed that ACLF progressively ameliorated liver injury by significantly reducing the amount of TLR4 and RAGE receptors and significantly reducing the amount of HMGB1 in both liver tissue and serum (Weixin Hou et al., 2021).

HMGB1 has extensively been released from hepatocytes in acute liver failure, ischemia-reperfusion injury (sterile inflammation), and hepatitis B virus (HBV) infection patients (C. Brenner et al., 2013). In addition, HMGB1 seems to inhibit cellular death through the induce autophagy in pancreatic tumour cells via RAGE receptors. Still, we have not been able to show the similar processes in the liver yet (Kang et al., 2010).

According to Hernandez et al., the deficiency of hepatocellular specific HMGB1 does not significantly affect the development of fibrosis or chronic injury responses such as cellular regeneration and local inflammation. Still, it has been shown to inhibit ductal and progenitor cell expansion and hepatic cell metaplasia. According to this study, as a DAMP molecule, HMGB1 regulates specific cell death mechanisms in the liver, which is already damaged, i.e., chronically injured (Hernandez et al., 2018). The fact that the translocation of HMGB1 is not seen in the early phase but at a later time also supports these data (Dai et al., 2018).

Conclusion

The incidence of liver diseases increases with many factors such as excessive calorie consumption, low physical activity, and overdose of hepatotoxic drugs, especially in western countries. Although liver damage develops with many factors and etiologies, there are limited treatment options. Considering the pathophysiology of liver diseases, cell death and inflammation response can be seen in disease regression or liver failure. A viable therapeutic strategy for disease regression may include using hepatoprotective and anti-inflammatory agents in different combinations or alone. Studies on the modulatory roles or potential anti-inflammatory roles of DAMP molecules, especially HMGB1, have increased the systemic inflammatory response. In this section, we tried to compile studies investigating the effects of HMGB1 on various liver diseases. HMGB1 modulates the hepatic injury with an anti-inflammatory or inflammatory role depending on the cause of injury. And we assume that HMGB1 modulates injury response with other DAMP molecules, not alone.

References

Albayrak, A., Uyanik, M. H., Cerrah, S., Altas, S., Dursun, H., Demir, M., & Uslu, H. (2010). Is HMGB1 a new indirect marker for revealing fibrosis in chronic hepatitis and a new therapeutic target in treatment? *Viral Immunol, 23*(6), 633-638. https://doi.org/10.1089/vim.2010.0080.

Andersson, U., & Tracey, K. J. (2011). HMGB1 Is a Therapeutic Target for Sterile Inflammation and Infection. *Annual Review of Immunology, 29*(1), 139-162. https://doi.org/10.1146/annurev-immunol-030409-101323.

Antoine, D. J., Lewis, P. S., Goldring, C. E., & Park, B. K. (2013). Are we closer to finding biomarkers for identifying acute drug-induced liver injury? *Biomark Med, 7*(3), 383-386. https://doi.org/10.2217/bmm.13.42.

Antoine, D. J., Williams, D. P., Kipar, A., Jenkins, R. E., Regan, S. L., Sathish, J. G., Kitteringham, N. R., & Park, B. K. (2009). High-mobility group box-1 protein and keratin-18, circulating serum proteins informative of acetaminophen-induced necrosis and apoptosis in vivo. *Toxicol Sci, 112*(2), 521-531. https://doi.org/10.1093/toxsci/kfp235.

Apetoh, L., Ghiringhelli, F., Tesniere, A., Obeid, M., Ortiz, C., Criollo, A., Mignot, G., Maiuri, M. C., Ullrich, E., Saulnier, P., Yang, H., Amigorena, S., Ryffel, B., Barrat, F. J., Saftig, P., Levi, F., Lidereau, R., Nogues, C., Mira, J. P., . . . Zitvogel, L. (2007). Toll-like receptor 4-dependent contribution of the immune system to anticancer chemotherapy and radiotherapy. *Nat Med, 13*(9), 1050-1059. https://doi.org/10.1038/nm1622.

Arriazu, E., Ge, X., Leung, T.-M., Magdaleno, F., Lopategi, A., Lu, Y., Kitamura, N., Urtasun, R., Theise, N., Antoine, D. J., & Nieto, N. (2017). Signalling via the osteopontin and high mobility group box-1 axis drives the fibrogenic response to liver injury. *Gut, 66*(6), 1123-1137. https://doi.org/10.1136/gutjnl-2015-310752.

Bergsbaken, T., Fink, S. L., & Cookson, B. T. (2009). Pyroptosis: host cell death and inflammation. *Nature Reviews Microbiology, 7*(2), 99-109. https://doi.org/10.1038/nrmicro2070.

Bonaldi, T., Talamo, F., Scaffidi, P., Ferrera, D., Porto, A., Bachi, A., Rubartelli, A., Agresti, A., & Bianchi, M. E. (2003). Monocytic cells hyperacetylate chromatin protein HMGB1 to redirect it towards secretion. *Embo J, 22*(20), 5551-5560. https://doi.org/10.1093/emboj/cdg516.

Bortoluci, K. R., & Medzhitov, R. (2010). Control of infection by pyroptosis and autophagy: role of TLR and NLR. *Cell Mol Life Sci*, *67*(10), 1643-1651. https://doi.org/10.1007/s00018-010-0335-5.

Brenner, C., Galluzzi, L., Kepp, O., & Kroemer, G. (2013). Decoding cell death signals in liver inflammation. *Journal of Hepatology*, *59*(3), 583-594. https://doi.org/10.1016/j.jhep.2013.03.033.

Cao, Z., Li, F., Xiang, X., Liu, K., Liu, Y., Tang, W., Lin, L., Guo, Q., Bao, S., Xie, Q., & Wang, H. (2015). Circulating cell death biomarker: good candidates of prognostic indicator for patients with hepatitis B virus related acute-on-chronic liver failure. *Sci Rep*, *5*, 14240. https://doi.org/10.1038/srep14240.

Chen, L., Lu, Q., Deng, F., Peng, S., Yuan, J., Liu, C., & Du, X. (2020). miR-103a-3p Could Attenuate Sepsis-Induced Liver Injury by Targeting HMGB1. *Inflammation*, *43*(6), 2075-2086. https://doi.org/10.1007/s10753-020-01275-0.

Chen, M., Huang, W., Wang, C., Nie, H., Li, G., Sun, T., Yang, F., Zhang, Y., Shu, K., Wang, C., & Gong, Q. (2014). High-mobility group box 1 exacerbates CCl_4-induced acute liver injury in mice. *Clin Immunol*, *153*(1), 56-63. https://doi.org/10.1016/j.clim.2014.03.021.

Chen, R., Hou, W., Zhang, Q., Kang, R., Fan, X. G., & Tang, D. (2013). Emerging role of high-mobility group box 1 (HMGB1) in liver diseases. *Mol Med*, *19*(1), 357-366. https://doi.org/10.2119/molmed.2013.00099.

Cheng, B. Q., Jia, C. Q., Liu, C. T., Lu, X. F., Zhong, N., Zhang, Z. L., Fan, W., & Li, Y. Q. (2008). Serum high mobility group box chromosomal protein 1 is associated with clinicopathologic features in patients with hepatocellular carcinoma. *Dig Liver Dis*, *40*(6), 446-452. https://doi.org/10.1016/j.dld.2007.11.024.

Church, R. J., & Watkins, P. B. (2017). The transformation in biomarker detection and management of drug-induced liver injury. *Liver Int*, *37*(11), 1582-1590. https://doi.org/10.1111/liv.13441.

Cookson, B. T., & Brennan, M. A. (2001). Pro-inflammatory programmed cell death. *Trends Microbiol*, *9*(3), 113-114. https://doi.org/10.1016/s0966-842x(00)01936-3.

Corsini, A., Ganey, P., Ju, C., Kaplowitz, N., Pessayre, D., Roth, R., Watkins, P. B., Albassam, M., Liu, B., Stancic, S., Suter, L., & Bortolini, M. (2012). Current challenges and controversies in drug-induced liver injury. *Drug Saf*, *35*(12), 1099-1117. https://doi.org/10.1007/bf03261997.

Dai, C., Xiao, X., Li, D., Tun, S., Wang, Y., Velkov, T., & Tang, S. (2018). Chloroquine ameliorates carbon tetrachloride-induced acute liver injury in mice via the concomitant inhibition of inflammation and induction of apoptosis. *Cell Death Dis*, *9*(12), 1164. https://doi.org/10.1038/s41419-018-1136-2.

Day, C. P., & James, O. F. (1998). Steatohepatitis: a tale of two "hits"? *Gastroenterology*, *114*(4), 842-845. https://doi.org/10.1016/s0016-5085(98)70599-2.

Deng, J., Tan, W., Luo, Q., Lin, L., Zheng, L., & Yang, J. (2021). Long Non-coding RNA MEG3 Promotes Renal Tubular Epithelial Cell Pyroptosis by Regulating the miR-18a-3p/GSDMD Pathway in Lipopolysaccharide-Induced Acute Kidney Injury. *Front Physiol*, *12*, 663216. https://doi.org/10.3389/fphys.2021.663216.

Ding, J., Cui, X., & Liu, Q. (2017). Emerging role of HMGB1 in lung diseases: friend or foe. *J Cell Mol Med*, *21*(6), 1046-1057. https://doi.org/10.1111/jcmm.13048.

Dolasia, K., Bisht, M. K., Pradhan, G., Udgata, A., & Mukhopadhyay, S. (2018). TLRs/NLRs: Shaping the landscape of host immunity. *Int Rev Immunol*, *37*(1), 3-19. https://doi.org/10.1080/08830185.2017.1397656.

Du, X. X., Shi, Y., Yang, Y., Yu, Y., Lou, H. G., Lv, F. F., Chen, Z., & Yang, Q. (2018). DAMP molecular IL-33 augments monocytic inflammatory storm in hepatitis B-precipitated acute-on-chronic liver failure. *Liver Int*, *38*(2), 229-238. https://doi.org/10.1111/liv.13503.

Eguchi, A., Wree, A., & Feldstein, A. E. (2014). Biomarkers of liver cell death. *Journal of Hepatology*, *60*(5), 1063-1074. https://doi.org/10.1016/j.jhep.2013.12.026

El-Serag, H. B., & Rudolph, K. L. (2007). Hepatocellular carcinoma: epidemiology and molecular carcinogenesis. *Gastroenterology*, *132*(7), 2557-2576. https://doi.org/10.1053/j.gastro.2007.04.061.

Erlandsson Harris, H., & Andersson, U. (2004). Mini-review: The nuclear protein HMGB1 as a proinflammatory mediator. *European Journal of Immunology*, *34*(6), 1503-1512. https://doi.org/10.1002/eji.200424916.

Evankovich, J., Cho, S. W., Zhang, R., Cardinal, J., Dhupar, R., Zhang, L., Klune, J. R., Zlotnicki, J., Billiar, T., & Tsung, A. (2010). High Mobility Group Box 1 Release from Hepatocytes during Ischemia and Reperfusion Injury Is Mediated by Decreased Histone Deacetylase Activity. *Journal of Biological Chemistry*, *285*(51), 39888-39897. https://doi.org/10.1074/jbc.m110.128348.

Fang, P., Dou, B., Liang, J., Hou, W., Ma, C., & Zhang, Q. (2021). Quercetin Reduces Oxidative Stress and Apoptosis by Inhibiting HMGB1 and Its Translocation, Thereby Alleviating Liver Injury in ACLF Rats. *Evid Based Complement Alternat Med*, *2021*, 2898995. https://doi.org/10.1155/2021/2898995.

Fang, P., Liang, J., Jiang, X., Fang, X., Wu, M., Wei, X., Yang, W., Hou, W., & Zhang, Q. (2020). Quercetin Attenuates d-GaLN-Induced L02 Cell Damage by Suppressing Oxidative Stress and Mitochondrial Apoptosis via Inhibition of HMGB1. *Front Pharmacol*, *11*, 608. https://doi.org/10.3389/fphar.2020.00608.

Fink, S. L., & Cookson, B. T. (2006). Caspase-1-dependent pore formation during pyroptosis leads to osmotic lysis of infected host macrophages. *Cellular Microbiology*, *8*(11), 1812-1825. https://doi.org/10.1111/j.1462-5822.2006.00751.x.

Fiuza, C., Bustin, M., Talwar, S., Tropea, M., Gerstenberger, E., Shelhamer, J. H., & Suffredini, A. F. (2003). Inflammation-promoting activity of HMGB1 on human microvascular endothelial cells. *Blood*, *101*(7), 2652-2660. https://doi.org/10.1182/blood-2002-05-1300.

Gardella, S., Andrei, C., Ferrera, D., Lotti, L. V., Torrisi, M. R., Bianchi, M. E., & Rubartelli, A. (2002). The nuclear protein HMGB1 is secreted by monocytes via a non-classical, vesicle-mediated secretory pathway. *EMBO Rep*, *3*(10), 995-1001. https://doi.org/10.1093/embo-reports/kvf198.

Ge, X., Antoine, D. J., Lu, Y., Arriazu, E., Leung, T.-M., Klepper, A. L., Branch, A. D., Fiel, M. I., & Nieto, N. (2014). High Mobility Group Box-1 (HMGB1) Participates in the Pathogenesis of Alcoholic Liver Disease (ALD). *Journal of Biological Chemistry*, *289*(33), 22672-22691. https://doi.org/10.1074/jbc.m114.552141

Habash, N. W., Sehrawat, T. S., Shah, V. H., & Cao, S. (2022). Epigenetics of alcohol-related liver diseases. *JHEP Rep*, *4*(5), 100466. https://doi.org/10.1016/j.jhepr.2022.100466.

Hernaez, R., Solà, E., Moreau, R., & Ginès, P. (2017). Acute-on-chronic liver failure: an update. *Gut*, *66*(3), 541-553. https://doi.org/10.1136/gutjnl-2016-312670.

Hernandez, C., Huebener, P., Pradere, J. P., Antoine, D. J., Friedman, R. A., & Schwabe, R. F. (2018). HMGB1 links chronic liver injury to progenitor responses and hepatocarcinogenesis. *J Clin Invest*, *128*(6), 2436-2451. https://doi.org/10.1172/jci91786.

Heymann, F., Peusquens, J., Ludwig-Portugall, I., Kohlhepp, M., Ergen, C., Niemietz, P., Martin, C., Van Rooijen, N., Ochando, J. C., Randolph, G. J., Luedde, T., Ginhoux, F., Kurts, C., Trautwein, C., & Tacke, F. (2015). Liver inflammation abrogates immunological tolerance induced by Kupffer cells. *Hepatology*, *62*(1), 279-291. https://doi.org/10.1002/hep.27793.

Hotchkiss, R. S., & Karl, I. E. (2003). The pathophysiology and treatment of sepsis. *N Engl J Med*, *348*(2), 138-150. https://doi.org/10.1056/NEJMra021333.

Hou, W., Wei, X., Liang, J., Fang, P., Ma, C., Zhang, Q., & Gao, Y. (2021). HMGB1-Induced Hepatocyte Pyroptosis Expanding Inflammatory Responses Contributes to the Pathogenesis of Acute-on-Chronic Liver Failure (ACLF). *J Inflamm Res*, *14*, 7295-7313. https://doi.org/10.2147/jir.S336626.

Hou, W., Wei, X., Liang, J., Fang, P., Ma, C., Zhang, Q., & Gao, Y. (2021). HMGB1-Induced Hepatocyte Pyroptosis Expanding Inflammatory Responses Contributes to the Pathogenesis of Acute-on-Chronic Liver Failure (ACLF). *Journal of Inflammation Research*, *Volume 14*, 7295-7313. https://doi.org/10.2147/jir.s336626.

Hu, X., Chen, H., Xu, H., Wu, Y., Wu, C., Jia, C., Li, Y., Sheng, S., Xu, C., Xu, H., Ni, W., & Zhou, K. (2020). Role of Pyroptosis in Traumatic Brain and Spinal Cord Injuries. *Int J Biol Sci*, *16*(12), 2042-2050. https://doi.org/10.7150/ijbs.45467.

Huang, C., Yu, K. K., Zheng, J. M., & Li, N. (2019). Steroid treatment in patients with acute-on-chronic liver failure precipitated by hepatitis B: A 10-year cohort study in a university hospital in East China. *J Dig Dis*, *20*(1), 38-44. https://doi.org/10.1111/1751-2980.12691.

Huang, Z.-B., Dai, X.-H., Xiao, M.-F., Zhou, R.-R., Zhao, S.-S., Zhang, B.-X., Yi, P.-P., Chen, R.-C., Li, W.-T., Yaser, A.-M., Huang, Y., & Fan, X.-G. (2013). HMGB1 release by human liver L02 and HepG2 cells induced by lipopolysaccharide. *Molecular Medicine Reports*, *8*(1), 103-112. https://doi.org/10.3892/mmr.2013.1482.

Huebener, P., Pradere, J.-P., Hernandez, C., Gwak, G.-Y., Caviglia, J. M., Mu, X., Loike, J. D., Jenkins, R. E., Antoine, D. J., & Schwabe, R. F. (2015). The HMGB1/RAGE axis triggers neutrophil-mediated injury amplification following necrosis. *Journal of Clinical Investigation*, *125*(2), 539-550. https://doi.org/10.1172/jci76887.

Iorga, A., Dara, L., & Kaplowitz, N. (2017). Drug-Induced Liver Injury: Cascade of Events Leading to Cell Death, Apoptosis or Necrosis. *Int J Mol Sci*, *18*(5). https://doi.org/10.3390/ijms18051018.

Jaeschke, H., McGill, M. R., & Ramachandran, A. (2012). Oxidant stress, mitochondria, and cell death mechanisms in drug-induced liver injury: Lessons learned from acetaminophen hepatotoxicity. *Drug Metabolism Reviews*, *44*(1), 88-106. https://doi.org/10.3109/03602532.2011.602688.

Jung, J. H., Park, J. H., Jee, M. H., Keum, S. J., Cho, M. S., Yoon, S. K., & Jang, S. K. (2011). Hepatitis C virus infection is blocked by HMGB1 released from virus-infected cells. *J Virol*, *85*(18), 9359-9368. https://doi.org/10.1128/jvi.00682-11.

Kang, R., Tang, D., Schapiro, N. E., Livesey, K. M., Farkas, A., Loughran, P., Bierhaus, A., Lotze, M. T., & Zeh, H. J. (2010). The receptor for advanced glycation end products (RAGE) sustains autophagy and limits apoptosis, promoting pancreatic tumor cell survival. *Cell Death Differ*, *17*(4), 666-676. https://doi.org/10.1038/cdd.2009.149.

Karlsson, S., Pettilä, V., Tenhunen, J., Laru-Sompa, R., Hynninen, M., & Ruokonen, E. (2008). HMGB1 as a predictor of organ dysfunction and outcome in patients with severe sepsis. *Intensive Care Med*, *34*(6), 1046-1053. https://doi.org/10.1007/s00134-008-1032-9.

Khambu, B., Huda, N., Chen, X., Antoine, D. J., Li, Y., Dai, G., Köhler, U. A., Zong, W.-X., Waguri, S., Werner, S., Oury, T. D., Dong, Z., & Yin, X.-M. (2018). HMGB1 promotes ductular reaction and tumorigenesis in autophagy-deficient livers. *Journal of Clinical Investigation*, *128*(6), 2419-2435. https://doi.org/10.1172/jci91814.

Khambu, B., Yan, S., Huda, N., & Yin, X. M. (2019). Role of High-Mobility Group Box-1 in Liver Pathogenesis. *Int J Mol Sci*, *20*(21). https://doi.org/10.3390/ijms20215314.

Komatsu, M., Kurokawa, H., Waguri, S., Taguchi, K., Kobayashi, A., Ichimura, Y., Sou, Y. S., Ueno, I., Sakamoto, A., Tong, K. I., Kim, M., Nishito, Y., Iemura, S., Natsume, T., Ueno, T., Kominami, E., Motohashi, H., Tanaka, K., & Yamamoto, M. (2010). The selective autophagy substrate p62 activates the stress responsive transcription factor Nrf2 through inactivation of Keap1. *Nat Cell Biol*, *12*(3), 213-223. https://doi.org/10.1038/ncb2021.

Kubes, P., & Mehal, W. Z. (2012). Sterile inflammation in the liver. *Gastroenterology*, *143*(5), 1158-1172. https://doi.org/10.1053/j.gastro.2012.09.008.

Kullak-Ublick, G. A., Andrade, R. J., Merz, M., End, P., Benesic, A., Gerbes, A. L., & Aithal, G. P. (2017). Drug-induced liver injury: recent advances in diagnosis and risk assessment. *Gut*, *66*(6), 1154-1164. https://doi.org/10.1136/gutjnl-2016-313369.

Kuraishy, A., Karin, M., & Sergei. (2011). Tumor Promotion via Injury- and Death-Induced Inflammation. *Immunity*, *35*(4), 467-477. https://doi.org/10.1016/j.immuni.2011.09.006.

Labbé, K., & Saleh, M. (2008). Cell death in the host response to infection. *Cell Death Differ*, *15*(9), 1339-1349. https://doi.org/10.1038/cdd.2008.91.

Lacey, C. A., Mitchell, W. J., Dadelahi, A. S., & Skyberg, J. A. (2018). Caspase-1 and Caspase-11 Mediate Pyroptosis, Inflammation, and Control of Brucella Joint Infection. *Infect Immun*, *86*(9). https://doi.org/10.1128/iai.00361-18.

Lamkanfi, M., Sarkar, A., Vande Walle, L., Vitari, A. C., Amer, A. O., Wewers, M. D., Tracey, K. J., Kanneganti, T. D., & Dixit, V. M. (2010). Inflammasome-dependent release of the alarmin HMGB1 in endotoxemia. *J Immunol*, *185*(7), 4385-4392. https://doi.org/10.4049/jimmunol.1000803.

Lee, W. M. (2012). Acute liver failure. *Semin Respir Crit Care Med*, *33*(1), 36-45. https://doi.org/10.1055/s-0032-1301733.

Levesque, E., Winter, A., Noorah, Z., Daurès, J. P., Landais, P., Feray, C., & Azoulay, D. (2017). Impact of acute-on-chronic liver failure on 90-day mortality following a first liver transplantation. *Liver Int*, *37*(5), 684-693. https://doi.org/10.1111/liv.13355.

Li, F., Li, H., Jin, X., Zhang, Y., Kang, X., Zhang, Z., Xu, M., Qian, Z., Ma, Z., Gao, X., Zhao, L., Wu, S., & Sun, H. (2019). Adipose-specific knockdown of Sirt1 results in obesity and insulin resistance by promoting exosomes release. *Cell Cycle*, *18*(17), 2067-2082. https://doi.org/10.1080/15384101.2019.1638694.

Li, H., Zhao, X.-K., Cheng, Y.-J., Zhang, Q., Wu, J., Lu, S., Zhang, W., Liu, Y., Zhou, M.-Y., Wang, Y., Yang, J., & Cheng, M.-L. (2019). Gasdermin D-mediated hepatocyte pyroptosis expands inflammatory responses that aggravate acute liver failure by upregulating monocyte chemotactic protein 1/CC chemokine receptor-2 to recruit macrophages. *World Journal of Gastroenterology*, *25*(44), 6527-6540. https://doi.org/10.3748/wjg.v25.i44.6527.

Li, L., Chen, L., Hu, L., Liu, Y., Sun, H. Y., Tang, J., Hou, Y. J., Chang, Y. X., Tu, Q. Q., Feng, G. S., Shen, F., Wu, M. C., & Wang, H. Y. (2011). Nuclear factor high-mobility group box1 mediating the activation of Toll-like receptor 4 signaling in hepatocytes in the early stage of nonalcoholic fatty liver disease in mice. *Hepatology*, *54*(5), 1620-1630. https://doi.org/10.1002/hep.24552.

Li, S., Tan, H.-Y., Wang, N., Zhang, Z.-J., Lao, L., Wong, C.-W., & Feng, Y. (2015). The Role of Oxidative Stress and Antioxidants in Liver Diseases. *International Journal of Molecular Sciences*, *16*(11), 26087-26124. https://doi.org/10.3390/ijms161125942.

Li, X., Jin, Q., Yao, Q., Xu, B., Li, Z., & Tu, C. (2016). Quercetin attenuates the activation of hepatic stellate cells and liver fibrosis in mice through modulation of HMGB1-TLR2/4-NF-κB signaling pathways. *Toxicol Lett*, *261*, 1-12. https://doi.org/10.1016/j.toxlet.2016.09.002.

Liu, H. B., Fan, X. G., Huang, J. J., Li, N., Peng, J. P., Li, S. L., & Wang, H. (2007). [Serum level of HMGB1 in patients with hepatitis B and its clinical significance]. *Zhonghua Gan Zang Bing Za Zhi*, *15*(11), 812-815.

Lotze, M. T., & Tracey, K. J. (2005). High-mobility group box 1 protein (HMGB1): nuclear weapon in the immune arsenal. *Nat Rev Immunol*, *5*(4), 331-342. https://doi.org/10.1038/nri1594.

Lu, B., Nakamura, T., Inouye, K., Li, J., Tang, Y., Lundbäck, P., Valdes-Ferrer, S. I., Olofsson, P. S., Kalb, T., Roth, J., Zou, Y., Erlandsson-Harris, H., Yang, H., Ting, J. P., Wang, H., Andersson, U., Antoine, D. J., Chavan, S. S., Hotamisligil, G. S., & Tracey, K. J. (2012). Novel role of PKR in inflammasome activation and HMGB1 release. *Nature*, *488*(7413), 670-674. https://doi.org/10.1038/nature11290.

Lu, B., Wang, C., Wang, M., Li, W., Chen, F., Tracey, K. J., & Wang, H. (2014). Molecular mechanism and therapeutic modulation of high mobility group box 1 release and action: an updated review. *Expert Rev Clin Immunol*, *10*(6), 713-727. https://doi.org/10.1586/1744666x.2014.909730.

Lu, W.-Y., Bird, T. G., Boulter, L., Tsuchiya, A., Cole, A. M., Hay, T., Guest, R. V., Wojtacha, D., Man, T. Y., Mackinnon, A., Ridgway, R. A., Kendall, T., Williams, M. J., Jamieson, T., Raven, A., Hay, D. C., Iredale, J. P., Clarke, A. R., Sansom, O. J., & Forbes, S. J. (2015). Hepatic progenitor cells of biliary origin with liver repopulation capacity. *Nature Cell Biology*, *17*(8), 971-983. https://doi.org/10.1038/ncb3203.

Lundbäck, P., Lea, J. D., Sowinska, A., Ottosson, L., Fürst, C. M., Steen, J., Aulin, C., Clarke, J. I., Kipar, A., Klevenvall, L., Yang, H., Palmblad, K., Park, B. K., Tracey, K. J., Blom, A. M., Andersson, U., Antoine, D. J., & Erlandsson Harris, H. (2016). A novel high mobility group box 1 neutralizing chimeric antibody attenuates drug-induced liver injury and postinjury inflammation in mice. *Hepatology*, *64*(5), 1699-1710. https://doi.org/10.1002/hep.28736.

Mihm, S. (2018). Danger-Associated Molecular Patterns (DAMPs): Molecular Triggers for Sterile Inflammation in the Liver. *International Journal of Molecular Sciences*, *19*(10), 3104. https://doi.org/10.3390/ijms19103104.

Miura, K., Kodama, Y., Inokuchi, S., Schnabl, B., Aoyama, T., Ohnishi, H., Olefsky, J. M., Brenner, D. A., & Seki, E. (2010). Toll-like receptor 9 promotes steatohepatitis by induction of interleukin-1beta in mice. *Gastroenterology*, *139*(1), 323-334.e327. https://doi.org/10.1053/j.gastro.2010.03.052.

Mofrad, P., Contos, M. J., Haque, M., Sargeant, C., Fisher, R. A., Luketic, V. A., Sterling, R. K., Shiffman, M. L., Stravitz, R. T., & Sanyal, A. J. (2003). Clinical and histologic spectrum of nonalcoholic fatty liver disease associated with normal ALT values. *Hepatology*, *37*(6), 1286-1292. https://doi.org/10.1053/jhep.2003.50229.

Mokdad, A. A., Lopez, A. D., Shahraz, S., Lozano, R., Mokdad, A. H., Stanaway, J., Murray, C. J., & Naghavi, M. (2014). Liver cirrhosis mortality in 187 countries between 1980 and 2010: a systematic analysis. *BMC Med*, *12*, 145. https://doi.org/10.1186/s12916-014-0145-y.

Nardo, A. D., Schneeweiss-Gleixner, M., Bakail, M., Dixon, E. D., Lax, S. F., & Trauner, M. (2021). Pathophysiological mechanisms of liver injury in COVID-19. *Liver Int*, *41*(1), 20-32. https://doi.org/10.1111/liv.14730.

Nathwani, R. A., Pais, S., Reynolds, T. B., & Kaplowitz, N. (2005). Serum alanine aminotransferase in skeletal muscle diseases. *Hepatology*, *41*(2), 380-382. https://doi.org/10.1002/hep.20548.

Park, J. S., Svetkauskaite, D., He, Q., Kim, J. Y., Strassheim, D., Ishizaka, A., & Abraham, E. (2004). Involvement of toll-like receptors 2 and 4 in cellular activation by high mobility group box 1 protein. *J Biol Chem*, *279*(9), 7370-7377. https://doi.org/10.1074/jbc.M306793200.

Parlesak, A., Schäfer, C., Schütz, T., Bode, J. C., & Bode, C. (2000). Increased intestinal permeability to macromolecules and endotoxemia in patients with chronic alcohol abuse in different stages of alcohol-induced liver disease. *J Hepatol*, *32*(5), 742-747. https://doi.org/10.1016/s0168-8278(00)80242-1.

Peralta, C., Jiménez-Castro, M. B., & Gracia-Sancho, J. (2013). Hepatic ischemia and reperfusion injury: Effects on the liver sinusoidal milieu. *Journal of Hepatology*, *59*(5), 1094-1106. https://doi.org/10.1016/j.jhep.2013.06.017.

Petrasek, J., Mandrekar, P., & Szabo, G. (2010). Toll-like receptors in the pathogenesis of alcoholic liver disease. *Gastroenterol Res Pract*, *2010*. https://doi.org/10.1155/2010/710381.

Rabinowitz, J. D., & White, E. (2010). Autophagy and Metabolism. *Science*, *330*(6009), 1344-1348. https://doi.org/10.1126/science.1193497.

Raven, A., Lu, W.-Y., Man, T. Y., Ferreira-Gonzalez, S., O'Duibhir, E., Dwyer, B. J., Thomson, J. P., Meehan, R. R., Bogorad, R., Koteliansky, V., Kotelevtsev, Y.,

Ffrench-Constant, C., Boulter, L., & Forbes, S. J. (2017). Cholangiocytes act as facultative liver stem cells during impaired hepatocyte regeneration. *Nature*, *547*(7663), 350-354. https://doi.org/10.1038/nature23015.

Rehm, J., Taylor, B., Mohapatra, S., Irving, H., Baliunas, D., Patra, J., & Roerecke, M. (2010). Alcohol as a risk factor for liver cirrhosis: a systematic review and meta-analysis. *Drug Alcohol Rev*, *29*(4), 437-445. https://doi.org/10.1111/j.1465-3362.2009.00153.x.

Rivera, C. A., Gaskin, L., Allman, M., Pang, J., Brady, K., Adegboyega, P., & Pruitt, K. (2010). Toll-like receptor-2 deficiency enhances non-alcoholic steatohepatitis. *BMC Gastroenterol*, *10*, 52. https://doi.org/10.1186/1471-230x-10-52.

Rockey, D. C., Seeff, L. B., Rochon, J., Freston, J., Chalasani, N., Bonacini, M., Fontana, R. J., & Hayashi, P. H. (2010). Causality assessment in drug-induced liver injury using a structured expert opinion process: comparison to the Roussel-Uclaf causality assessment method. *Hepatology*, *51*(6), 2117-2126. https://doi.org/10.1002/hep.23577.

Rutherford, A., King, L. Y., Hynan, L. S., Vedvyas, C., Lin, W., Lee, W. M., & Chung, R. T. (2012). Development of an accurate index for predicting outcomes of patients with acute liver failure. *Gastroenterology*, *143*(5), 1237-1243. https://doi.org/10.1053/j.gastro.2012.07.113.

Sarin, S. K., Choudhury, A., Sharma, M. K., Maiwall, R., Al Mahtab, M., Rahman, S., Saigal, S., Saraf, N., Soin, A. S., Devarbhavi, H., Kim, D. J., Dhiman, R. K., Duseja, A., Taneja, S., Eapen, C. E., Goel, A., Ning, Q., Chen, T., Ma, K., . . . Yoon, E. L. (2019). Acute-on-chronic liver failure: consensus recommendations of the Asian Pacific association for the study of the liver (APASL): an update. *Hepatol Int*, *13*(4), 353-390. https://doi.org/10.1007/s12072-019-09946-3.

Scaffidi, P., Misteli, T., & Bianchi, M. E. (2002). Release of chromatin protein HMGB1 by necrotic cells triggers inflammation. *Nature*, *418*(6894), 191-195. https://doi.org/10.1038/nature00858.

Schiraldi, M., Raucci, A., Muñoz, L. M., Livoti, E., Celona, B., Venereau, E., Apuzzo, T., De Marchis, F., Pedotti, M., Bachi, A., Thelen, M., Varani, L., Mellado, M., Proudfoot, A., Bianchi, M. E., & Uguccioni, M. (2012). HMGB1 promotes recruitment of inflammatory cells to damaged tissues by forming a complex with CXCL12 and signaling via CXCR4. *J Exp Med*, *209*(3), 551-563. https://doi.org/10.1084/jem.20111739.

Seitz, H. K., Bataller, R., Cortez-Pinto, H., Gao, B., Gual, A., Lackner, C., Mathurin, P., Mueller, S., Szabo, G., & Tsukamoto, H. (2018). Alcoholic liver disease. *Nat Rev Dis Primers*, *4*(1), 16. https://doi.org/10.1038/s41572-018-0014-7.

Shi, H., Han, W., Shi, H., Ren, F., Chen, D., Chen, Y., & Duan, Z. (2017). Augmenter of liver regeneration protects against carbon tetrachloride-induced liver injury by promoting autophagy in mice. *Oncotarget*, *8*(8), 12637-12648. https://doi.org/10.18632/oncotarget.14478.

Stauffer, J. K., Scarzello, A. J., Jiang, Q., & Wiltrout, R. H. (2012). Chronic inflammation, immune escape, and oncogenesis in the liver: A unique neighborhood for novel intersections. *Hepatology*, *56*(4), 1567-1574. https://doi.org/10.1002/hep.25674.

Sun, Z., Nyanzu, M., Yang, S., Zhu, X., Wang, K., Ru, J., Yu, E., Zhang, H., Wang, Z., Shen, J., Zhuge, Q., & Huang, L. (2020). VX765 Attenuates Pyroptosis and HMGB1/TLR4/NF-κB Pathways to Improve Functional Outcomes in TBI Mice. *Oxid Med Cell Longev*, *2020*, 7879629. https://doi.org/10.1155/2020/7879629.

Szabo, G., & Bala, S. (2010). Alcoholic liver disease and the gut-liver axis. *World J Gastroenterol*, *16*(11), 1321-1329. https://doi.org/10.3748/wjg.v16.i11.1321.

Tang, D., Kang, R., Zeh, H. J., & Lotze, M. T. (2011). High-Mobility Group Box 1, Oxidative Stress, and Disease. *Antioxidants & Redox Signaling*, *14*(7), 1315-1335. https://doi.org/10.1089/ars.2010.3356.

Tsung, A., Klune, J. R., Zhang, X., Jeyabalan, G., Cao, Z., Peng, X., Stolz, D. B., Geller, D. A., Rosengart, M. R., & Billiar, T. R. (2007). HMGB1 release induced by liver ischemia involves Toll-like receptor 4 dependent reactive oxygen species production and calcium-mediated signaling. *Journal of Experimental Medicine*, *204*(12), 2913-2923. https://doi.org/10.1084/jem.20070247.

Tsung, A., Sahai, R., Tanaka, H., Nakao, A., Fink, M. P., Lotze, M. T., Yang, H., Li, J., Tracey, K. J., Geller, D. A., & Billiar, T. R. (2005). The nuclear factor HMGB1 mediates hepatic injury after murine liver ischemia-reperfusion. *Journal of Experimental Medicine*, *201*(7), 1135-1143. https://doi.org/10.1084/jem.20042614.

Ueno, T., & Komatsu, M. (2017). Autophagy in the liver: functions in health and disease. *Nat Rev Gastroenterol Hepatol*, *14*(3), 170-184. https://doi.org/10.1038/nrgastro.2016.185.

Van Golen, R. F., Reiniers, M. J., Marsman, G., Alles, L. K., Van Rooyen, D. M., Petri, B., Van Der Mark, V. A., Van Beek, A. A., Meijer, B., Maas, M. A., Zeerleder, S., Verheij, J., Farrell, G. C., Luken, B. M., Teoh, N. C., Van Gulik, T. M., Murphy, M. P., & Heger, M. (2019). The damage-associated molecular pattern HMGB1 is released early after clinical hepatic ischemia/reperfusion. *Biochimica et Biophysica Acta (BBA) - Molecular Basis of Disease*, *1865*(6), 1192-1200. https://doi.org/10.1016/j.bbadis.2019.01.014.

Vénéreau, E., Ceriotti, C., & Bianchi, M. E. (2015). DAMPs from Cell Death to New Life. *Front Immunol*, *6*, 422. https://doi.org/10.3389/fimmu.2015.00422.

Wang, H., Bloom, O., Zhang, M., Vishnubhakat, J. M., Ombrellino, M., Che, J., Frazier, A., Yang, H., Ivanova, S., Borovikova, L., Manogue, K. R., Faist, E., Abraham, E., Andersson, J., Andersson, U., Molina, P. E., Abumrad, N. N., Sama, A., & Tracey, K. J. (1999). HMG-1 as a late mediator of endotoxin lethality in mice. *Science*, *285* (5425), 248-251. https://doi.org/10.1126/science.285.5425.248.

Wang, H. B., O. Zhang, M. et al. (1999). Wang 1999 [Research]. *Science*, *285*, 248-251.

Wang, K. (2015). Autophagy and apoptosis in liver injury. *Cell Cycle*, *14*(11), 1631-1642. https://doi.org/10.1080/15384101.2015.1038685.

Wang, L. W., Chen, H., & Gong, Z. J. (2010). High mobility group box-1 protein inhibits regulatory T cell immune activity in liver failure in patients with chronic hepatitis B. *Hepatobiliary Pancreat Dis Int*, *9*(5), 499-507.

Wang, S., Pacher, P., De Lisle, R. C., Huang, H., & Ding, W.-X. (2016). A Mechanistic Review of Cell Death in Alcohol-Induced Liver Injury. *Alcoholism: Clinical and Experimental Research*, *40*(6), 1215-1223. https://doi.org/10.1111/acer.13078.

Wang, Y., Zhang, H., Chen, Q., Jiao, F., Shi, C., Pei, M., Lv, J., Zhang, H., Wang, L., & Gong, Z. (2020). TNF-alpha/HMGB1 inflammation signalling pathway regulates pyroptosis during liver failure and acute kidney injury. *Cell Prolif*, *53*(6), e12829. https://doi.org/10.1111/cpr.12829.

Wen, S., Li, X., Ling, Y., Chen, S., Deng, Q., Yang, L., Li, Y., Shen, J., Qiu, Y., Zhan, Y., Lai, H., Zhang, X., Ke, Z., & Huang, W. (2020). HMGB1-associated necroptosis and Kupffer cells M1 polarization underlies remote liver injury induced by intestinal ischemia/reperfusion in rats. *FASEB J*, *34*(3), 4384-4402. https://doi.org/10.1096/fj.201900817R.

Woolbright, B. L., & Jaeschke, H. (2018). Is Keratin-18 only a Marker of Cell Death in Acute-On-Chronic Liver Failure? *J Lab Precis Med*, *3*. https://doi.org/10.21037/jlpm.2018.03.07.

Wu, J. J., Quijano, C., Wang, J., & Finkel, T. (2010). Metabolism meets autophagy. *Cell Cycle*, *9*(24), 4780-4781. https://doi.org/10.4161/cc.9.24.14273.

Xu, H., Yao, Y., Su, Z., Yang, Y., Kao, R., Martin, C. M., & Rui, T. (2011). Endogenous HMGB1 contributes to ischemia-reperfusion-induced myocardial apoptosis by potentiating the effect of TNF-α/JNK. *Am J Physiol Heart Circ Physiol*, *300*(3), H913-921. https://doi.org/10.1152/ajpheart.00703.2010.

Xu, J., Jiang, Y., Wang, J., Shi, X., Liu, Q., Liu, Z., Li, Y., Scott, M. J., Xiao, G., Li, S., Fan, L., Billiar, T. R., Wilson, M. A., & Fan, J. (2014). Macrophage endocytosis of high-mobility group box 1 triggers pyroptosis. *Cell Death Differ*, *21*(8), 1229-1239. https://doi.org/10.1038/cdd.2014.40.

Yamamoto, T., & Tajima, Y. (2017). HMGB1 is a promising therapeutic target for acute liver failure. *Expert Rev Gastroenterol Hepatol*, *11*(7), 673-682. https://doi.org/10.1080/17474124.2017.1345625.

Yang, H., Antoine, D. J., Andersson, U., & Tracey, K. J. (2013). The many faces of HMGB1: molecular structure-functional activity in inflammation, apoptosis, and chemotaxis. *J Leukoc Biol*, *93*(6), 865-873. https://doi.org/10.1189/jlb.1212662.

Yang, H., Hreggvidsdottir, H. S., Palmblad, K., Wang, H., Ochani, M., Li, J., Lu, B., Chavan, S., Rosas-Ballina, M., Al-Abed, Y., Akira, S., Bierhaus, A., Erlandsson-Harris, H., Andersson, U., & Tracey, K. J. (2010). A critical cysteine is required for HMGB1 binding to Toll-like receptor 4 and activation of macrophage cytokine release. *Proc Natl Acad Sci U S A*, *107*(26), 11942-11947. https://doi.org/10.1073/pnas.1003893107.

Yang, H., Wang, H., Chavan, S. S., & Andersson, U. (2015). High Mobility Group Box Protein 1 (HMGB1): The Prototypical Endogenous Danger Molecule. *Mol Med*, *21 Suppl 1*(Suppl 1), S6-s12. https://doi.org/10.2119/molmed.2015.00087.

Yang, L., Li, P., Fu, S., Calay, E. S., & Hotamisligil, G. S. (2010). Defective hepatic autophagy in obesity promotes ER stress and causes insulin resistance. *Cell Metab*, *11*(6), 467-478. https://doi.org/10.1016/j.cmet.2010.04.005.

Yang, M., Cao, L., Xie, M., Yu, Y., Kang, R., Yang, L., Zhao, M., & Tang, D. (2013). Chloroquine inhibits HMGB1 inflammatory signaling and protects mice from lethal sepsis. *Biochem Pharmacol*, *86*(3), 410-418. https://doi.org/10.1016/j.bcp.2013.05.013.

Zhang, F., He, Y., & Duan, Z. (2008). Changes of high mobility group box 1 in serum of pig acute hepatic failure model and significance. *J Huazhong Univ Sci Technolog Med Sci*, *28*(1), 52-55. https://doi.org/10.1007/s11596-008-0113-x.

Zhao, X., Xue, X., Wang, C., Wang, J., Peng, C., & Li, Y. (2022). Emerging roles of Sirtuins in alleviating alcoholic liver Disease: A comprehensive review. *Int Immunopharmacol*, *108*, 108712. https://doi.org/10.1016/j.intimp.2022.108712.

Zhou, R. R., Zhao, S. S., Zou, M. X., Zhang, P., Zhang, B. X., Dai, X. H., Li, N., Liu, H. B., Wang, H., & Fan, X. G. (2011). HMGB1 cytoplasmic translocation in patients with acute liver failure. *BMC Gastroenterol*, *11*, 21. https://doi.org/10.1186/1471-230x-11-21.

Section III. HMGB1 and Neurodegenerative Diseases

Chapter 5

The Role of HMGB-1 and Its Inhibitors in Brain Disorders

Onur Cagin Gurlek[1], MD, Anisa Dehghani[2], PhD and Hulya Karatas[1], MD, PhD

[1]Hacettepe University,
Institute of Neurological Sciences and Psychiatry,
Ankara, Turkey
[2]Leiden University Medical Centre,
Department of Human Genetics,
Migraine Research Group, Leiden, Netherlands

Abstract

High Mobility Group Box-1 (HMGB-1) plays a crucial role in several physiological and pathological processes in the body depending on its localization, post-translational modification status, the environment of the compartment it occupies, and essential for life. In the basal state, HMGB-1 is concentrated 30-fold in the nucleus compared to the cytoplasm, but depending on the environment and stimulus, it can be localized to the cytoplasm, cell membrane, and extracellular compartments. Its most studied and well-known receptors are Receptor for Advanced Glycation End-products (RAGE) and Toll-like Receptors (TLR), specifically TLR4. HMGB-1 can bind to RAGE and TLR4 and act upon cell differentiation, migration, and, most importantly, inflammation. This chapter aims to summarize new data on the experimental and clinical research of HMGB1, and its current inhibitors in brain disorders, specifically epilepsy, headache, neuroimmunological disorders, neurodegenerative disorders, and stroke.

In: HMGB1
Editors: Eylem Taskin, Celal Guven and Salih Tunc Kaya
ISBN: 979-8-88697-408-9

Keywords: HMGB-1, RAGE, TLR-4, neurological diseases, monoclonal antibodies

Introduction to High Mobility Group Box-1 Protein

High Mobility Group Box-1 (HMGB-1) is the most abundant and well-studied member of the High Mobility Group (HMG) proteins which are a family of non-histone, DNA-binding nuclear proteins and member of the alarmin family (Hock et al. 2007; Chikhirzhina et al. 2020). HMGB-1 plays a crucial role in several different physiological and pathological processes in the body depending on its localization, post-translational modification status, and the environment of the compartment it occupies, so it's essential for life (Fang, Schachner, and Shen 2012). HMGB-1 by the HMGB1 gene on 13q12 (Ferrari et al. 1996) and its transcription is regulated by a silencer region upstream to its promoter and an enhancer region in intron 1 (Lum and Lee 2001). In addition, the transcription level is regulated in opposing ways by p53 and its homolog p73, decreasing and increasing its transcription rate, respectively (Uramoto et al. 2003). Other transcription factors affecting HMGB-1 include but are not limited to Signal transducer and activator of transcription 3 (STAT3) (Wolfson, Mapes, and Garcia 2014), estrogen (Borrmann et al. 2001), Kruppel-like factor 4 (KLF4) (Liu et al. 2008), miR181 (Hutchison et al. 2013) and miR129 (Zhu, Chen, and Zhang 2020). It is a highly conserved protein, and 90% percent of DNA sequence is identical between humans, bovines, rats, and hamsters (Wen et al. 1989). It has two DNA-binding domains named A box and B box, respectively, an acidic C-terminal tail that acts as a regulatory region, two nuclear export signals, and two nuclear localization signals. Interestingly, the A-box domain, by itself, acts as an inhibitor to HMGB-1, and this inhibitory effect is increased when the A-box domain is given together with the C-terminal tail (Bonaldi et al. 2003).

Depending on the cell's redox state, HMGB-1 can be found in three different isoforms according to the oxidation status of three major cysteine residues: C23, C45, and C106 (Antoine et al. 2014). "Fully reduced HMGB1" only has thiols and is the primary isoform in the nucleus; it acts as a chemoattractant when released extracellularly and in high numbers when the cell is damaged due to injury or necrosis (Antoine et al. 2014). "Disulfide HMGB1" has a disulfide bond between C23 and C45, giving the HMGB-1 its pro-inflammatory cytokine-like properties when released extracellularly. Lastly, "Sulfonyl HMGB-1" is fully oxidized to sulfonyl groups on C23, C45, and C106 and is inert regarding cell migration and cytokine-like activity (Antoine et al. 2014). In general, reactive oxygen species and oxidation levels rise with inflammation, so other isoforms of HMGB-1 turn into sulfonyl HMGB-1, creating a negative feedback loop (Antoine et al. 2014). Furthermore, HMGB-1 is not only modified via oxidation but also is modified with post-translational mechanisms such as methylation (Ito, Fukazawa, and Yoshida 2007), phosphorylation (Youn and Shin 2006), and acetylation. These modifications determine whether HMGB-1 is imported or exported from the nucleus (Andersson, Antoine, and Tracey, 2014; Lu et al., 2014).

In the basal state, HMGB-1 is concentrated 30-fold in the nucleus compared to the cytoplasm. However, depending on the environment and stimulus, it can be localized to the cytoplasm, cell membrane, and extracellular compartments. It has different functions in each compartment (Kuehl, Salmond, and Tran 1984; Andersson, Antoine, and Tracey 2014; Lu et al. 2014). In the nucleus, it can act on nucleosome assembly and number (Celona et al. 2011), regulate the rate of transcription and act as a transcription factor (Cato et al. 2008; Singh and Dixon 1990), work as a DNA chaperone or contribute to DNA repair mechanisms (Kang et al. 2014). In the cytoplasm, it mainly works as a regulator of autophagy (Tang et al. 2010). When membrane-bound, it has a crucial role in neutrophil extracellular traps (NET) formation, (NETosis) platelet activation, and clot formation (Kim and Lee 2020; Tadie et al. 2013). HMGB-1 is released into extracellular space in two different major pathways. The first one is its active secretion by living cells upon encountering different stimuli, including lipopolysaccharide (LPS) (Li et al. 2020), tumor necrosis factor α (TNF-α) (Yang et al. 2016), Interferon- γ (IFN-γ) (Rendon-Mitchell et al. 2003), free DNA (Jiang and Pisetsky 2008). HMGB-1 lacks a leader signal sequence needed for active secretion through the classical endoplasmic reticulum-Golgi secretory pathway, so it is thought to be secreted through an unconventional secretory pathway likely involving lysosomes (Daniels and Brough 2017) exosomes, or pyroptotic proteins like inflammasomes such as NLRP3 and pore-forming proteins like gasdermins (Li et al. 2020). However, the exact mechanism remains unclear (Volchuk et al., 2020). The second major pathway is its passive release after cell death and injury, which is regulated by multiple mechanisms (Tang et al., 2010). It has also been shown that HMGB-1 is directly secreted by the neurons via hyperexcitation (Fujita et al., 2016) or direct optogenetic stimulation (Yang et al., 2021).

Upon release, HMGB-1 acts as a Damage-Associated Molecular Pattern (DAMP), which is a part of the innate immune system and is generally released from cells under stress and signals to other cells about potential injuries (Venereau, Ceriotti, and Bianchi 2015). Acting as a DAMP, HMGB-1 induces numerous different cell types such as macrophages (He et al. 2012), neutrophils (Park et al. 2003), astrocytes (Pedrazzi et al. 2007), microglia (Shi et al. 2018) to produce various pro-inflammatory molecules like cytokines (He et al. 2012; Park et al. 2003; Pedrazzi et al. 2007; Shi et al. 2018), chemokines (Lin et al. 2016) and inducible nitric oxide synthase (Su et al. 2016). During the development of the central nervous system (CNS), it also has physiological roles in neurogenesis, cell differentiation, neurite extension, and neuronal migration (Fang, Schachner, and Shen 2012). It also modulates and is modulated by different neurotransmitter systems, such as the cholinergic or glutaminergic systems (Bonanno et al., 2007; Wang et al., 2004).

It interacts with several different receptors and proteins when released, including integrins (Orlova et al., 2007), α-synuclein (Lindersson et al., 2004), C-X-C chemokine receptor type 4 (CXCR4) (Schiraldi et al., 2012) and N-methyl-D-aspartate receptor (NMDAR) (Pedrazzi et al., 2012). However, its most studied and well-known receptors are Receptor for Advanced Glycation End-products (RAGE) and Toll-like Receptors (TLR), specifically TLR4. HMGB-1 can bind to RAGE and TLR4 and act upon cell differentiation, migration, and, most importantly, inflammation (Yang et al., 2015). They show a synergistic interaction during inflammation, and without RAGE and TLR4, immune cells secrete substantially fewer cytokines. It has been shown that when HMGB-1 binds to RAGE through dynamin-dependent signaling, it is taken into the cell via endocytosis, and this triggers a pyroptotic cascade, most likely through TLR4 (Kokkola et al., 2005; Xu et al., 2014).

HMGB-1 and Epilepsy

Epilepsy is a complex, chronic and common neurological disorder characterized by a predisposition to unprovoked and recurring seizures, affecting around %1 of the population. Pathophysiology of the disease remains elusive, and the fact that around a third of the patients remain unresponsive to anti-epileptic drugs shows a need for new therapeutic approaches (Devinsky et al., 2018). In the context of epilepsy, HMGB1 has been researched extensively on its contributions to the emergence of epileptic circuits (epileptogenesis) from physiologic circuits, the generation of spontaneous seizures (ictogenesis), effects on the severity of the disease and comorbidities, and its potential as a biomarker.

HMGB-1 can directly influence both epileptogenesis and ictogenesis via modifying neuronal excitability. It directly regulates the NMDAR via a specific binding zone and increases the Ca2+ current passing through the channel (Pedrazzi et al., 2012). Especially, its disulfide form increases the phosphorylation of the N-methyl D-aspartate receptor subtype 2B (NR2B) subunit, increasing the channel's permeability to Ca2+ ions (Balosso et al., 2014; Maroso et al., 2010). These effects are abolished when either NMDAR antagonist MK-801, HMGB1 competitive inhibitor Box-A of HMGB-1, or TLR4 antagonist Rhodobacter sphaeroides LPS (LPS-RS) is given (Balosso et al., 2014; Pedrazzi et al., 2012). A natural inhibitor of HMGB-1, Glycyrrhizic Acid (GLA), showed anti-excitotoxic effects when administered together with excitotoxic agents like NMDA and kainite, supporting the hyperexcitability-inducing effect of HMGB-1 (Luo et al., 2013). Since overexcited neurons can secrete HMGB1 (Fujita et al., 2016; Yang et al., 2021), an insult might create a vicious cycle of hyperexcitability and neuroinflammation involving hyperexcitable neurons, secretion of cytokines by neurons, astrocytes, and microglia and reduced intake of glutamates by the astrocytes in the synaptic cleft which could explain the transformation of physiologic circuits into epileptogenic ones (Crews and Vetreno 2016).

The relationship between HMGB-1 and epilepsy has been studied extensively. The levels of HMGB-1 were shown to increase in the serum of adult epilepsy patients (Huang et al. 2020; Kan et al. 2019; Walker et al. 2014; Walker et al. 2022; Wang, Liu, et al. 2021), the serum of children with epilepsy (Kamasak et al. 2020; Wu et al. 2021; Zhu et al. 2018), the cerebrospinal fluid (CSF) of adult epilepsy patients (Wang, Liu, et al. 2021), the CSF of children after febrile seizures (Kaya et al. 2021), both the serum and the CSF of elderly patients with suspected autoimmune epilepsy (Han et al. 2020) and the CSF of patients with anti-NMDAR encephalitis (Ai et al. 2018). Its use as a biomarker of prognosis, drug-resistance status, or severity was suggested (Huang et al. 2020; Kamasak et al. 2020; Kan et al. 2019; Walker et al. 2022; Wang, Liu, et al. 2021; Wu et al. 2021; Zhu et al. 2018) but conflicting results show the need for further studies before committing (Wang, Liu, et al. 2021). Interestingly, in a population of children with temporal lobe epilepsy, levels of miRNA-29a in serum were lower than in the control group, and its levels were correlated negatively with HMGB-1 levels. When miRNA-29a was studied further, it was found to target the HMGB1 gene and lower its expression (Wu et al., 2021).

This relationship was further studied in both animal models and human samples. In 2010 Maroso et al. used the kainate-induced acute seizure model, bicuculline-induced acute seizure model, and kainate-induced chronic seizures modeling temporal lobe epilepsy (TLE) to study the effects of any changes in HMGB-1 and TLR4. An increase was observed in the expression of HMGB1 in the hippocampus, and translocation from the nucleus to the cytoplasm was seen in both neurons and astrocytes of all models used compared to control groups (Maroso et al., 2010). There was also significant TLR4 expression in neurons in all models, while there was no observable signal in controls. Changes in the expression of HMGB-1 and TLR4 in surgical hippocampus samples obtained from human subjects with TLE were also parallel to what was observed in the animal models. In order to show how HMGB-1 modifies ictogenesis, C57BL/6 mice were injected with HMGB-1 before kainate-induced seizures. It was seen that HMGB-1 increased both the frequency and the total duration of the seizures (Maroso et al. 2010). In order to determine the role of TLR4 in this process, experiments were repeated using C3H/HeJ mice with functionally inactive TLR4. While HMGB-1 did not affect seizures in the C3H/HeJ mice, it caused an increase in the frequency and total duration of seizures in their wild-type counterparts (Maroso et al., 2010). An interesting observation was that the C3H/HeJ mice showed a natural resistance to seizures, determined by the longer latency measured before a seizure and a reduction in the frequency and the total duration of seizures. NR2B-containing NMDAR inhibitor, ifenprodil, overturned the pro-epileptic effects of HMGB-1 (Maroso et al. 2010). Applying Box-A, LPS-RS, or another inhibitor of TLR4 cyanobacterial LPS was sufficient to reduce the frequency and total time spent in the seizure in all models. However, ifenprodil was only effective in reducing seizure parameters in the chronic model, showing its effect through modulating the changes in the excitability caused by HMGB-1 increase in the natural progression of the disease (Maroso et al., 2010).

In 2011, the translocation of HMGB-1 from the nucleus to the cytoplasm and an increase in the expression of TLR4 and RAGE were shown in neurons and astrocytes of samples of people with focal cortical dysplasia, cortical tubers from patients with tuberous sclerosis complex and in samples taken from gangliogliomas, which are known to cause seizures (Zurolo et al. 2011). In another study, an increase in HMGB-1, its hyperacetylation, and its subsequent translocation from the nucleus to the cytoplasm was shown again in the hippocampus and cortex of kainate-induced status epilepticus maximal electroshock models in mice (Walker et al., 2014). The increase in HMGB-1 and TLR4 and the translocation of HMGB-1 were also shown in different models of epilepsy like Sombati's cell model (Huang et al., 2015), coriaria lactone stimulated human microglia cells (Shi et al., 2018), and lithium-chloride-and-pilocarpine model (Liu, Wu, and Wang, 2017) and in brain samples from patients with seizure-related conditions such as Rasmussen's encephalitis (Luan et al., 2016) and focal cortical dysplasia (Zhang et al., 2018). In a study from 2017, long-term exposure to HMGB-1 was shown to cause an increase in the soma size, dendrite length, and Synaptosomal-Associated Protein, 25kDa (SNAP-25) (Yang et al., 2017), which can be related to structural changes seen in epilepsy brains.

Various approaches have been tried to target HMGB-1 for treatment. One popular approach is the use of anti-HMGB-1 monoclonal antibodies (mAb). Anti-HMGB-1 mAb was shown to reduce neuronal death, microglial activation, and proinflammatory molecule synthesis in a kainate-induced acute seizure model (Li et al., 2013). A similar study showed its anti-seizure effects in maximal electroshock seizure, pentylenetetrazole-induced, and kindling-induced acute seizure models and a kainate-induced chronic epilepsy model. Anti-seizure effects were not observed in the TLR4 absent mice. Its safety profile was also tested, and there were no detectable side effects, even in long-term use in very high doses (Zhao et al. 2017). Most excitingly, anti-HMGB-1 mAb was shown to partially inhibit the development of chronic epilepsy after induction of status epilepticus (Zhao et al., 2017). In another study on pilocarpine-and-methylscopolamine treated mice, anti-HMGB-1 mAb was effective in reducing the blood-brain barrier (BBB) damage and proinflammatory molecule synthesis as well as protecting against neural apoptosis and prolonging the seizure latency (Fu et al., 2017). In a pilocarpine-induced status epilepticus model, anti-HMGB-1 mAb increased the 30-minute therapeutic time window to 180-minute, which is the time frame where it's possible to treat the seizure and stop it from becoming benzodiazepine-refractory (Zhao, Zheng, et al. 2020). Again, in a pilocarpine-induced epilepsy model, long-term use of the anti-HMGB-1 mAb could correct the learning and memory deficits seen in mice models. In addition, on the microscopic level, it was able to reduce the hippocampal neuronal damage caused by seizures (Cui et al., 2021).

Another common HMGB-1 inhibitor used in epilepsy research is the naturally occurring Glycyrrhizin or Glycyrrhizic acid (GLA) which is found naturally in licorice (Glycyrrhiza glabra) root. GLA's neuroprotective role was shown in a mice kainate-induced epilepsy model. Administration of GLA 30 minutes before seizure induction reduced the neuronal loss, lowered the HMGB-1 level in the hippocampus, and inhibited HMGB-1 release into the serum (Luo et al. 2013, 2014).

Table 1. Interventional studies targeting HMGB-1 on animals and their results

Disease	Intervention	Results
Epilepsy	Anti-HMGB-1 monoclonal antibodies	Decreased the seizure threshold, total seizure duration, seizure frequency and seizure stage (Zhao et al. 2017) Decreased cognitive impairments (Cui et al. 2021; Li et al. 2013) Decreased the neuronal loss (Fu et al. 2017; Li et al. 2013; Zhao et al. 2017) Decreased the microglial activation, proinflammatory cytokine levels (Fu et al. 2017; Li et al. 2013) Increased the therapeutic time window (Zhao, Zheng, et al. 2020) Increased the latency to seizure (Fu et al. 2017; Zhao et al. 2017) Protected the BBB integrity (Fu et al. 2017)
	Box-A	Decreased the seizure frequency and total seizure duration (Maroso et al. 2010) Protected the BBB integrity (Zaben et al. 2021) Decreased neuronal loss and microglial activation (Zaben et al. 2021)
	Glycyrrhizic Acid	Prevented excitotoxicity (Luo et al. 2013) Decreased neuronal loss, BBB integrity and mortality rate (Li et al. 2019; Luo et al. 2013, 2014) Improved the seizure parameters and cognitive impairment (Paudel et al. 2021)
	Rhodobacter sphaeroides LPS	Decreased the seizure frequency and total seizure duration (Maroso et al. 2010)
	Cyanobacterial LPS	Decreased the seizure frequency and total seizure duration (Maroso et al. 2010)
	TLR4 knock out	Increased the latency to seizure (Maroso et al. 2010) Decreased the seizure frequency and total seizure duration (Maroso et al. 2010)
	Ifendropil	Inhibited the ictogenic effects of HMGB-1 (Maroso et al. 2010)
	Celecoxib	Increased the latency to seizure (Alsaegh et al. 2021; Morales-Sosa et al. 2018) Decreased the seizure severity, seizure frequency, proinflammatory cytokines and ROS activity (Alsaegh et al. 2021; Morales-Sosa et al. 2018)
	MicroRNA-129-5p	Increased the latency to seizure (Liu, Wu, and Wang 2017) Decreased the proinflammatory cytokine levels, neuronal loss and microglial activation (Liu, Wu, and Wang 2017)

Disease	Intervention	Results
	HMGB-1 siRNA	Increased the latency to seizure (Liu, Wu, and Wang 2017) Decreased the proinflammatory cytokine levels, neuronal loss and microglial activation (Liu, Wu, and Wang 2017)
	N-acetylcysteine + Sulforaphane	Increased the latency to seizure (Pauletti et al. 2019) Decreased seizure frequency, neuronal loss and cognitive impairments (Pauletti et al. 2019)
	Pentoxifylline	Decreased seizure severity, neuronal loss, proinflammatory cytokine levels and learning deficits (Badawi et al. 2021)
	TAK-242	Increased the seizure threshold (Zaniani et al. 2022) Decreased the seizure duration, proinflammatory cytokine levels and memory deficits (Zaniani et al. 2022)
Migraine	Box-A	Decreased neuroinflammation and mast cell degranulation (Karatas et al. 2013)
	Anti-HMGB-1 monoclonal antibodies	Decreased neuroinflammation, mast cell degranulation and microglial activation (Karatas et al. 2013; Takizawa et al. 2016)
	HMGB-1 shRNA	Decreased neuroinflammation and mast cell degranulation (Karatas et al. 2013)
	Glycyrrhizic Acid	Decreased microglial activation (Takizawa et al. 2017; Takizawa et al. 2016)
	TAK-242	Decreased neuroinflammation, microglial activation and hyperalgesia (Su et al. 2018)
Multiple Sclerosis	Anti-HMGB-1 monoclonal antibodies	Improved the clinical outcome (Robinson et al. 2013) Decreased cell infiltration into CNS, oligodendrocyte loss, neuroinflammation, D387 increase and demyelinization (Robinson et al. 2013; Uzawa, Mori, Taniguchi, et al. 2013)
	Glycyrrhizic Acid	Increased the latency to the onset of symptoms (Sun et al. 2018) Decreased the severity of the symptoms, gliosis, neuronal loss, demyelination, and cell infiltration into CNS (Sun et al. 2018)
	Inflachromene	Increased the latency to the onset of symptoms (Lee et al. 2014) Decreased the severity of the symptoms, microglial activation, neuronal loss, and demyelination (Lee et al. 2014)
	Recombinant Thrombomodulin	Decreased the disease severity and IL-17 increase (Uzawa et al. 2018)
	Matrine	Decreased the disease severity and proinflammatory cytokine levels (Chu et al. 2021)
	Recombinant Sonic Hedgehog	Decreased disease severity and demyelination (Xiao et al. 2021)

Table 1. (Continued)

Disease	Intervention	Results
Frontotemporal Dementia	Anti-HMGB-1 monoclonal antibodies	Decreased memory impairment, neuronal loss, DNA damage, protein aggregation, synapse abnormalities and tau phosphorylation (Jin et al. 2021)
Alzheimer's Disease	Anti-HMGB-1 monoclonal antibodies	Decreased cognitive impairment, DNA damage and neurite degeneration (Fujita et al. 2016) No effect on Aβ accumulation (Fujita et al. 2016)
	Glycyrrhizic Acid	Decreased the tau phosphorylation, neurofibrillary tangles and cognitive impairment (Zhao, Xu, et al. 2021)
	Ethyl Pyruvate + Glycyrrhizic Acid	Decreased the tau phosphorylation, neurofibrillary tangles, and cognitive impairment (Gaikwad et al. 2021)
Parkinson Disease	Anti-HMGB-1 monoclonal antibodies	Increased dopamine levels (Santoro et al. 2016) Decreased gliosis, neurodegeneration, functional impairments and proinflammatory cytokine levels (Masai et al. 2021; Santoro et al. 2016; Sasaki et al. 2016) Protected the BBB integrity (Sasaki et al. 2016)
	Glycyrrhizic Acid	Decreased microglial and astrocytic activation and neurodegeneration (Santoro et al. 2016) Increased dopamine levels (Santoro et al. 2016)
Huntington Disease	Transgenic HMGB-1 overexpression	Decreased neurodegeneration (Qi et al. 2007)
	TLR4 knock out TLR2 knock out TLR3 knock out	Increased the lifespan (Griffioen, Mattson, and Okun 2018)
Spinocerebellar Ataxia 1	Transgenic HMGB-1 overexpression	No effect on neuroinflammation and protein aggregation (Ito et al. 2015) Decreased mitochondrial fragmentation, mtDNA damage, cerebellar neurodegeneration, and DNA double strand break damage (Ito et al. 2015; Qi et al. 2007) Increased the cortical thickness, expression of mtDNA repair genes, mitochondria numbers and lifespan (Ito et al. 2015) Improved motor activity (Ito et al. 2015) Normalized gene expression patterns (Ito et al. 2015)

Disease	Intervention	Results
	Transfective HMGB-1 overexpression	No effect on neuroinflammation (Ito et al. 2015) Improved motor activity (Ito et al. 2015) Increased the lifespan and cortical thickness (Ito et al. 2015) Decreased cerebellar neurodegeneration, DNA double strand break damage and mtDNA damage (Ito et al. 2015)
Amyotrophic lateral sclerosis	Anti-HMGB-1 monoclonal antibodies	Decreased the motor symptoms short-term (Lee et al. 2019) No effect on survival, gliosis, or later symptoms (Lee et al. 2019)
	TLR4 knock out	Decreased the motor symptoms short-term (Lee et al. 2020) Increased the lifespan (Lee et al. 2020)
	RAGE knock out	Decreased the progression speed (Lee et al. 2015) Increased the lifespan (Lee et al. 2015)
Ischemic Stroke	Anti-HMGB-1 monoclonal antibodies	Decreased the NETosis, infarct volume, neuronal loss, microglial activity, neuroinflammation and neurological deficits (Kim, Lee, et al. 2019; Liu et al. 2007; Muhammad et al. 2008) Protected the BBB integrity (Liu et al. 2007; Zhang et al. 2011)
	Box-A	Decreased the NETosis, infarct volume, neuroinflammation and neurological deficits (Denorme et al. 2021; Jin et al. 2011; Kim, Ryu, et al. 2019; Muhammad et al. 2008)
	Glycyrrhizic Acid	Decreased neuronal loss, infarct volume, cell infiltration into CNS, neuroinflammation, free radical formation, hemorrhagic transformation, neurological deficits, and mortality (Kim, Jin, et al. 2012; Xiong et al. 2016)
	Ethyl Pyruvate	Decreased the infarct volume (Shin et al. 2014)
	sRAGE	Decreased the infarct volume and neuroinflammation (Muhammad et al. 2008)
	RAGE knock out	Decreased the infarct volume (Muhammad et al. 2008)
	HMGB-1 binding heptamer peptide	Decreased the infarct volume, neuroinflammation, neurological deficits and mortality (Kim, Shin, et al. 2012; Li et al. 2018) Protected the BBB integrity (Li et al. 2018) Prevented hemorrhagic transformation (Li et al. 2018)
	Hypothermia	Decreased the infarct volume and neuroinflammation (Lee et al. 2016)
	Hyperbaric oxygen treatment	Decreased the infarct volume, neuronal loss and neuroinflammation (Zhao, Xu, et al. 2021)

Table 1. (Continued)

Disease	Intervention	Results
Intracerebral Hemorrhage	Anti-HMGB-1 monoclonal antibodies	Decreased the neuronal loss, neurological deficits, cerebral edema and proinflammatory cytokine levels (Wang et al. 2017; Xu et al. 2022) Increased the expression of protective proteins and anti-inflammatory cytokine levels (Xu et al. 2022)
Intracerebral Hemorrhage	Ethyl Pyruvate	Decreased neuronal loss, neurological deficits, microglial activation and cerebral edema (Lei et al. 2013b)
	Glycyrrhizic Acid	Decreased neuronal loss, cerebral edema, and neurological deficits (Ohnishi et al. 2011)
Subarachnoidal Hemorrhage	Anti-HMGB-1 monoclonal antibodies	Decreased the delayed cerebral vasospasm, neurological deficits, proinflammatory cytokine levels, weight loss and susceptibility of endothelia to delayed cerebral vasospasm (Haruma et al. 2016; Wang et al. 2019)
	Glycyrrhizic Acid	Decreased neurological deficits, neuroinflammation, neuronal loss and BBB integrity (Ieong et al. 2018)
	Oleanolic acid	Decreased neuronal loss, lesion severity, neurological deficits, proinflammatory cytokine levels, and weight loss (Han et al. 2021)
	Bone marrow mesenchymal stem cell derived exosomes	Protected the BBB integrity (Xiong et al. 2020) Decreased the proinflammatory cytokine, neurological deficits, and mortality (Xiong et al. 2020)

GLA was also effective in a rat lithium-chloride-and-pilocarpine-induced status epilepticus model, and it reduced both HMGB-1 expression and translocation, BBB disruption, and hippocampal neuron loss. Moreover, it significantly lowered the mortality rate from 25% to 9% (Li et al., 2019). In a later study done by a different group, this neuroprotective effect was overturned by the inhibition of Sirtuin3, which is known to modulate HMGB-1 through its effect on cellular oxidation levels (Wu et al., 2020). GLA administration before seizures was successfully used to lower seizure parameters and reduce cognitive impairment in a zebrafish chronic epilepsy model (Paudel et al., 2021). Unfortunately, an important limitation of GLA use in humans is its inhibitory effect on 11β-hydroxysteroid dehydrogenase, which may cause a mineralocorticoid excess in the body, leading to hypertension, hypernatremia, and hypokalemia (O'Connell et al. 2016).

The isolated form of HMGB-1 Box-A (shortly Box-A) also acts as an inhibitor. It has been shown to be effective in downregulating pro-inflammatory protein synthesis, decreasing BBB dysfunction, and reducing microglial activation and hippocampal neuronal loss (Balosso et al., 2014; Yu et al., 2019). Furthermore, Zaben et al. showed in 3D cultures, 2D cultures, and neurospheres made from surgically resected sclerotic hippocampi that the microenvironment in these cultures was anti-neurogenic due to high levels of HMGB-1 and IL-1B and neurogenesis could be rescued by the application of Box-A (Zaben et al., 2021). Other molecules that inhibited the HMGB-1 axis and were found to be effective in reducing seizures or the damage caused by them are specific cyclooxygenase-2 inhibitor celecoxib (Alsaegh et al., 2021; Morales-Sosa et al., 2018), miR-129-5p (Liu, Wu, and Wang 2017), an antioxidant combination of N-acetylcysteine and sulforaphane (Pauletti et al. 2019), pentoxifylline (Badawi et al. 2021), specific TLR4 inhibitor TAK-242 (Zaniani et al. 2022) and Carbenoxolone which targets both HMGB-1 and gap junctions of the BBB (Table 1) (Gareri et al. 2005; Gareri et al. 2004; Volnova et al. 2022).

HMGB-1 and Headaches

Migraine is a primary headache disorder caused by the activation of the trigeminocervical complex and involves episodes of moderate to severe headaches, photophobia, phonophobia, nausea, and other debilitating symptoms. It affects approximately 15% of the world population, and a third of those affected experience transient neurological deficits (most commonly visual) called an aura 5 to 60 minutes before the onset of the headache, which usually correlates with a wave of depolarization across the brain called Cortical Spreading Depolarization (CSD or SD).

Prevalence of migraine and aura are three times more common in adult women than in adult men. However, the prevalence of migraine is equal in men and women both before puberty and after menopause (Ferrari et al., 2022). Interestingly, there are reports of increased headache frequency and aura prevalence in transgender women receiving gender-affirming hormone therapy (GAHT) (MacGregor and van den Brink, 2019). Conversely, transgender men with existing headache conditions who are receiving GAHT described a decrease in the frequency and duration of their headaches (MacGregor and van den Brink 2019). All of these show that there is a solid biological relationship between sex and migraines.

The relationship between migraine and HMGB-1 was first shown in 2013. Using a mechanical stimulation-based cortico-dural pinprick model to induce a single CSD and application of KCl over dura mater to induce multiple CSDs, it was shown that a single CSD was sufficient to open neuronal Pannexin-1 mega-channels, which in turn caused the cleavage of caspase-1 and release of HMGB-1 with IL1-B into extracellular space from neurons and translocation of NF-kB from the cytoplasm to the nucleus in astrocytes (Karatas et al., 2013). Activation of trigeminocervical and the presence of headache were confirmed by visualization of mast cell degranulation in dura and scoring of face grimacing, respectively (Karatas et al. 2013). Carbenoxolone, which is used as an inhibitor of both Pannexin-1 (Panx1) and HMGB-1, was shown to abrogate caspase-1 cleavage, HMGB-1 translocation and release, NF-kB translocation, and trigeminocervical activation. It also alleviated the pain measured through grimace scoring. HMGB-1 inhibition by Box-A, anti-HMGB-1 mAb, or HMGB-1 shRNA is also effective in preventing NF-kB translocation and dural mast cell degranulation (Karatas et al., 2013). Multiple CSD's induced by KCl application were also shown to increase the mRNA of HMGB-1, its receptors TLR2 and TLR4, and their adaptor proteins IRAK4 and MyD88 (Takizawa et al., 2017; Takizawa et al., 2016). Enlargement and activation of microglia after multiple CSDs were also observed, and these changes could be abolished by application of GLA, anti-HMGB-1 mAb, or double knocking-out of TLR2/4 (Takizawa et al. 2017; Takizawa et al. 2016).

Another study showed that sleep deprivation and insufficient glycogen supply to the brain, which are known risk factors of a migraine attack, caused Panx1 channel opening, caspase-1 cleavage, and HMGB-1 release (Kilic et al.

2018). More recently, neuroinflammation and migraine-HMGB-1 relationship were investigated in two transgenic mice models of familial hemiplegic migraine type 1. These mice carried two different mutations in the α1A subunit of neuronal voltage-gated CaV2.1 channels causing increased cortical excitatory transmission. It was shown that naïve FHM1 mice showed a baseline level of HMGB-1 release that was not present in their WT counterparts which could explain their known susceptibility to CSD (Dehghani et al., 2021). CSD induced in one hemisphere caused an increase in HMGB1 release and astrocytic nuclear NF-kB translocation in both hemispheres in FHM1 mice in contrast to exclusively ipsilateral activation seen in WT mice, which could be related to widespread neuroinflammation known to be present in migraineurs. Application of NMDAR inhibitor MK-801 to the contralateral hemisphere abolished the propagation of the CSD to the contralateral hemisphere without affecting its induction in the ipsilateral one and a single systemic application of ifenprodil, which is selective for NR2B-subunit carrying NMDAR known to interact reciprocally with HMGB-1, reduced the basal level of HMGB-1 release seen in mutant FHM1 (Dehghani et al., 2021) NR2B-subunit carrying NMDAR also contributes to excitotoxicity in familial hemiplegic migraine type 2 (Crivellaro et al., 2021).

Taken together with the knowledge that HMGB-1 increases NR2B subunit containing NMDAR activity selectively by increasing NR2B tyrosine phosphorylation (Balosso et al., 2014; Maroso et al., 2010) and overexcitation of neurons present during CSD propagation causes them to release HMGB-1 (Fujita et al., 2016; Yang et al., 2021), NR2B phosphorylation and inhibition could be an attractive therapeutic target. In models of migraine that did not use CSD induction, there was no direct examination of HMGB-1, but several studies showed the importance of proteins working together with HMGB-1. To model chronic migraine, rats were subjected to several inflammatory soup injections containing histamine, serotonin, bradykinin, and prostaglandin E2 to induce trigeminocervical and dural activation. An increase in NR2B phosphorylation was seen as proportional to the level of synaptic activity, which in turn increased CGRP expression. When NR2B phosphorylation was inhibited by tyrosine kinase inhibitors genistein or PP2, CGRP expression, synaptic activity, and frequency of migraine attacks were reduced, while the mechanical threshold that was found deceased in the tactile sensory test was increased (Liang et al. 2017; Wang et al. 2018). Another study using the same model found that pretreatment with a specific inhibitor of TLR4 TAK-242 abolished hyperalgesia, prevented microglial activation, and blocked the IL-1B and NF-kB increase (Table 1) (Su et al. 2018).

Several studies also showed that sex hormones might affect migraines via HMGB-1. It is known that HMGB-1 has estrogen binding sites (Borrmann et al. 2001), and it increases with exposure to estrogen (Chau, Lam, and Lee 1998; He, Liang, and Lippard 2000). Considering that CSD is significantly enhanced by estrogen and progesterone (Sachs et al., 2007) application of estrogen to ovariectomized rats is sufficient to induce CSD episodes (Sandweiss et al., 2017). CSD susceptibility arises after orchiectomy in FHM1 mice models, which can be treated with testosterone replacement (Eikermann-Haerter et al., 2009), and a relationship between sex hormones and migraine was epidemiologically shown (Ferrari et al., 2022; MacGregor and van den Brink 2019). Further research on sex hormones and HMGB-1 interactions in the context of migraine and CSD is sorely needed.

A recent study demonstrated that HMGB-1 levels were increased in the serum of Covid-19 patients with headaches (Bolay et al. 2021), which could point to HMGB-1 and its common partner proteins being an effector in headaches seen secondary to infections and other inflammatory conditions.

HMGB-1 and Neuroimmunological Disorders

Neuroimmunological disorders are a group of disorders causing inflammation, neurodegeneration, and/or demyelination in the nervous system that might lead to a combination of optic neuritis, visual defects, cerebellar dysfunction, and sensorimotor deficits, usually caused by the interaction between immune system elements and neural structures of the body.

The most common disorders are Multiple Sclerosis (MS) and Neuromyelitis Optica Spectrum Disorders (NMOSD) (Filippi et al., 2018). Multiple Sclerosis is the prototypical neuro-immunological disorder. Clinical manifestations occur due to demyelinated lesions in both the white and gray matter of the brain and the spinal cord. Patients are grouped according to the time course of their symptoms. Remitting-Relapsing MS (RRMS) is characterized by transient episodes of moderate to severe neurological deficits (relapses) and a relatively symptom-free remission phase between relapses. Much the less common Primary Progressive MS (PPMS) is characterized by progressive and permanent deficits, acute relapse periods where symptoms increase transiently, and no symptoms free remission phase but a steady progression between relapses (Filippi et al., 2018). Most RRMS patients will develop Secondary Progressive MS (SPMS), which, like PPMS, has no symptom-free periods as the burden of the disease cumulates (Filippi et al.,

2018). It is more common in women than men and is much more common in Europe and North America, especially as the latitude increases (Filippi et al. 2018). MS is managed by disease-modifying treatments such as glatiramer acetate, interferon B, teriflunomide, dimethyl fumarate, fingolimod, natalizumab, ocrelizumab for long-term control or short-term high-dose corticosteroids for acute relapses (Filippi et al., 2018).

The second most common group of neuroimmunological disorders is Neuromyelitis Optica Spectrum Disorders. NMOSD is a group of disorders sharing a common etiology and clinical presentation. Neuromyelitis Optica (NMO) is the most common form, which is characterized by acute attacks of optic neuritis and transverse myelitis (Jarius et al., 2020). Most patients have autoreactive antibodies targeting aquaporin-4 (AQP4) channels which primarily cause inflammatory damage to and loss of astrocytes, and sometimes secondary damage to and loss of other CNS cells like neurons and oligodendrocytes causing neurodegeneration and demyelination (Jarius et al., 2020). Some patients, who are not positive for the AQP4 antibodies test, may be positive for autoantibodies targeting Myelin Oligodendrocyte Glycoprotein (MOG) on the myelin sheaths (Jarius et al., 2020). NMOSD usually shows a similar course to RRMS. Acute attacks are managed with short-term high-dose corticosteroids, while long-term management is done through eculizumab, satralizumab, inebilizumab, or other immunosuppressants (Jarius et al., 2020).

The Association of HMGB-1 with neuroimmunological disorders was first demonstrated by Andersson et al. (2008). They showed that in active lesions from autopsy samples taken from MS patients, there were cytoplasmic translocations of HMGB-1 in both microglia and macrophages of acute MS and RRMS patients. In contrast, inactive lesions and lesions from PPMS and SPMS patients showed nuclear staining. To compare the levels of expression of different molecules from the HMGB-1 cascade, they compared the HMGB-1, RAGE, TLR2, and TLR4 mRNA levels of peripheral blood mononuclear cells (PBMC) and CSF macrophages (Andersson et al. 2008). There was an increase in the expression of TLR2 and RAGE in PBMC samples but no difference in HMGB-1 and TLR4. However, in the CSF macrophages, all of them were upregulated.

Furthermore, the level of HMGB-1 expression was correlated with the expression of each receptor. HMGB-1 levels in the CSF of MS patients were increased compared to the control group (Andersson et al., 2008). These findings were tested in an animal model of MS, experimental autoimmune encephalomyelitis (EAE). After rats were immunized to MOG to induce the pathology, the level of HMGB-1 mRNA was measured and found to be increased in the active lesion, reflecting the histopathological damage and clinical scoring (Andersson et al., 2008). HMGB-1 expression remained elevated through the active phase and decreased in the remission phase, confirming the studies done on autopsy samples (Andersson et al. 2008). Immunostaining was used to demonstrate the translocation of HMGB-1 in microglia and macrophages and the diffuse staining within the active lesions. RAGE, TLR2, and TLR4 expression levels were also increased (Andersson et al., 2008). Several clinical and animal studies have been performed since then that further explored this association between HMGB-1 and neuro-immunological diseases.

Several studies showed a relationship between the HMGB-1 receptor RAGE polymorphisms and MS and/or NMO (Kolonaite et al., 2021; Tiszlavicz et al., 2009). However, only one study directly examined the effect of HMGB-1 polymorphisms on MS. 8 polymorphisms of the gene were screened among a population of Iranian MS patients, and there was no significant difference between patients and healthy controls (Asouri et al. 2020). More studies are needed on different populations before determining the role of genetic differences of the HMGB-1 cascade on MS and/or NMO.

There has been an abundance of studies on levels of HMGB-1 and its receptors, specifically RAGE, and how they change in MS, their diagnostic and prognostic value. In MS, several studies found an increase in HMGB-1 levels in peripheral blood (Bucova et al., 2020; D'Angelo et al., 2018; Malhotra et al., 2015; Sternberg et al., 2016; Uzawa, Mori, Masuda, et al., 2013), but in other studies, such an increase could not be found (Glasnovic et al., 2014; Nasi et al., 2020). Studies also disagreed on the relationship between the levels of HMGB-1 and the course of the disease; while one found higher serum levels of HMGB-1 in the RRMS subtype (Malhotra et al., 2015), the other found higher serum levels of HMGB-1 in SPMS and PPMS subtypes (Sternberg et al., 2016). Studies on CSF levels of HMGB-1 also contested each other (Glasnovic et al., 2014; Uzawa, Mori, Masuda, et al., 2013). A meta-analysis performed in 2018 included data from seven studies that show that, overall, there was no increase in serum levels of HMGB-1 in MS patients. However, the RRMS subtype displayed a higher serum level of HMGB-1 compared to PPMS and healthy controls. The level of HMGB-1 in the CSF was also increased (Zhen et al., 2019). Various studies also measured the soluble RAGE (sRAGE) levels in MS patients, which acts as an inhibitor by sequestering the ligands of RAGE and halting its activation. One study found a decreased serum level of sRAGE correlating with their relapse rate (Sternberg et al., 2008), while another found it unchanged in serum but decreased in the CSF (Glasnovic et al., 2014). Two studies measured the effects of disease-modifying treatment options on the HMGB-1 and sRAGE levels. It was shown that fingolimod treatment increased the serum levels of sRAGE while decreasing the HMGB-1 levels (Sternberg et al. 2018), and interferon-β was displayed to normalize the serum levels of HMGB-1 from the increased levels they found before treatment (D'Angelo et al. 2018). In NMO, serum levels and CSF levels of HMGB-1 were found to be increased with CSF levels correlating with the amount of astrocyte damage (Uzawa, Mori, Masuda, et al. 2013; Wang et al. 2013; Wang et al. 2012).

Animal studies also have shown the relationship between EAE and HMGB-1 repeatedly. In 2013, Robinson et al. found that EAE mice had higher levels of HMGB-1 in their serum and recombinant HMGB-1 application increased the lymphocytic response to the myelin epitope in a dose-dependent way. Both prophylactic and post-immunization applications of anti-HMGB-1 mAb significantly mitigated the clinical disease, although the effect was less pronounced and transient in the post-immunization group (Robinson et al. 2013). In the same year, Uzawa et al. showed similar changes in EAE models upon post-immunization treatment with the anti-HGMB-1 mAb, but they reported a more prolonged effect, up to 45 days compared to return to baseline levels in 7d days seen in the study done by Robinson et al. (Robinson et al. 2013; Uzawa, Mori, Taniguchi, et al. 2013). The application of anti-HMGB-1 mAb also inhibited the release of HMGB-1 and IL-17 (Uzawa, Mori, Taniguchi, et al., 2013). Effects of GLA were also examined in other studies. In a study done in 2018 by Sun et al. EAE induced mice were divided into three groups, and GLA was applicated in 3 different stages of the EAE; dubbed early, midterm-late, and late. In the naïve group, they saw an increase in HMGB-1 levels in CSF, spinal cord homogenate, and serum. They also showed increased immunostaining in the lateral column, dorsal column, ventral horn, and dorsal horn of the spinal cord. GLA treatment significantly inhibited these increases. Furthermore, it delayed the onset of symptoms and lowered the severity of the symptoms, decreased the activation of microglia and astrocytes, neuronal loss, demyelination, and mononuclear cell infiltration both in the brain and the spinal cord (Sun et al., 2018). In 2014, a new inhibitor of microglial activation that targeted HMGB-1 and HMGB-2 was found, called inflachromene (ICM), that works via binding the DNA-binding domain in Box-A and inhibiting its translocation to the cytoplasm from the nucleus. They also showed that ICM was effective in alleviating EAE when given post-immunization. Furthermore, ICM treatment delayed disease onset, decreased the clinical score, and lowered demyelination and neurite damage (Lee et al.,

2014). Another study examined the effects of recombinant thrombomodulin (rTM) since it is a known inhibitor of HMGB-1 that works by sequestering and cleaving it (Ito et al., 2008; Uzawa et al., 2018). They found that rTM effectively decreases the disease severity, suppresses the HMGB-1 translocation, and inhibits the IL-17 increase. However, since the plasma levels of HMGB-1 were unchanged, they speculated that this effect might not depend on its effect on HMGB-1 (Uzawa et al., 2018). In 2021 another novel inhibitor of HMGB-1 was found to be effective in EAE. Matrine, a compound derived from the plants in the genus Sophora, was effective in diminishing the severity of ongoing EAE, reducing the levels of HMGB-1 and other proinflammatory markers, demyelination, and inflammatory infiltration. Its inhibitory effect on HMGB-1/TLR4/NF-κB was also confirmed in in-vitro experiments (Chu et al., 2021). Recently it was shown that miRNA levels targeting HMGB-1 also change in MS. Expression levels of miR-218-5p, which downregulate HMGB-1 (Mosarrezaii Aghdam et al., 2022; Yu et al., 2018) was decreased in both the PBMC's and B lymphocytes of MS patients (Mosarrezaii Aghdam et al. 2022; Sievers et al. 2012). Another miRNA targeting HMGB-1, miR-129 expression, was also measured in the serum of MS patients, but there was no significant change (Montazeri, Eskandari, and Mansouri, 2021).

HMGB-1 may also affect neuroimmunological disorders through several other pathways. HMGB-1 is a known positive regulator of autophagy through the interactions of its extracellular form with RAGE and its intracellular form with Beclin1 (Kang et al., 2014). While autophagy seems to increase and Beclin1 seems to decrease in MS patients, this may be another pathway HMGB-1 works through (Misrielal et al., 2020; Zhen et al., 2015). In another study, selective inhibitors of NR2B subunit-containing NMDARs regulated by HMGB-1 were found to be effective in decreasing the neurological deficits seen in EAE mice compared to nonselective NMDAR inhibitor memantine and vehicle (Farjam et al., 2014). Another study published recently revealed a completely new and beneficial effect of HMGB-1 in EAE. It has been known that sonic hedgehog (shh) expression increases during all stages of EAE but is found in only remyelinating lesions, reflecting the levels of remyelination (Seifert et al., 2005). Recombinant HMGB-1 application increased the shh expression levels in primary astrocyte culture, and anti-HMGB-1 mAb and GLA inhibited the increase of shh in vitro and in vivo, respectively, through JNK/STAT3 signaling. This effect was suppressed after RAGE was inhibited genetically or pharmacologically. When recombinant shh was given to EAE mice receiving GLA, the disease's progress was inhibited, and demyelination was ameliorated (Table 1) (Xiao et al., 2021).

HMGB-1 and Neurodegenerative Diseases

Neurodegenerative diseases are a group of diseases that affect the nervous system and are characterized by a progressive loss of neurons, sometimes accompanied by glial loss and BBB dysfunction. Every disease in this group is characterized by a specific and defined loss order regarding the neuron type and affected regions. In almost every neurodegenerative disease, there is an aggregation of different protein types such as amyloid-β (Aβ), tau, TAR DNA-binding protein 43 (TDP-43), α-Synuclein (α-syn), or Huntingtin (Htt) (Wolfe 2018). Aging is a shared risk factor for various neurodegenerative diseases (Enokido et al., 2008). HMGB-1 level changes as the brain ages, depending on the cell type (Basisty et al., 2020; Enokido et al., 2008). Both overexpression and depletion of the HMGB-1 may cause neurodegeneration (Chavan et al., 2012; Davalos et al., 2013; Zhao, Rouhiainen, et al., 2020) due to its effects on neurogenesis, differentiation, autophagy, DNA repair, neuroinflammation, and synaptic plasticity (Chavan et al., 2012; Davalos et al., 2013; Enokido et al., 2008; Misrielal et al., 2020; Tang et al., 2010; Zaben et al., 2021; Zhao, Rouhiainen, et al., 2020).

The most common neurodegenerative disease is Alzheimer's disease (Wolfe 2018). Alzheimer's Disease (AD) is a progressive neurodegenerative disorder, and the accumulation of extracellular Aβ plaques, intracellular tau neurofibrillary tangles, and neuronal loss prominent in frontotemporal lobes, limbic system, and hippocampal formation are the hallmarks of the disease (Knopman et al., 2021). Clinically AD starts to manifest as memory problems, and the deficits encompass other cognitive, emotional, and social functions as the pathology progress (Knopman et al., 2021; Wolfe, 2018). Risk factors for AD include aging, diabetes mellitus, hypertension, obesity, depression, and mutations in lipid-binding protein apolipoprotein E (Knopman et al., 2021). In addition, a subgroup of patients has autosomal dominant inherited AD (DIAD), which is characterized by earlier onset, but a similar course, and these patients carry mutations in genes encoding the amyloid-β-protein precursor (APP), presenilin-1 (PSEN1), and presenilin-2 (PSEN2) (Wolfe 2018). APP is the precursor of Aβ, while PSEN1 and PSEN2 are part of a group of enzymes called gamma secretases which process APP into Aβ monomers together with beta secretases (Knopman et al., 2021).

HMGB-1 is increased in the CSF and brain homogenates of AD patients (Fujita et al., 2016; Takata et al., 2003). Furthermore, it interacts with several different processes related to the pathological mechanism of the disease, such as tau, Aβ, or DNA repair. Regarding Aβ accumulation, HMGB-1 directly inhibits the microglial clearance of Aβ40 and Aβ42 via phagocytosis (Takata et al., 2003; Takata et al., 2004; Takata et al., 2012). HMGB-1 also colocalizes with Aβ plaques and surrounds activated microglia (Takata et al. 2003; Takata et al. 2004; Takata et al. 2012). When exposed to Aβ in vitro, there is an acute increase of HMGB-1 in microglia (Caldeira et al., 2017). Similarly, when neurons are exposed to Aβ monomers in vitro, there is an increase in the expression and translocation of HMGB-1, which causes the activation of RAGE, TLR4, and NF-kB. An increase in HMGB-1 and subsequent activation of RAGE, TLR4, and NF-kB can be abolished by shRNA application targeting HMGB-1 (Nan et al., 2019).

In a recent study on the 3xTg-AD mice model, there was an increase in APP processing and mir-146a levels in both hippocampus and cortex at three months when mice were in early stages and pre-symptomatic, with an accompanying decrease in HMGB-1 in both areas. In the animals' cortex, but not the hippocampus, there was also an increase in mir-155 and mir-124 levels (Fernandes et al., 2022). When the mice were symptomatic at nine months, HMGB-1 and mir-155 were elevated in both regions, while mir-124 and mir-146a levels were decreased (Fernandes et al., 2022). Another recent study also showed the changes in RNA regulation of HMGB-1 expression in AD. In in-vitro and in-vivo models of AD, there was an elevation of Long non-coding RNA H19 (H19) levels and a decrease in mir-129 levels. H19 binds and inhibits mir-129, which in turn suppresses HMGB-1. Normalizing the levels of H19 or mir-129 increased neuronal survival and prevented neurodegeneration (Zhang et al., 2021). In addition, several studies showed that inhibiting HMGB-1 via anti-HMGB-1 mAb suppressed Aβ and tau deposition and neurodegeneration and promoted cell survival in the brain by opsonization of Aβ (Fujita et al., 2016; Hei et al., 2018).

The relationship between tau and HMGB-1 was also shown repeatedly. In both tauopathy models and the brain samples from the AD patients, upregulation and translocation of HMGB-1 were shown (Nilson et al., 2017; Zhao, Tan, et al., 2021). Furthermore, inhibition of HMGB-1 by GLA or the genetic knockdown of NLRP3 decreased the tau phosphorylation, p-tau deposition and improved cognitive functions (Zhao, Tan, et al., 2021). Tau oligomers also directly promoted astrocyte senescence through HMGB-1 promoted neuroinflammation and cognitive deterioration. Effects of tau oligomers could be counteracted with anti-HMGB-1 mAb, GLA, or ethyl pyruvate (another HMGB-1 inhibitor) (Gaikwad et al., 2021). HMGB-1 also disturbs DNA double-strand break repair mechanisms in AD via promoting the abnormal phosphorylation of the 70 kDa subunit of Ku antigen, which is required for non-homologous end joining and known to decrease with age and AD (Tanaka et al. 2021).

Another group of neurodegenerative diseases closely related to AD is Frontotemporal Dementias (FTD), which targets frontal and temporal regions. It is identified with the deposition of tau and TDP-43 pathologically and disturbance of personality, language, and other higher cognitive functions (Wolfe 2018). HMGB-1 increases tau deposition in FTD patients and animal models (Gaikwad et al., 2021; Nilson et al., 2017). It also colocalizes with phosphorylated TDP-43 in atherosclerotic plaques of patients (Umahara et al., 2020). Moreover, in 2021 using dynamical network analysis, Jin et al. showed that both FTD and AD models share a common pathological network involving HMGB-1/TLR4 signaling, which was replicated in postmortem brains of patients (Jin et al. 2021). Furthermore, anti-HMGB-1 mAb therapy in 4 different transgenic FTD mice models decreased memory impairment, neuronal loss, DNA damage, protein aggregation, synapse abnormalities, and tau phosphorylation (Table 1) (Jin et al., 2021).

Parkinson's Disease (PD) is the second most common neurodegenerative disease and the most common neurodegenerative disease affecting motor systems. PD is characterized by a progressive loss of dopaminergic neurons and the presence of Lewy bodies which are formed by the intracellular aggregation of α-syn. Defects of the dopaminergic system cause motor symptoms such as resting tremors, bradykinesia, and postural instability clinically (Poewe et al., 2017; Wolfe, 2018). Impairments in autophagy and mitochondrial dysfunctions also contribute to the pathology (Poewe et al., 2017). Several studies demonstrated that HMGB-1 levels were elevated in serum (Baran et al. 2019; Santoro et al. 2016; Yang et al. 2018), CSF (Santoro et al. 2016), and substantia nigra (Santoro et al. 2016) samples of the PD patients compared to the healthy controls. Poor treatment response, higher staging, and disease duration were also related to higher HMGB-1 and TLR4 levels (Yang et al., 2018). HMGB-1 was also shown to bind aggregated α-syn in Lewy bodies procured from brain samples of patients with PD or dementia with Lewy bodies (Lindersson et al. 2004). In 1-Methyl-4-phenylpyridinium (Gao et al. 2011; Santoro et al. 2016), 6-hydroxydopamine (Sasaki et al. 2016), methamphetamine (Masai et al. 2021), and rotenone (Gao et al. 2011; Huang et al. 2017) induced models of dopaminergic neurodegeneration, HMGB-1 levels were increased. HMGB-1 could also directly activate microglia in some of these models (Gao et al., 2011; Masai et al., 2021).

In all these studies, neurodegeneration and other detrimental effects induced by the toxins could be suppressed by anti-HMGB-1 mAb (Gao et al., 2011; Huang et al., 2017; Masai et al. 2021; Santoro et al., 2016; Sasaki et al. 2016). HMGB-1 was also partly responsible for the impairment in autophagy via its interactions with Beclin-1 and α-syn (Huang et al. 2017; Song et al. 2014; Wang et al. 2016), but the results on the exact effect of HMGB-1 were conflicting, which may be due to use of different models, indiscrimination of different isoforms and different oxidative states of the cells. Interestingly two studies by Kim et al. showed that HMGB-1 could promote tyrosine hydroxylase activity and support the dopaminergic system via RAGE using both in vitro (Kim et al., 2017) and in vivo (Kim, Ryu, et al., 2019). However, this effect may result from the stereotaxic application of HMGB-1 rather than systematic application and bypassing its peripheral effects, timing, or dosage (Table 1).

Polyglutamine diseases are a group of neurodegenerative repeat expansion diseases characterized by the increased number of CAG repeats in the coding region of specific genes, which are translated as a long chain of glutamine (PolyQ) in the protein encoded by the affected gene (Bunting, Hamilton, and Tabrizi 2022). HMGB-1 has been shown to affect three of these diseases. First is Huntington's Disease (HD), in which CAG repeat is located in Huntingtin (Htt) gene. As a result, HD patients show striatal and cortical neurodegeneration, which causes motor, psychiatric, and cognitive symptoms (Bunting, Hamilton, and Tabrizi, 2022). Second is Spinocerebellar Ataxia type 1 (SCA1), caused by repeats of the ataxin-1 gene, which produce cerebellar, spinal, and bulbar symptoms due to neurodegeneration in the brainstem and cerebellum (Bunting, Hamilton, and Tabrizi 2022). The last one is Spinocerebellar Ataxia type 17 (SCA17), which arises due to repeats on TATA-binding protein (TBP) and is characterized by cortical, cerebellar, and striatal neurodegeneration pathologically and cerebellar ataxia, dementia, psychiatric symptoms, and motor abnormalities (Bunting, Hamilton, and Tabrizi 2022).

In 2007, Qi et al. showed via the coprecipitation method that both mutant Htt and ataxin-1 decreased the levels of HMGB-1 and HMGB-2 in neurons. Furthermore, this reduction was more pronounced in the cortex for Htt and the cerebellum for ataxin-1, which are more susceptible to HD and SCA1, respectively (Qi et al., 2007). In a mice model of HD, this tissue-specific reduction was shown to be accompanied by a reduction in levels of the proteins participating in base excision repair (Goula et al. 2009), which could mean HMGB-1 might have a disease-modifying role in the temporal and spatial distribution of neuron loss. A similar reduction in HMGB-1 was also demonstrated in vitro for SCA17 due to coprecipitation of HMGB-1 with TBP (Liu et al. 2009). Overexpression or exogenous application of HMGB-1 lowered protein aggregation, increased cell viability, lifespan, and expression of both mitochondrial and nuclear DNA repair genes (Ito et al., 2015; Min et al., 2013; Qi et al., 2007). Overexpression of HMGB-1 after the emergence of symptoms also improved motor performance and drastically increased the lifespan of the affected mice without causing an increase in neuro-inflammation (Ito et al., 2015).

A possible explanation for this positive effect could be the chaperone-like activity of HMGB-1 protein. It can directly interact with PolyQ repeats and decrease their accumulation (Min et al., 2013). However, when the concentration of HMGB-1 reaches a threshold, it can have detrimental effects and aggravate the accumulation of the PolyQ repeats rather than attenuate it (Min et al., 2013). It can also bind the nuclear localization signal on Htt and mediate its entry into the nucleus (Son et al., 2019). Various studies also showed the possibility of disease-modifying effects of HMGB-1 receptors RAGE and TLRs on HD (Angelopoulou, Paudel, and Piperi 2020; Anzilotti et al. 2012; Griffioen, Mattson, and Okun 2018; Shi et al. 2018). Based on all these data, there are efforts to prepare viral vectors targeting HMGB-1 and increasing its expression suitable for human use, and clinical trials are being designed (Table 1) (Fujita and Okazawa 2017).

Human Immunodeficiency Virus (HIV) Associated Neurocognitive Disorder (HAND) is the neurocognitive impairments seen after a long duration of infection in people living with HIV due to neurodegeneration driven by HIV (Clifford and Ances 2013). Although severe forms of disease became much less prominent with the introduction of combination antiretroviral therapy, around half of the people living with HIV show milder forms of HAND, probably due to the status of CNS as a reservoir for HIV (Clifford and Ances 2013). HMGB-1 is closely related to HAND; it can induce reactivation in latently infected cells (Thierry et al. 2007) and is increased in both CSF (Gougeon et al. 2017) and neuron-derived exosomes, especially in people showing neurocognitive decline (Pulliam et al. 2020; Sun et al. 2017; Sun, Fernandes, and Pulliam 2019). Naturally occurring anti-HMGB-1 antibodies were found in the serum of patients with HAND, and the level of autoantibodies targeting HMGB-1 could discriminate patients with cognitive impairment even in very early stages (Gougeon et al. 2017). Furthermore,

HMGB-1 levels in neuron-derived exosomes combined with CD4+ cell count and neurofilament light chain levels were found to be the best predictor of HAND using machine learning (Pulliam et al., 2020).

Amyotrophic lateral sclerosis (ALS) is a rare neurodegenerative disease affecting both upper cortical motor neurons and lower spinal motor neurons leading to prominent motor dysfunction like muscle weakness and atrophy, dysphagia, dysarthria, or respiratory insufficiency, and nonmotor symptoms like cognitive impairment or mood disorders (Hardiman et al. 2017). Neuropathological changes include protein aggregation comprised of TDP-43 in motor neurons, neuroinflammation, mitochondrial dysfunction, glial dysfunction, and DNA repair impairments (Hardiman et al., 2017; Wolfe, 2018).

HMGB-1 levels in neurons were usually unchanged in the pre-symptomatic stage and reduced or unchanged in a symptomatic stage in a model of ALS expressing mutant Superoxide dismutase-1. (SOD1 mice) (Gomes et al. 2019; Lo Coco et al. 2007). In contrast, the HMGB-1 signal was increased in the reactive astrocytes after the symptomatic stage (Gomes et al., 2019). HMGB-1, RAGE, TLR2, and TLR4 levels were also shown to increase in the astrocytes taken from the spinal cords of a population of sporadic ALS patients (Casula et al., 2011). In another in vivo model of ALS, the Wobbler mice HMGB-1 and TLR4 expression were also increased (Meyer et al. 2020). Interestingly, anti-HMGB-1 autoantibodies were also found in the serum of ALS patients, and levels of autoantibodies correlated with the disease severity significantly (Hwang et al., 2013).

HMGB-1 blockage by the anti-HMGB-1 mAb caused a short-term relief from the motor deficits but did not affect the survival, the glial activation status, or motor performance in the later stages in the SOD1 mice (Lee et al. 2019). In contrast, TLR4 or RAGE deletion extended survival and increased motor performance (Lee et al., 2020; Lee et al., 2015). Furthermore, RAGE deletion also reduced HMGB-1 and TLR4 expression by half (Lee et al., 2015). These results suggest that HMGB-1 has potential as both a biomarker and a therapeutic target in neurodegenerative diseases (Table 1).

HMGB-1 and Strokes

Stroke is the disturbance of the blood flow to a part of the CNS, which can be classified into two main categories (Campbell et al., 2019; Unnithan, J, and Mehta, 2022). First is ischemic stroke, which happens when blood supply to an area of the CNS is cut off, most often due to a thromboembolic event (Campbell et al. 2019). The resulting infarctions are usually divided into two areas, an ischemic core where the cells are irreversibly damaged and an ischemic penumbra where the cells can be saved with rapid reperfusion, however when the ischemia persists for a long time, reperfusion might increase the damage (Campbell et al. 2019; Ikhlas and Atherton 2022). The second category is hemorrhagic stroke which is bleeding into the brain due to a vessel rupture and can be further divided into subarachnoid hemorrhage (SAH) and intracerebral hemorrhage (ICH) based on the location of the event (Unnithan, J, and Mehta 2022). Some of the risk factors of having a stroke include hypertension, dyslipidemia, smoking, atherosclerosis, and diabetes mellitus (Campbell et al., 2019), all of which can affect the levels of HMGB-1 in the brain to influence the pathology and the prognosis (Kikuchi et al. 2013).

HMGB-1 increases in both the acute and the chronic phases of ischemic stroke (Goldstein et al. 2006; Le et al. 2018; Schulze et al. 2013; Tsukagawa et al. 2017) and after reperfusion (Li et al. 2018; Wang, Jiang, et al. 2020). A serum level of HMGB-1 after an ischemic stroke is correlated with infarct volume and deficit severity (Le et al., 2018; Tsukagawa et al., 2017). A higher level of serum HMGB-1 after the reperfusion increased the risk of a poor outcome three months and one year after the event (Tsukagawa et al., 2017; Wang, Jiang, et al., 2020). Moreover, it also increases the mortality risk in the first three months (Wang, Jiang, et al., 2020). In a study to determine the predisposing genetic variances in a Chinese Han population, RAGE and HMGB-1 variants were shown to modify ischemic stroke susceptibility (Li et al., 2017). Using a transient middle cerebral artery occlusion (MCAO) model in mice, the release of HMGB-1 started as early as 30 minutes after the occlusive event in the ischemic core (Qiu et al., 2008). After the reperfusion, HMGB-1 was measurably higher in the CSF and was translocated into the cytoplasm of neurons in the peri-infarct area (Qiu et al. 2008). In another transient MCAO study, after leaving the nucleus, HMGB-1 was shown to colocalize with peroxisomes and mitochondria, specifically after the reperfusion. It also modulated mitochondrial fission activity via dynamin-related protein one, which is essential for mitochondrial fission (Wang, Liu, et al. 2020).

Based on its physiological role in platelet activation and NETosis (Kang et al. 2014; Kim and Lee 2020), HMGB-1 was investigated directly in cerebral thromboembolism. In a rat permanent MCAO model, levels of NETosis marker citrullinated histone H3 (CitH3) (Pan et al. 2017) and HMGB-1 were increased in the blood, CSF, and infarct region (Kim, Lee, et al. 2019). Using recombinant HMGB-1 on isolated neutrophils, it was shown that both fully reduced and disulfide isoforms of HMGB-1 could induce the NET formation and upregulate CitH3 via CXCr4 and TLR4, specific receptors for fully reduced and disulfide forms, respectively (Kim, Lee, et al. 2019). Anti-HMGB-1 mAb and Box-A suppressed the CitH3 upregulation and NETosis in permanent MCAO. HMGB-1 or NETosis inhibition also decreased neuronal loss and blood vessel damage seen during the delayed damage period of ischemic stroke (Kim, Lee, et al., 2019). HMGB-1 was also closely related to neutrophils, NETs, and platelets, forming around 1.5% of the total area in human thrombectomy samples (Essig et al., 2021). HMGB-1 also contributes to excitotoxicity seen after ischemic stroke, which is the leading cause of secondary cell death in the penumbra after ischemia/reperfusion (Kim et al., 2006; Lin, Chen, and Wei, 2020).

In primary cortical cultures, there was a vast HMGB-1 release after treatment with glutamate or NMDA to induce excitotoxicity, but no such release was detected after staurosporine treatment to induce apoptosis. HMGB-1 then activated microglia and increased the proinflammatory cytokine levels (Lin, Chen, and Wei 2020). Furthermore, oxygen-glucose deprivation/reoxygenation (OGD/R) model of ischemia/reperfusion hindered the glutamate clearance in primary astrocyte culture, an essential contributor to post-ischemic excitotoxicity (Barreto et al. 2011; Kim et al. 2006). OGD/R caused an increase in the HMGB-1 and reduced glutamate aspartate transporter expression, impairing glutamate clearance. Applying recombinant HMGB-1 to primary astrocytes was enough to hinder glutamate clearance, while treatment with siRNAs targeting HMGB-1 or TLR4 before OGD/R or GLA treatment concurrent with OGD/R protected the glutamate clearance system of the astrocytes (Kim et al. 2006). HMGB-1 expression was also directly induced by reactive nitrogen species (RNS) like peroxynitrite both in vivo and in vitro (Chen et al. 2020). RNS marker nitrotyrosine level elevation and HMGB-1 level elevation in the serum after ischemic stroke correlated strongly. They increased further if the patient had undergone a hemorrhagic transformation after delayed tissue plasminogen activator treatment (Chen et al., 2020).

In several animal stroke studies, HMGB-1 was tried as a potential therapeutic target and the results were promising. Treatment with mAb reduced the infarct volume (Liu et al. 2007; Muhammad et al. 2008), preserved BBB integrity (Liu et al. 2007; Zhang et al. 2011), prevented microglia activation (Liu et al. 2007), decreased neuroinflammation (Liu et al. 2007), and improved the outcome (Liu et al. 2007). Box-A was also effective in reducing the infarct volume (Jin et al. 2011; Muhammad et al. 2008), microglial activation (Jin et al. 2011), neuroinflammation (Jin et al. 2011) and NETosis (Denorme et al. 2021) while improving the functional outcome (Jin et al. 2011). Similarly, GLA also improved the outcome (Chen et al. 2020; Kim, Jin, et al. 2012) and reduced infarct volume (Kim, Shin, et al. 2012), neuroinflammation (Chen et al. 2020; Kim, Jin, et al. 2012), peripheral immune cell infiltration (Xiong et al. 2016), RNS formation (Chen et al. 2020), hemorrhagic transformation neuroinflammation (Kim, Shin, et al. 2012), peripheral immune cell infiltration, RNS formation (Chen et al. 2020), and mortality neuroinflammation (Chen et al. 2020; Kim, Jin, et al. 2012), peripheral immune cell infiltration (Xiong et al. 2016), RNS formation (Chen et al. 2020).

Other inhibitors of HMGB-1 like ethyl pyruvate (EP) (Shin et al. 2014), sRAGE (Muhammad et al. 2008), and HMGB1-binding heptamer peptide (Kim, Jin, et al. 2012; Li et al. 2018) were also tried and found to be protective or potentially therapeutic against ischemic stroke. Interestingly, several treatments being researched for their therapeutic effect on ischemic stroke influenced HMGB-1. Therapeutic hypothermia decreases the rate of neural loss in the penumbra and increases the therapeutic window, reducing the infarct volume via inhibiting the HMGB-1 release and the subsequent neuroinflammatory cytokine expression (Lee et al. 2016). Hyperbaric oxygen treatment increases the level of deacetylase Sirtuin 1 (SIRT1), which deacetylases HMGB-1, suppressing its translocation to cytoplasm, which in turn inhibits cell apoptosis, decreases inflammation, and promotes cell proliferation both in vitro and in vivo (Zhao, Xu et al. 2021). Uric acid was shown to be effective in different animal models in reducing infarct volume and neurological deficits (Paul and Candelario-Jalil, 2021).

Furthermore, URICO-ICTUS, a multicentric double-blind, randomized clinical trial, showed that it reduced the incidence of early ischemic worsening, improved neurological deficits, and decreased disability rate without any major side effects (Amaro et al. 2016). Researchers demonstrated that uric acid inhibits the OGD/R induced translocation and release of HMGB-1, reduces the infarct volume, and decreases microglial activation. These effects were inhibited by probenecid treatment which inhibits the entrance of uric acid into the cell via inhibiting Glucose transporter nine and urate transporter 1 (Wang, Zhao, et al., 2021).

HMGB-1 also has some positive effects during ischemic stroke recovery. In primary astrocyte cultures, low doses of HMGB-1 (1-10 ng/mL) increased the rate of endothelial progenitor cell (EPC) proliferation, and this effect was abolished when HMGB-1 was suppressed with siRNA, RAGE was suppressed with anti-RAGE antibodies or with a high concentration HMGB-1 (100-1000 ng/mL) (Hayakawa et al. 2012). In addition, HMGB-1 released from astrocytes was shown to increase the number of EPCs in the peri-infarct area (Hayakawa et al. 2012), and HMGB-1 released from microglia could induce EPCs to promote neovascularization via paracrine signaling (Chen et al. 2014). Astrocytic HMGB-1 release also promoted recovery in white matter stroke by binding to RAGE on EPCs to increase brain-derived neurotrophic factor (BDNF) and basic fibroblast growth factor (FGF) release (Chen et al. 2014). Choi et al. found that lower concentrations of HMGB-1 (0.1-1 ng/mL) reduced oligodendrocyte death through TLR2 in vitro, and blockage of HMGB-1 with GLA increased the lesion severity and aggravated the neurological deficits (Choi et al. 2017). Hei et al. showed that HMGB-1 blockage with anti-HMGB-1 mAb was protective in a chronic cerebral hypoperfusion optic tract white matter lesion model. It decreased the severity of the lesion, increased neurological outcome, and decreased neuro-inflammation (Hei et al. 2019). These conflicting findings may be due to the timing of the intervention since the first two studies focused on the first three days, the acute phase of white matter stroke, while the last study specifically inhibited HMGB-1 at day 28 to not affect the protective effects in the acute phase (Table 1) (Choi et al. 2017; Hayakawa et al. 2013; Hei et al. 2019).

Intracerebral hemorrhage (ICH) is bleeding into the brain and/or the ventricles (Unnithan, J, and Mehta, 2022). Patients have higher levels of circulating HMGB-1 and sRAGE after ICH (Lei, Geng, and Zhong 2020; Zhou et al. 2010), and HMGB-1 levels correlate with lesion severity and neurological deficit severity (Lei, Geng, and Zhong 2020; Zhou et al. 2010). Hemoglobin, which is released into the brain in massive amounts after ICH, can directly induce HMGB-1 translocation and its subsequent release in microglia and neurons (Sun et al., 2014; Zhou et al., 2010). In the acute phase of in vivo ICH models, HMGB-1 translocates from the nucleus to the cytoplasm and is released extracellularly (Lei et al. 2013b, a; Lei et al. 2015; Ohnishi et al. 2011). Inhibition of HMGB-1 translocation and release by GLA, EP, or anti-HMGB-1 mAb reduced HMGB-1 levels, brain edema, neuroinflammation, and neuronal loss (Lei et al. 2013b; Ohnishi et al. 2011; Wang et al. 2017). They also improved the functional outcome and lessened the neurological deficits (Lei et al. 2013b; Ohnishi et al. 2011; Wang et al. 2017). GLA also effectively blocked the direct induction of HMGB-1 release from neurons by hemoglobin (Sun et al., 2014). Anti-HMGB-1 mAb also upregulated nuclear factor erythroid 2-related factor (Nrf2) and heme oxygenase-1 (HO-1) (Xu et al. 2022).

Nfr2 is a transcription factor that upregulates the expression of proteins that protect the cell against various insults, and HO-1 is one of its targets that breakdowns hemoglobin and regulates inflammation, apoptosis, and angiogenesis (Loboda et al. 2016). Contrary to its effects on the acute phase, in the chronic or late phase of the ICH models, HMGB-1 takes a recuperative role. HMGB-1 levels are even more elevated than in the acute phase in the chronic phase, together with RAGE, TLR2, TLR4, and trophic factors like BDNF, vascular endothelial growth factor (VEGF), and nerve growth factor (NGF) levels (Lei et al. 2013b; Lei et al. 2015). Neurogenesis was also more pronounced in that time frame (Lei et al. 2013b; Lei et al. 2015). When HMGB-1 release was decreased via EP or RAGE activity was inhibited by its antagonist FPS-ZM, the levels of BDNF, NGF, and VEGF were reduced drastically. Furthermore, neurogenesis was diminished, and clinical symptom severity increased (Lei et al. 2013b; Lei et al. 2015).

These results suggest that HMGB-1 also has both negative and positive effects during the ICH recovery period that could depend on the time frame or the metabolic state of the tissue. Another study on miR-129-5p's role on ICH highlighted this even further. When miR-129-5p was upregulated, HMGB-1 increase was suppressed, and subsequent neuroinflammation and brain edema were decreased. However, VEGF levels and microvessel density, which mark ongoing angiogenesis, were also decreased, again suggesting a biphasic role for HMGB-1 (Table 1) (Ma, Li, and Shang 2017).

The other type of hemorrhagic stroke is subarachnoid hemorrhage (SAH) which is bleeding into the subarachnoid space between the pia mater and arachnoid membrane (Unnithan, J, and Mehta 2022). HMGB-1 levels rise in the CSF (Nakahara et al. 2009; Sokol et al. 2015) and the serum (Chaudhry et al. 2018; Hemmer et al. 2022; Zhu et al. 2012) of the SAH patients, and higher levels of HMGB-1 in CSF (Nakahara et al. 2009; Sokol et al. 2015) or serum (Zhu et al. 2012) predict a worse outcome. In animal studies, HMGB-1 protein levels rose as early as 2 hours and peaked at day one. Furthermore, its expression was still upregulated for up to a day (Murakami et al., 2011; Sun et al., 2014). HMGB-1 rapidly translocated into the cytoplasm, mainly in neurons in the early stages, but translocation also started in microglia around day one and overtook neurons (Murakami et al., 2011; Sun et al., 2014). On day five, most HMGB-1 expressions were in microglia (Murakami et al., 2011). There was an increase in the levels of cleaved caspase 3, an apoptosis marker, in neurons (Sun et al. 2014) and levels of pro-inflammatory markers and astrogliosis after SAH (Murakami et al. 2011).

HMGB-1 was also closely related to secondary damage mechanisms in SAH. After SAH majority of patients develop a complication called "delayed cerebral vasospasm" (DCV) or shortly "vasospasm," which is the contraction of the cerebral arteries several days after SAH (Findlay, Nisar, and Darsaut 2016; Tsiskaridze, Lindgren, and Qureshi 2016). Another complication is "delayed cerebral ischemia" (DCI) which can be diagnosed with the sudden appearance of global or focal ischemic symptoms or new ischemic lesion on radiological imaging that cannot be explained by other means (Tsiskaridze, Lindgren, and Qureshi 2016). While DCI and DCV are closely related and DCV is a major cause of DCI, both DCI and DCV can happen without the other (Findlay, Nisar, and Darsaut 2016; Tsiskaridze, Lindgren, and Qureshi 2016).

The G allele of rs2249825 in the HMGB-1 gene increases the risk for delayed cerebral ischemia after SAH (Hendrix et al., 2017). A recent study reported that a higher serum level of HMGB-1 was a good predictor of DCI (Loboda et al. 2016), but earlier studies could not find a correlation (Chaudhry et al. 2018), so further research with larger sample sizes is needed to detect the role of HMGB-1 in DCI. The relationship between DCV and HMGB-1 is easier to study due to more reliable animal models. Clinical studies show that HMGB-1 levels are higher in patients who develop DCV than in those who do not (Chaudhry et al., 2018; Zhu et al., 2012). It was shown in animal models that in the smooth muscle cells of the basilar artery, HMGB-1 was translocated to the cytoplasm in vascular cells and HMGB-1 levels increased after SAH (Haruma et al., 2016; Zhao et al., 2016). HMGB-1 peaked on days 3 and 5, coinciding with the highest amount of vasoconstriction (Zhao et al. 2016). HMGB-1 can directly induce vascular smooth muscle cells to change their expression patterns and become more susceptible to DCV (Wang et al., 2019). Anti-HMGB-1 mAb treatment counters the induction of vascular smooth muscle cells (Wang et al. 2019) and decreases DCV, inhibits HMGB-1 translocation, alleviates neurological deficits, reduces weight loss, and lowers the concentration of inflammatory cytokines and vasoconstrictor molecules even 3 hours after SAH (Haruma et al. 2016).

GLA can also significantly improve the functional outcome, reduce HMGB1 expression, decrease HMGB1 levels, inhibit BBB permeability, reduce neuroinflammation, and diminish neural loss after SAH (Ieong et al. 2018). Two recent studies tried two novel methods to counter the HMGB-1 influence. In 2021, Han et al. used Oleanolic acid (OA), which is found in various plants and foods. OA was shown to reduce weight loss, neurological deficits, neuroinflammation, apoptosis levels, and lesion severity when administered systematically 1 hour after SAH. The protective effect of OA was due to increased deacetylation of HMGB-1 by SIRT1, which inhibited HMGB-1 translocation into the cytoplasm (Han et al., 2021). Bone marrow mesenchymal stem cell use is another promising therapeutic agent, showing its effect mainly through paracrine signaling by the stem cells (Liang et al., 2014). In 2020, Xiong et al. showed that treatment after SAH using exosomes taken from bone marrow mesenchymal stem cells improved the functional outcome, protected BBB integrity, and lowered the increase in proinflammatory and proapoptotic proteins via upregulating miR-129-5p which suppresses HMGB-1 expression (Xiong et al. 2020).

Lastly, HMGB-1 showed some beneficial effects in the chronic phase of SAH. HMGB-1 and RAGE levels were increased during the chronic stage, together with an increase in NGF, VEGF, BDNF, and the rate of neurogenesis. When EP or GLA inhibited HMGB-1, a reduction in the amounts of trophic factors and the rate of neurogenesis was seen (Tian et al., 2017). Moreover, EP and GLA treatment also lowered the number of surviving neurons, increased proinflammatory marker levels, increased brain edema, and worsened the functional outcome. Intriguingly, when SAH mice were given only fully reduced or sulfonyl HMGB-1 during the first 14 days after SAH was induced, they showed opposing effects. Fully reduced HMGB-1 decreased the neurogenesis, increased the edema, lowered the trophic factor levels, escalated neuroinflammation, and hindered recovery. In contrast, sulfonyl HMGB-1 increased neurogenesis, upregulated the trophic factors, and ameliorated neurological deficits (Tian et al., 2017). RAGE antagonist FPS-ZM1 suppressed the rehabilitative effects of physiological HMGB-1 and supplemented sulfonyl HMGB-1 (Table 1) (Tian et al. 2017).

Conclusion

HMGB-1 plays an essential role in many pathophysiological conditions, such as epilepsy, headache, neuroimmunological diseases, neurodegeneration, ischemia/reperfusion, and infection-related disorders like COVID-19. Moreover, HMGB-1 inhibitors either target the molecule or its receptors directly by natural products, synthetic molecules, or genetic manipulations that have beneficial effects on diseases' pathophysiology (Figure 1). As a result, HMGB-1 can be regarded as a good candidate for a biomarker and a treatment target for neurological diseases in the near future.

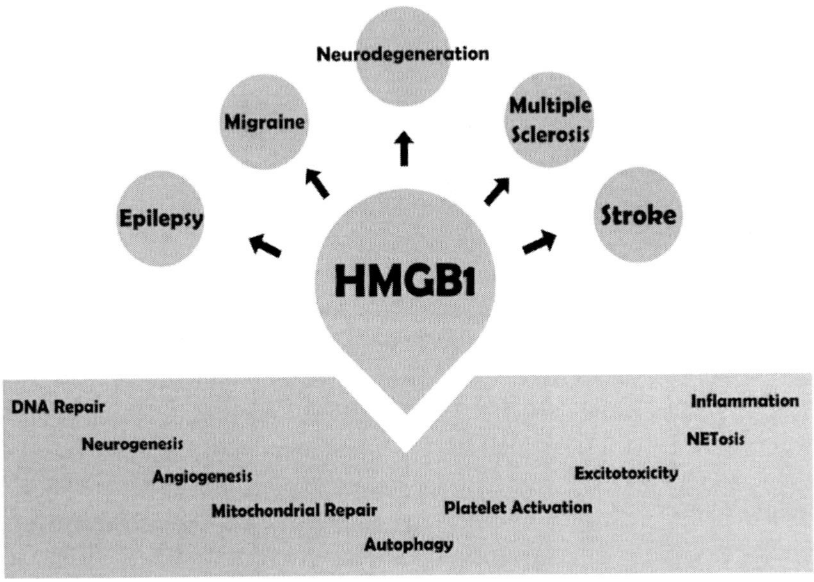

Figure 1. The role of HMGB-1 in neurological diseases.

References

Ai, P., X. Zhang, Z. Xie, G. Liu, X. Liu, S. Pan, and H. Wang. 2018. "The HMGB1 is increased in CSF of patients with an Anti-NMDAR encephalitis." *Acta Neurol Scand* 137 (2):277-282. https://doi.org/10.1111/ane.12850.

Alsaegh, H., H. Eweis, F. Kamal, and A. Alrafiah. 2021. "Celecoxib Decrease Seizures Susceptibility in a Rat Model of Inflammation by Inhibiting HMGB1 Translocation." *Pharmaceuticals (Basel)* 14 (4). https://doi.org/10.3390/ph14040380.

Amaro, S., C. Laredo, A. Renu, L. Llull, S. Rudilosso, V. Obach, X. Urra, A. M. Planas, A. Chamorro, and Urico-Ictus Investigators. 2016. "Uric Acid Therapy Prevents Early Ischemic Stroke Progression: A Tertiary Analysis of the URICO-ICTUS Trial (Efficacy Study of Combined Treatment With Uric Acid and r-tPA in Acute Ischemic Stroke)." *Stroke* 47 (11):2874-2876. https://doi.org/10.1161/STROKEAHA.116.014672.

Andersson, A., R. Covacu, D. Sunnemark, A. I. Danilov, A. Dal Bianco, M. Khademi, E. Wallstrom, A. Lobell, L. Brundin, H. Lassmann, and R. A. Harris. 2008. "Pivotal advance: HMGB1 expression in active lesions of human and experimental multiple sclerosis." *J Leukoc Biol* 84 (5):1248-55. https://doi.org/10.1189/jlb.1207844.

Andersson, U., D. J. Antoine, and K. J. Tracey. 2014. "The functions of HMGB1 depend on molecular localization and post-translational modifications." *J Intern Med* 276 (5):420-4. https://doi.org/10.1111/joim.12309.

Angelopoulou, E., Y. N. Paudel, and C. Piperi. 2020. "Exploring the role of high-mobility group box 1 (HMGB1) protein in the pathogenesis of Huntington's disease." *J Mol Med (Berl)* 98 (3):325-334. https://doi.org/10.1007/s00109-020-01885-z.

Antoine, D. J., H. E. Harris, U. Andersson, K. J. Tracey, and M. E. Bianchi. 2014. "A systematic nomenclature for the redox states of high mobility group box (HMGB) proteins." *Mol Med* 20:135-7. https://doi.org/10.2119/molmed.2014.00022.

Anzilotti, S., C. Giampa, D. Laurenti, L. Perrone, G. Bernardi, M. A. Melone, and F. R. Fusco. 2012. "Immunohistochemical localization of receptor for advanced glycation end (RAGE) products in the R6/2 mouse model of Huntington's disease." *Brain Res Bull* 87 (2-3):350-8. https://doi.org/10.1016/j.brainresbull.2011.01.009.

Asouri, M., H. Alinejad Rokni, M. A. Sahraian, S. Fattahi, N. Motamed, R. Doosti, H. Rahimi, M. Lotfi, A. Moslemi, M. Karimpoor, F. Mahboudi, and H. Akhavan-Niaki. 2020. "Analysis of Single Nucleotide Polymorphisms in HLA-DRA, IL2RA, and HMGB1 Genes in Multiple Sclerosis." *Rep Biochem Mol Biol* 9 (2):198-208. https://doi.org/10.29252/rbmb.9.2.199.

Badawi, G. A., M. M. Shokr, H. F. Zaki, and A. F. Mohamed. 2021. "Pentoxifylline prevents epileptic seizure via modulating HMGB1/RAGE/TLR4 signalling pathway and improves memory in pentylenetetrazol kindling rats." *Clin Exp Pharmacol Physiol* 48 (8):1111-1124. https://doi.org/10.1111/1440-1681.13508.

Balosso, S., J. Liu, M. E. Bianchi, and A. Vezzani. 2014. "Disulfide-containing high mobility group box-1 promotes N-methyl-D-aspartate receptor function and excitotoxicity by activating Toll-like receptor 4-dependent signaling in hippocampal neurons." *Antioxid Redox Signal* 21 (12):1726-40. https://doi.org/10.1089/ars.2013.5349.

Baran, A., M. Bulut, M. C. Kaya, O. Demirpence, B. Sevim, E. Akil, and S. Varol. 2019. "High-sensitivity C-reactive protein and high mobility group box-1 levels in Parkinson's disease." *Neurol Sci* 40 (1):167-173. https://doi.org/10.1007/s10072-018-3611-z.

Barreto, G., R. E. White, Y. Ouyang, L. Xu, and R. G. Giffard. 2011. "Astrocytes: targets for neuroprotection in stroke." *Cent Nerv Syst Agents Med Chem* 11 (2):164-73. https://doi.org/10.2174/187152411796011303.

Basisty, N., A. Kale, O. H. Jeon, C. Kuehnemann, T. Payne, C. Rao, A. Holtz, S. Shah, V. Sharma, L. Ferrucci, J. Campisi, and B. Schilling. 2020. "A proteomic atlas of senescence-associated secretomes for aging biomarker development." *PLoS Biol* 18 (1):e3000599. https://doi.org/10.1371/journal.pbio.3000599.

Bolay, H., O. Karadas, B. Ozturk, R. Sonkaya, B. Tasdelen, T. D. S. Bulut, O. Gulbahar, A. Ozge, and B. Baykan. 2021. "HMGB1, NLRP3, IL-6 and ACE2 levels are elevated in COVID-19 with headache: a window to the infection-related headache mechanism." *J Headache Pain* 22 (1):94. https://doi.org/10.1186/s10194-021-01306-7.

Bonaldi, T., F. Talamo, P. Scaffidi, D. Ferrera, A. Porto, A. Bachi, A. Rubartelli, A. Agresti, and M. E. Bianchi. 2003. "Monocytic cells hyperacetylate chromatin protein HMGB1 to redirect it towards secretion." *EMBO J* 22 (20):5551-60. https://doi.org/10.1093/emboj/cdg516.

Bonanno, G., L. Raiteri, M. Milanese, S. Zappettini, E. Melloni, M. Pedrazzi, M. Passalacqua, C. Tacchetti, C. Usai, and B. Sparatore. 2007. "The high-mobility group box 1 cytokine induces transporter-mediated release of glutamate from glial subcellular particles (gliosomes) prepared from in situ-matured astrocytes." *Int Rev Neurobiol* 82:73-93. https://doi.org/10.1016/S0074-7742(07)82004-6.

Borrmann, L., I. Kim, D. Schultheiss, P. Rogalla, and J. Bullerdiek. 2001. "Regulation of the expression of HMG1, a co-activator of the estrogen receptor." *Anticancer Res* 21 (1A):301-5.

Bucova, M., B. Majernikova, V. Durmanova, D. Cudrakova, K. Gmitterova, I. Lisa, E. Klimova, K. Kluckova, and M. Buc. 2020. "HMGB1 as a potential new marker of disease activity in patients with multiple sclerosis." *Neurol Sci* 41 (3):599-604. https://doi.org/10.1007/s10072-019-04136-3.

Bunting, E. L., J. Hamilton, and S. J. Tabrizi. 2022. "Polyglutamine diseases." *Curr Opin Neurobiol* 72:39-47. https://doi.org/10.1016/j.conb.2021.07.001.

Caldeira, C., C. Cunha, A. R. Vaz, A. S. Falcao, A. Barateiro, E. Seixas, A. Fernandes, and D. Brites. 2017. "Key Aging-Associated Alterations in Primary Microglia Response to Beta-Amyloid Stimulation." *Front Aging Neurosci* 9:277. https://doi.org/10.3389/fnagi.2017.00277.

Campbell, B. C. V., D. A. De Silva, M. R. Macleod, S. B. Coutts, L. H. Schwamm, S. M. Davis, and G. A. Donnan. 2019. "Ischaemic stroke." *Nat Rev Dis Primers* 5 (1):70. https://doi.org/10.1038/s41572-019-0118-8.

Casula, M., A. M. Iyer, W. G. Spliet, J. J. Anink, K. Steentjes, M. Sta, D. Troost, and E. Aronica. 2011. "Toll-like receptor signaling in amyotrophic lateral sclerosis spinal cord tissue." *Neuroscience* 179:233-43. https://doi.org/10.1016/j.neuroscience.2011.02.001.

Cato, L., K. Stott, M. Watson, and J. O. Thomas. 2008. "The interaction of HMGB1 and linker histones occurs through their acidic and basic tails." *J Mol Biol* 384 (5):1262-72. https://doi.org/10.1016/j.jmb.2008.10.001.

Celona, B., A. Weiner, F. Di Felice, F. M. Mancuso, E. Cesarini, R. L. Rossi, L. Gregory, D. Baban, G. Rossetti, P. Grianti, M. Pagani, T. Bonaldi, J. Ragoussis, N. Friedman, G. Camilloni, M. E. Bianchi, and A. Agresti. 2011. "Substantial histone reduction modulates genomewide nucleosomal occupancy and global transcriptional output." *PLoS Biol* 9 (6):e1001086. https://doi.org/10.1371/journal.pbio.1001086.

Chau, K. Y., H. Y. Lam, and K. L. Lee. 1998. "Estrogen treatment induces elevated expression of HMG1 in MCF-7 cells." *Exp Cell Res* 241 (1):269-72. https://doi.org/10.1006/excr.1998.4052.

Chaudhry, S. R., A. Guresir, B. Stoffel-Wagner, R. Fimmers, T. M. Kinfe, D. Dietrich, A. Lamprecht, H. Vatter, E. Guresir, and S. Muhammad. 2018. "Systemic High-Mobility Group Box-1: A Novel Predictive Biomarker for Cerebral Vasospasm in Aneurysmal Subarachnoid Hemorrhage." *Crit Care Med* 46 (11):e1023-e1028. https://doi.org/10.1097/CCM.0000000000003319.

Chavan, S. S., P. T. Huerta, S. Robbiati, S. I. Valdes-Ferrer, M. Ochani, M. Dancho, M. Frankfurt, B. T. Volpe, K. J. Tracey, and B. Diamond. 2012. "HMGB1 mediates cognitive impairment in sepsis survivors." *Mol Med* 18:930-7. https://doi.org/10.2119/molmed.2012.00195.

Chen, C., X. Lin, J. Wang, G. Tang, Z. Mu, X. Chen, J. Xu, Y. Wang, Z. Zhang, and G. Y. Yang. 2014. "Effect of HMGB1 on the paracrine action of EPC promotes post-ischemic neovascularization in mice." *Stem Cells* 32 (10):2679-89. https://doi.org/10.1002/stem.1754.

Chen, H., B. Guan, B. Wang, H. Pu, X. Bai, X. Chen, J. Liu, C. Li, J. Qiu, D. Yang, K. Liu, Q. Wang, S. Qi, and J. Shen. 2020. "Glycyrrhizin Prevents Hemorrhagic Transformation and Improves Neurological Outcome in Ischemic Stroke with Delayed Thrombolysis Through Targeting Peroxynitrite-Mediated HMGB1 Signaling." *Transl Stroke Res* 11 (5):967-982. https://doi.org/10.1007/s12975-019-00772-1.

Chikhirzhina, E., T. Starkova, A. Beljajev, A. Polyanichko, and A. Tomilin. 2020. "Functional Diversity of Non-Histone Chromosomal Protein HmgB1." *Int J Mol Sci* 21 (21). https://doi.org/10.3390/ijms21217948.

Choi, J. Y., Y. Cui, S. T. Chowdhury, and B. G. Kim. 2017. "High-mobility group box-1 as an autocrine trophic factor in white matter stroke." *Proc Natl Acad Sci USA* 114 (25):E4987-E4995. https://doi.org/10.1073/pnas.1702035114.

Chu, Y., Y. Jing, X. Zhao, M. Wang, M. Zhang, R. Ma, W. Ma, Y. Lv, and L. Zhu. 2021. "Modulation of the HMGB1/TLR4/NF-kappaB signaling pathway in the CNS by matrine in experimental autoimmune encephalomyelitis." *J Neuroimmunol* 352:577480. https://doi.org/10.1016/j.jneuroim.2021.577480.

Clifford, D. B., and B. M. Ances. 2013. "HIV-associated neurocognitive disorder." *Lancet Infect Dis* 13 (11):976-86. doi: 10.1016/S1473-3099(13)70269-X.

Crews, F. T., and R. P. Vetreno. 2016. "Mechanisms of neuroimmune gene induction in alcoholism." *Psychopharmacology (Berl)* 233 (9):1543-57. https://doi.org/10.1007/s00213-015-3906-1.

Crivellaro, G., A. Tottene, M. Vitale, M. Melone, G. Casari, F. Conti, M. Santello, and D. Pietrobon. 2021. "Specific activation of GluN1-N2B NMDA receptors underlies facilitation of cortical spreading depression in a genetic mouse model of migraine with reduced astrocytic glutamate clearance." *Neurobiol Dis* 156:105419. https://doi.org/10.1016/j.nbd.2021.105419.

Cui, Y., Y. Liang, Y. Liu, L. Wang, and L. Cao. 2021. "Corrigendum to 'High mobility group box 1 antibody represses autophagy and alleviates hippocampus damage in pilocarpine-induced mouse epilepsy model' [Acta Histochemica 122 (2020) 151485]." *Acta Histochem* 123 (8):151815. https://doi.org/10.1016/j.acthis.2021.151815.

D'Angelo, C., M. Reale, E. Costantini, M. Di Nicola, I. Porfilio, C. de Andres, L. Fernandez-Paredes, S. Sanchez-Ramon, and L. Pasquali. 2018. "Profiling of Canonical and Non-Traditional Cytokine Levels in Interferon-beta-Treated Relapsing-Remitting-Multiple Sclerosis Patients." *Front Immunol* 9:1240. https://doi.org/10.3389/fimmu.2018.01240.

Daniels, M. J., and D. Brough. 2017. "Unconventional Pathways of Secretion Contribute to Inflammation." *Int J Mol Sci* 18 (1). https://doi.org/10.3390/ijms18010102.

Davalos, A. R., M. Kawahara, G. K. Malhotra, N. Schaum, J. Huang, U. Ved, C. M. Beausejour, J. P. Coppe, F. Rodier, and J. Campisi. 2013. "p53-dependent release of

Alarmin HMGB1 is a central mediator of senescent phenotypes." *J Cell Biol* 201 (4):613-29. https://doi.org/10.1083/jcb.201206006.

Dehghani, A., T. Phisonkunkasem, S. Yilmaz Ozcan, T. Dalkara, Amjm van den Maagdenberg, E. A. Tolner, and H. Karatas. 2021. "Widespread brain parenchymal HMGB1 and NF-kappaB neuroinflammatory responses upon cortical spreading depolarization in familial hemiplegic migraine type 1 mice." *Neurobiol Dis* 156:105424. https://doi.org/10.1016/j.nbd.2021.105424.

Denorme, Frederik, Irina Portier, Mark Cody, Ramesh Grandhi, Matthew D Neal, Jennifer J Majersik, Christian Yost, and Robert A. Campbell. 2021. "Platelet-Mediated NET Formation Exacerbates Ischemic Stroke Brain Injury." *Blood* 138 (Supplement 1):437-437. https://doi.org/10.1182/blood-2021-151423.

Devinsky, O., A. Vezzani, T. J. O'Brien, N. Jette, I. E. Scheffer, M. de Curtis, and P. Perucca. 2018. "Epilepsy." *Nat Rev Dis Primers* 4:18024. https://doi.org/10.1038/nrdp.2018.24.

Eikermann-Haerter, K., M. J. Baum, M. D. Ferrari, A. M. van den Maagdenberg, M. A. Moskowitz, and C. Ayata. 2009. "Androgenic suppression of spreading depression in familial hemiplegic migraine type 1 mutant mice." *Ann Neurol* 66 (4):564-8. https://doi.org/10.1002/ana.21779.

Enokido, Y., A. Yoshitake, H. Ito, and H. Okazawa. 2008. "Age-dependent change of HMGB1 and DNA double-strand break accumulation in mouse brain." *Biochem Biophys Res Commun* 376 (1):128-33. https://doi.org/10.1016/j.bbrc.2008.08.108.

Essig, F., L. Babilon, C. Vollmuth, A. M. Kollikowski, M. Pham, L. Solymosi, K. G. Haeusler, P. Kraft, G. Stoll, and M. K. Schuhmann. 2021. "High Mobility Group Box 1 Protein in Cerebral Thromboemboli." *Int J Mol Sci* 22 (20). https://doi.org/10.3390/ijms222011276.

Fang, P., M. Schachner, and Y. Q. Shen. 2012. "HMGB1 in development and diseases of the central nervous system." *Mol Neurobiol* 45 (3):499-506. https://doi.org/10.1007/s12035-012-8264-y.

Farjam, M., F. B. Beigi Zarandi, S. Farjadian, B. Geramizadeh, A. R. Nikseresht, and M. R. Panjehshahin. 2014. "Inhibition of NR2B-Containing N-methyl-D-Aspartate Receptors (NMDARs) in Experimental Autoimmune Encephalomyelitis, a Model of Multiple Sclerosis." *Iran J Pharm Res* 13 (2):695-705.

Fernandes, A., C. Caldeira, C. Cunha, E. Ferreiro, A. R. Vaz, and D. Brites. 2022. "Differences in Immune-Related Genes Underlie Temporal and Regional Pathological Progression in 3xTg-AD Mice." *Cells* 11 (1). https://doi.org/10.3390/cells11010137.

Ferrari, M. D., P. J. Goadsby, R. Burstein, T. Kurth, C. Ayata, A. Charles, M. Ashina, Amjm van den Maagdenberg, and D. W. Dodick. 2022. "Migraine." *Nat Rev Dis Primers* 8 (1):2. https://doi.org/10.1038/s41572-021-00328-4.

Ferrari, S., P. Finelli, M. Rocchi, and M. E. Bianchi. 1996. "The active gene that encodes human high mobility group 1 protein (HMG1) contains introns and maps to chromosome 13." *Genomics* 35 (2):367-71. https://doi.org/10.1006/geno.1996.0369.

Filippi, M., A. Bar-Or, F. Piehl, P. Preziosa, A. Solari, S. Vukusic, and M. A. Rocca. 2018. "Multiple sclerosis." *Nat Rev Dis Primers* 4 (1):43. https://doi.org/10.1038/s41572-018-0041-4.

Findlay, J. M., J. Nisar, and T. Darsaut. 2016. "Cerebral Vasospasm: A Review." *Can J Neurol Sci* 43 (1):15-32. https://doi.org/10.1017/cjn.2015.288.

Fu, L., K. Liu, H. Wake, K. Teshigawara, T. Yoshino, H. Takahashi, S. Mori, and M. Nishibori. 2017. "Therapeutic effects of anti-HMGB1 monoclonal antibody on pilocarpine-induced status epilepticus in mice." *Sci Rep* 7 (1):1179. https://doi.org/10.1038/s41598-017-01325-y.

Fujita, K., and H. Okazawa. 2017. "Molecularly-Targeted Therapy of Spinocerebellar Ataxia Type 1 by HMGB1 " *BRAIN and NERVE* 69 (8):925-932. https://doi.org/10.11477/mf.1416200844.

Fujita, K., K. Motoki, K. Tagawa, X. Chen, H. Hama, K. Nakajima, H. Homma, T. Tamura, H. Watanabe, M. Katsuno, C. Matsumi, M. Kajikawa, T. Saito, T. Saido, G. Sobue, A. Miyawaki, and H. Okazawa. 2016. "HMGB1, a pathogenic molecule that induces neurite degeneration via TLR4-MARCKS, is a potential therapeutic target for Alzheimer's disease." *Sci Rep* 6:31895. https://doi.org/10.1038/srep31895.

Gaikwad, S., N. Puangmalai, A. Bittar, M. Montalbano, S. Garcia, S. McAllen, N. Bhatt, M. Sonawane, U. Sengupta, and R. Kayed. 2021. "Tau oligomer induced HMGB1 release contributes to cellular senescence and neuropathology linked to Alzheimer's disease and frontotemporal dementia." *Cell Rep* 36 (3):109419. https://doi.org/10.1016/j.celrep.2021.109419.

Gao, H. M., H. Zhou, F. Zhang, B. C. Wilson, W. Kam, and J. S. Hong. 2011. "HMGB1 acts on microglia Mac1 to mediate chronic neuroinflammation that drives progressive neurodegeneration." *J Neurosci* 31 (3):1081-92. https://doi.org/10.1523/JNEUROSCI.3732-10.2011.

Gareri, P., D. Condorelli, N. Belluardo, E. Russo, A. Loiacono, V. Barresi, A. Trovato-Salinaro, M. B. Mirone, G. Ferreri Ibbadu, and G. De Sarro. 2004. "Anticonvulsant effects of carbenoxolone in genetically epilepsy prone rats (GEPRs)." *Neuropharmacology* 47 (8):1205-16. https://doi.org/10.1016/j.neuropharm.2004.08.021.

Gareri, P., D. Condorelli, N. Belluardo, R. Citraro, V. Barresi, A. Trovato-Salinaro, G. Mudo, G. F. Ibbadu, E. Russo, and G. De Sarro. 2005. "Antiabsence effects of carbenoxolone in two genetic animal models of absence epilepsy (WAG/Rij rats and lh/lh mice)." *Neuropharmacology* 49 (4):551-63. https://doi.org/10.1016/j.neuropharm.2005.04.012.

Glasnovic, A., H. Cvija, M. Stojic, I. Tudoric-Deno, S. Ivcevic, D. Romic, N. Ticinovic, V. Vuletic, I. Lazibat, and D. Grcevic. 2014. "Decreased level of sRAGE in the cerebrospinal fluid of multiple sclerosis patients at clinical onset." *Neuroimmunomodulation* 21 (5):226-33. https://doi.org/10.1159/000357002.

Goldstein, R. S., M. Gallowitsch-Puerta, L. Yang, M. Rosas-Ballina, J. M. Huston, C. J. Czura, D. C. Lee, M. F. Ward, A. N. Bruchfeld, H. Wang, M. L. Lesser, A. L. Church, A. H. Litroff, A. E. Sama, and K. J. Tracey. 2006. "Elevated high-mobility group box 1 levels in patients with cerebral and myocardial ischemia." *Shock* 25 (6):571-4. https://doi.org/10.1097/01.shk.0000209540.99176.72.

Gomes, C., C. Cunha, F. Nascimento, J. A. Ribeiro, A. R. Vaz, and D. Brites. 2019. "Cortical Neurotoxic Astrocytes with Early ALS Pathology and miR-146a Deficit

Replicate Gliosis Markers of Symptomatic SOD1G93A Mouse Model." *Mol Neurobiol* 56 (3):2137-2158. https://doi.org/10.1007/s12035-018-1220-8.

Gougeon, M. L., B. Poirier-Beaudouin, J. Durant, C. Lebrun-Frenay, H. Saidi, V. Seffer, M. Ticchioni, S. Chanalet, H. Carsenti, A. Harvey-Langton, M. Laffon, J. Cottalorda, C. Pradier, P. Dellamonica, and M. Vassallo. 2017. "HMGB1/anti-HMGB1 antibodies define a molecular signature of early stages of HIV-Associated Neurocognitive Isorders (HAND)." *Heliyon* 3 (2):e00245. https://doi.org/10.1016/j.heliyon.2017.e00245.

Goula, A. V., B. R. Berquist, D. M. Wilson, 3rd, V. C. Wheeler, Y. Trottier, and K. Merienne. 2009. "Stoichiometry of base excision repair proteins correlates with increased somatic CAG instability in striatum over cerebellum in Huntington's disease transgenic mice." *PLoS Genet* 5 (12):e1000749. https://doi.org/10.1371/journal.pgen.1000749.

Griffioen, K., M. P. Mattson, and E. Okun. 2018. "Deficiency of Toll-like receptors 2, 3 or 4 extends life expectancy in Huntington's disease mice." *Heliyon* 4 (1):e00508. https://doi.org/10.1016/j.heliyon.2018.e00508.

Han, Y., Z. Tong, C. Wang, X. Li, and G. Liang. 2021. "Oleanolic acid exerts neuroprotective effects in subarachnoid hemorrhage rats through SIRT1-mediated HMGB1 deacetylation." *Eur J Pharmacol* 893:173811. https://doi.org/10.1016/j.ejphar.2020.173811.

Han, Y., L. Yang, X. Liu, Y. Feng, Z. Pang, and Y. Lin. 2020. "HMGB1/CXCL12-Mediated Immunity and Th17 Cells Might Underlie Highly Suspected Autoimmune Epilepsy in Elderly Individuals." *Neuropsychiatr Dis Treat* 16:1285-1293. https://doi.org/10.2147/NDT.S242766.

Hardiman, O., A. Al-Chalabi, A. Chio, E. M. Corr, G. Logroscino, W. Robberecht, P. J. Shaw, Z. Simmons, and L. H. van den Berg. 2017. "Amyotrophic lateral sclerosis." *Nat Rev Dis Primers* 3:17071. https://doi.org/10.1038/nrdp.2017.71.

Haruma, J., K. Teshigawara, T. Hishikawa, D. Wang, K. Liu, H. Wake, S. Mori, H. K. Takahashi, K. Sugiu, I. Date, and M. Nishibori. 2016. "Anti-high mobility group box-1 (HMGB1) antibody attenuates delayed cerebral vasospasm and brain injury after subarachnoid hemorrhage in rats." *Sci Rep* 6:37755. https://doi.org/10.1038/srep37755.

Hayakawa, K., L. D. Pham, Z. S. Katusic, K. Arai, and E. H. Lo. 2012. "Astrocytic high-mobility group box 1 promotes endothelial progenitor cell-mediated neurovascular remodeling during stroke recovery." *Proc Natl Acad Sci USA* 109 (19):7505-10. https://doi.org/10.1073/pnas.1121146109.

Hayakawa, K., N. Miyamoto, J. H. Seo, L. D. Pham, K. W. Kim, E. H. Lo, and K. Arai. 2013. "High-mobility group box 1 from reactive astrocytes enhances the accumulation of endothelial progenitor cells in damaged white matter." *J Neurochem* 125 (2):273-80. https://doi.org/10.1111/jnc.12120.

He, Q., C. H. Liang, and S. J. Lippard. 2000. "Steroid hormones induce HMG1 overexpression and sensitize breast cancer cells to cisplatin and carboplatin." *Proc Natl Acad Sci USA* 97 (11):5768-72. https://doi.org/10.1073/pnas.100108697.

He, Q., H. You, X. M. Li, T. H. Liu, P. Wang, and B. E. Wang. 2012. "HMGB1 promotes the synthesis of pro-IL-1beta and pro-IL-18 by activation of p38 MAPK and NF-kappaB through receptors for advanced glycation end-products in macrophages." *Asian Pac J Cancer Prev* 13 (4):1365-70. https://doi.org/10.7314/apjcp.2012.13.4.1365.

Hei, Y., R. Chen, X. Yi, Q. Long, D. Gao, and W. Liu. 2018. "HMGB1 Neutralization Attenuates Hippocampal Neuronal Death and Cognitive Impairment in Rats with Chronic Cerebral Hypoperfusion via Suppressing Inflammatory Responses and Oxidative Stress." *Neuroscience* 383:150-159. https://doi.org/10.1016/j.neuroscience.2018.05.010.

Hei, Y., X. Zhang, R. Chen, Y. Zhou, D. Gao, and W. Liu. 2019. "High-Mobility Group Box 1 Neutralization Prevents Chronic Cerebral Hypoperfusion-Induced Optic Tract Injuries in the White Matter Associated with Down-regulation of Inflammatory Responses." *Cell Mol Neurobiol* 39 (7):1051-1060. https://doi.org/10.1007/s10571-019-00702-7.

Hemmer, S., S. Senger, C. J. Griessenauer, A. Simgen, J. Oertel, J. Geisel, and P. Hendrix. 2022. "Admission serum high mobility group box 1 (HMGB1) protein predicts delayed cerebral ischemia following aneurysmal subarachnoid hemorrhage." *Neurosurg Rev* 45 (1):807-817. https://doi.org/10.1007/s10143-021-01607-0.

Hendrix, P., P. M. Foreman, M. R. Harrigan, W. S. Rd Fisher, N. A. Vyas, R. H. Lipsky, M. Lin, B. C. Walters, R. S. Tubbs, M. M. Shoja, J. F. Pittet, M. Mathru, and C. J. Griessenauer. 2017. "Impact of High-Mobility Group Box 1 Polymorphism on Delayed Cerebral Ischemia After Aneurysmal Subarachnoid Hemorrhage." *World Neurosurg* 101:325-330. https://doi.org/10.1016/j.wneu.2017.01.121.

Hock, R., T. Furusawa, T. Ueda, and M. Bustin. 2007. "HMG chromosomal proteins in development and disease." *Trends Cell Biol* 17 (2):72-9. https://doi.org/10.1016/j.tcb.2006.12.001.

Huang, J., J. Yang, Y. Shen, H. Jiang, C. Han, G. Zhang, L. Liu, X. Xu, J. Li, Z. Lin, N. Xiong, Z. Zhang, J. Xiong, and T. Wang. 2017. "HMGB1 Mediates Autophagy Dysfunction via Perturbing Beclin1-Vps34 Complex in Dopaminergic Cell Model." *Front Mol Neurosci* 10:13. https://doi.org/10.3389/fnmol.2017.00013.

Huang, J. S., Y. Wu, Q. Huang, S. J. Li, J. M. Ye, X. Wei, Q. D. Liu, Y. Liu, and M. G. Ma. 2015. "Expression level and distribution of HMGB1 in Sombati's cell model and kainic acid-induced epilepsy model." *Eur Rev Med Pharmacol Sci* 19 (15):2928-33.

Huang, Q., J. Liu, Z. Shi, and X. Zhu. 2020. "Correlation of MMP-9 and HMGB1 expression with the cognitive function in patients with epilepsy and factors affecting the prognosis." *Cell Mol Biol (Noisy-le-grand)* 66 (3):39-47.

Hutchison, E. R., E. M. Kawamoto, D. D. Taub, A. Lal, K. Abdelmohsen, Y. Zhang, W. H. Wood, 3rd, E. Lehrmann, S. Camandola, K. G. Becker, M. Gorospe, and M. P. Mattson. 2013. "Evidence for miR-181 involvement in neuroinflammatory responses of astrocytes." *Glia* 61 (7):1018-28. https://doi.org/10.1002/glia.22483.

Hwang, C. S., G. T. Liu, M. D. Chang, I. L. Liao, and H. T. Chang. 2013. "Elevated serum autoantibody against high mobility group box 1 as a potent surrogate biomarker for amyotrophic lateral sclerosis." *Neurobiol Dis* 58:13-8. https://doi.org/10.1016/j.nbd.2013.04.013.

Ieong, C., H. Sun, Q. Wang, and J. Ma. 2018. "Glycyrrhizin suppresses the expressions of HMGB1 and ameliorates inflammative effect after acute subarachnoid hemorrhage in rat model." *J Clin Neurosci* 47:278-284. https://doi.org/10.1016/j.jocn.2017.10.034.

Ikhlas, M., and N. S. Atherton. 2022. "Vascular Reperfusion Injury." In *Stat Pearls*. Treasure Island (FL).

Ito, H., K. Fujita, K. Tagawa, X. Chen, H. Homma, T. Sasabe, J. Shimizu, S. Shimizu, T. Tamura, S. Muramatsu, and H. Okazawa. 2015. "HMGB1 facilitates repair of mitochondrial DNA damage and extends the lifespan of mutant ataxin-1 knock-in mice." *EMBO Mol Med* 7 (1):78-101. https://doi.org/10.15252/emmm.201404392.

Ito, I., J. Fukazawa, and M. Yoshida. 2007. "Post-translational methylation of high mobility group box 1 (HMGB1) causes its cytoplasmic localization in neutrophils." *J Biol Chem* 282 (22):16336-44. https://doi.org/10.1074/jbc.M608467200.

Ito, T., K. Kawahara, K. Okamoto, S. Yamada, M. Yasuda, H. Imaizumi, Y. Nawa, X. Meng, B. Shrestha, T. Hashiguchi, and I. Maruyama. 2008. "Proteolytic cleavage of high mobility group box 1 protein by thrombin-thrombomodulin complexes." *Arterioscler Thromb Vasc Biol* 28 (10):1825-30. https://doi.org/10.1161/ATVBA HA.107.150631.

Jarius, S., F. Paul, B. G. Weinshenker, M. Levy, H. J. Kim, and B. Wildemann. 2020. "Neuromyelitis optica." *Nat Rev Dis Primers* 6 (1):85. https://doi.org/10.1038/s41572-020-0214-9.

Jiang, W., and D. S. Pisetsky. 2008. "The induction of HMGB1 release from RAW 264.7 cells by transfected DNA." *Mol Immunol* 45 (7):2038-44. https://doi.org/10.1016/j.m olimm.2007.10.019.

Jin, M., X. Jin, H. Homma, K. Fujita, H. Tanaka, S. Murayama, H. Akatsu, K. Tagawa, and H. Okazawa. 2021. "Prediction and verification of the AD-FTLD common pathomechanism based on dynamic molecular network analysis." *Commun Biol* 4 (1):961. https://doi.org/10.1038/s42003-021-02475-6.

Jin, Y. C., S. W. Kim, F. Cheng, J. H. Shin, J. K. Park, S. Lee, J. E. Lee, P. L. Han, M. Lee, K. K. Kim, H. Choi, and J. K. Lee. 2011. "The effect of biodegradable gelatin microspheres on the neuroprotective effects of high mobility group box 1 A box in the postischemic brain." *Biomaterials* 32 (3):899-908. https://doi.org/10.1016/j.bioma terials.2010.09.054.

Kamasak, T., B. Dilber, S. O. Yaman, B. D. Durgut, T. Kurt, E. Coban, E. A. Arslan, S. Sahin, S. C. Karahan, and A. Cansu. 2020. "HMGB-1, TLR4, IL-1R1, TNF-alpha, and IL-1beta: novel epilepsy markers?" *Epileptic Disord* 22 (2):183-193. https://doi.org/ 10.1684/epd.2020.1155.

Kan, M., L. Song, X. Zhang, J. Zhang, and P. Fang. 2019. "Circulating high mobility group box-1 and toll-like receptor 4 expressions increase the risk and severity of epilepsy." *Braz J Med Biol Res* 52 (7):e7374. https://doi.org/10.1590/1414-431X20197374.

Kang, R., R. Chen, Q. Zhang, W. Hou, S. Wu, L. Cao, J. Huang, Y. Yu, X. G. Fan, Z. Yan, X. Sun, H. Wang, Q. Wang, A. Tsung, T. R. Billiar, H. J. Zeh, 3rd, M. T. Lotze, and D. Tang. 2014. "HMGB1 in health and disease." *Mol Aspects Med* 40:1-116. https://doi.org/10.1016/j.mam.2014.05.001.

Karatas, H., S. E. Erdener, Y. Gursoy-Ozdemir, S. Lule, E. Eren-Kocak, Z. D. Sen, and T. Dalkara. 2013. "Spreading depression triggers headache by activating neuronal Panx1 channels." *Science* 339 (6123):1092-5. https://doi.org/10.1126/science.1231897.

Kaya, M. A., N. Erin, O. Bozkurt, N. Erkek, O. Duman, and S. Haspolat. 2021. "Changes of HMGB-1 and sTLR4 levels in cerebrospinal fluid of patients with febrile seizures." *Epilepsy Res* 169:106516. https://doi.org/10.1016/j.eplepsyres.2020.106516.

Kikuchi, K., S. Tancharoen, T. Ito, Y. Morimoto-Yamashita, N. Miura, K. Kawahara, I. Maruyama, Y. Murai, and E. Tanaka. 2013. "Potential of the angiotensin receptor blockers (ARBs) telmisartan, irbesartan, and candesartan for inhibiting the HMGB1/RAGE axis in prevention and acute treatment of stroke." *Int J Mol Sci* 14 (9):18899-924. https://doi.org/10.3390/ijms140918899.

Kilic, K., H. Karatas, B. Donmez-Demir, E. Eren-Kocak, Y. Gursoy-Ozdemir, A. Can, J. M. Petit, P. J. Magistretti, and T. Dalkara. 2018. "Inadequate brain glycogen or sleep increases spreading depression susceptibility." *Ann Neurol* 83 (1):61-73. https://doi.org/10.1002/ana.25122.

Kim, I. D., J. H. Shin, H. K. Lee, Y. C. Jin, and J. K. Lee. 2012. "Intranasal delivery of HMGB1-binding heptamer peptide confers a robust neuroprotection in the postischemic brain." *Neurosci Lett* 525 (2):179-83. https://doi.org/10.1016/j.neulet.2012.07.040.

Kim, J. B., J. Sig Choi, Y. M. Yu, K. Nam, C. S. Piao, S. W. Kim, M. H. Lee, P. L. Han, J. S. Park, and J. K. Lee. 2006. "HMGB1, a novel cytokine-like mediator linking acute neuronal death and delayed neuroinflammation in the postischemic brain." *J Neurosci* 26 (24):6413-21. https://doi.org/10.1523/JNEUROSCI.3815-05.2006.

Kim, S. J., M. J. Ryu, J. Han, Y. Jang, J. Kim, M. J. Lee, I. Ryu, X. Ju, E. Oh, W. Chung, G. R. Kweon, and J. Y. Heo. 2017. "Activation of the HMGB1-RAGE axis upregulates TH expression in dopaminergic neurons via JNK phosphorylation." *Biochem Biophys Res Commun* 493 (1):358-364. https://doi.org/10.1016/j.bbrc.2017.09.017.

Kim, S. J., M. J. Ryu, J. Han, Y. Jang, M. J. Lee, X. Ju, I. Ryu, Y. L. Lee, E. Oh, W. Chung, J. Y. Heo, and G. R. Kweon. 2019. "Non-cell autonomous modulation of tyrosine hydroxylase by HMGB1 released from astrocytes in an acute MPTP-induced Parkinsonian mouse model." *Lab Invest* 99 (9):1389-1399. https://doi.org/10.1038/s41374-019-0254-5.

Kim, S. W., H. Lee, H. K. Lee, I. D. Kim, and J. K. Lee. 2019. "Neutrophil extracellular trap induced by HMGB1 exacerbates damages in the ischemic brain." *Acta Neuropathol Commun* 7 (1):94. https://doi.org/10.1186/s40478-019-0747-x.

Kim, S. W., and J. K. Lee. 2020. "Role of HMGB1 in the Interplay between NETosis and Thrombosis in Ischemic Stroke: A Review." *Cells* 9 (8). https://doi.org/10.3390/cells9081794.

Kim, S. W., Y. Jin, J. H. Shin, I. D. Kim, H. K. Lee, S. Park, P. L. Han, and J. K. Lee. 2012. "Glycyrrhizic acid affords robust neuroprotection in the postischemic brain via anti-inflammatory effect by inhibiting HMGB1 phosphorylation and secretion." *Neurobiol Dis* 46 (1):147-56. https://doi.org/10.1016/j.nbd.2011.12.056.

Knopman, D. S., H. Amieva, R. C. Petersen, G. Chetelat, D. M. Holtzman, B. T. Hyman, R. A. Nixon, and D. T. Jones. 2021. "Alzheimer disease." *Nat Rev Dis Primers* 7 (1):33. https://doi.org/10.1038/s41572-021-00269-y.

Kokkola, R., A. Andersson, G. Mullins, T. Ostberg, C. J. Treutiger, B. Arnold, P. Nawroth, U. Andersson, R. A. Harris, and H. E. Harris. 2005. "RAGE is the major receptor for the proinflammatory activity of HMGB1 in rodent macrophages." *Scand J Immunol* 61 (1):1-9. https://doi.org/10.1111/j.0300-9475.2005.01534.x.

Kolonaite, G., A. Vilkeviciute, L. Kriauciuniene, G. Gedvilaite, and R. Liutkeviciene. 2021. "Association of RAGE rs1800624 and rs1800625 gene polymorphisms with predisposition to optic neuritis and optic neuritis together with multiple sclerosis." *Ophthalmic Genet* 42 (6):685-690. https://doi.org/10.1080/13816810.2021.1952619.

Kuehl, L., B. Salmond, and L. Tran. 1984. "Concentrations of high-mobility-group proteins in the nucleus and cytoplasm of several rat tissues." *J Cell Biol* 99 (2):648-54. https://doi.org/10.1083/jcb.99.2.648.

Le, K., S. Mo, X. Lu, A. Idriss Ali, D. Yu, and Y. Guo. 2018. "Association of circulating blood HMGB1 levels with ischemic stroke: a systematic review and meta-analysis." *Neurol Res* 40 (11):907-916. https://doi.org/10.1080/01616412.2018.1497254.

Lee, J. D., N. Liu, S. C. Levin, L. Ottosson, U. Andersson, H. E. Harris, and T. M. Woodruff. 2019. "Therapeutic blockade of HMGB1 reduces early motor deficits, but not survival in the SOD1(G93A) mouse model of amyotrophic lateral sclerosis." *J Neuroinflammation* 16 (1):45. https://doi.org/10.1186/s12974-019-1435-2.

Lee, J. D., T. S. McDonald, J. N. T. Fung, and T. M. Woodruff. 2020. "Absence of Receptor for Advanced Glycation End Product (RAGE) Reduces Inflammation and Extends Survival in the hSOD1(G93A) Mouse Model of Amyotrophic Lateral Sclerosis." *Mol Neurobiol* 57 (10):4143-4155. https://doi.org/10.1007/s12035-020-02019-9.

Lee, J. H., E. J. Yoon, J. Seo, A. Kavoussi, Y. E. Chung, S. P. Chung, I. Park, C. H. Kim, and J. S. You. 2016. "Hypothermia inhibits the propagation of acute ischemic injury by inhibiting HMGB1." *Mol Brain* 9 (1):81. https://doi.org/10.1186/s13041-016-0260-0.

Lee, J. Y., J. D. Lee, S. Phipps, P. G. Noakes, and T. M. Woodruff. 2015. "Absence of toll-like receptor 4 (TLR4) extends survival in the hSOD1 G93A mouse model of amyotrophic lateral sclerosis." *J Neuroinflammation* 12:90. https://doi.org/10.1186/s12974-015-0310-z.

Lee, S., Y. Nam, J. Y. Koo, D. Lim, J. Park, J. Ock, J. Kim, K. Suk, and S. B. Park. 2014. "A small molecule binding HMGB1 and HMGB2 inhibits microglia-mediated neuroinflammation." *Nat Chem Biol* 10 (12):1055-60. https://doi.org/10.1038/nchembio.1669.

Lei, C., B. Wu, T. Cao, S. Zhang, and M. Liu. 2015. "Activation of the high-mobility group box 1 protein-receptor for advanced glycation end-products signaling pathway in rats during neurogenesis after intracerebral hemorrhage." *Stroke* 46 (2):500-6. https://doi.org/10.1016/j.neuroscience.2012.10.02310.1161/STROKEAHA.114.006825.

Lei, C., J. Geng, and L. Zhong. 2020. "The association between plasma HMGB1 and sRAGE and clinical outcome in intracerebral hemorrhage." *J Neuroimmunol* 345:577266. https://doi.org/10.1016/j.jneuroim.2020.577266.

Lei, C., S. Lin, C. Zhang, W. Tao, W. Dong, Z. Hao, M. Liu, and B. Wu. 2013a. "Effects of high-mobility group box1 on cerebral angiogenesis and neurogenesis after intracerebral hemorrhage." *Neuroscience* 229:12-9. https://doi.org/10.1016/j.neuroscience.2012.10.054.

Lei, C., S. Lin, C. Zhang, W. Tao, W. Dong, Z. Hao, M. Liu, and B. Wu. 2013b. "High-mobility group box1 protein promotes neuroinflammation after intracerebral hemorrhage in rats." *Neuroscience* 228:190-9. https://doi.org/10.1016/j.neuroscience.2012.10.023.

Li, M., S. Chen, X. Shi, C. Lyu, Y. Zhang, M. Tan, C. Wang, N. Zang, X. Liu, Y. Hu, J. Shen, L. Zhou, and Y. Gu. 2018. "Cell permeable HMGB1-binding heptamer peptide ameliorates neurovascular complications associated with thrombolytic therapy in rats with transient ischemic stroke." *J Neuroinflammation* 15 (1):237. https://doi.org/10.1186/s12974-018-1267-5.

Li, W., M. Deng, P. A. Loughran, M. Yang, M. Lin, C. Yang, W. Gao, S. Jin, S. Li, J. Cai, B. Lu, T. R. Billiar, and M. J. Scott. 2020. "LPS Induces Active HMGB1 Release From Hepatocytes Into Exosomes Through the Coordinated Activities of TLR4 and Caspase-11/GSDMD Signaling." *Front Immunol* 11:229. https://doi.org/10.3389/fimmu.2020.00229.

Li, Y. J., L. Wang, B. Zhang, F. Gao, and C. M. Yang. 2019. "Glycyrrhizin, an HMGB1 inhibitor, exhibits neuroprotective effects in rats after lithium-pilocarpine-induced status epilepticus." *J Pharm Pharmacol* 71 (3):390-399. https://doi.org/10.1111/jphp.13040.

Li, Y., J. Zhu, L. Chen, W. Hu, M. Wang, S. Li, X. Gu, H. Tao, B. Zhao, G. Ma, and K. Li. 2017. "Genetic predisposition to ischaemic stroke by RAGE and HMGB1 gene variants in Chinese Han population." *Oncotarget* 8 (59):100150-100164. https://doi.org/10.18632/oncotarget.22112.

Li, Z., B. Li, X. Zhu, P. Yin, J. Liu, S. Huang, and R. Sun. 2013. "Neuroprotective effects of anti-high-mobility group box 1 antibody in juvenile rat hippocampus after kainic acid-induced status epilepticus." *Neuroreport* 24 (14):785-90. https://doi.org/10.1097/WNR.0b013e328363fed3.

Liang, X., S. Wang, G. Qin, J. Xie, G. Tan, J. Zhou, D. W. McBride, and L. Chen. 2017. "Tyrosine Phosphorylation of NR2B Contributes to Chronic Migraines via Increased Expression of CGRP in Rats." *Biomed Res Int* 2017:7203458. https://doi.org/10.1155/2017/7203458.

Liang, X., Y. Ding, Y. Zhang, H. F. Tse, and Q. Lian. 2014. "Paracrine mechanisms of mesenchymal stem cell-based therapy: current status and perspectives." *Cell Transplant* 23 (9):1045-59. https://doi.org/10.3727/096368913X667709.

Lin, C. H., H. Y. Chen, and K. C. Wei. 2020. "Role of HMGB1/TLR4 Axis in Ischemia/Reperfusion-Impaired Extracellular Glutamate Clearance in Primary Astrocytes." *Cells* 9 (12). https://doi.org/10.3390/cells9122585.

Lin, F., D. Xue, T. Xie, and Z. Pan. 2016. "HMGB1 promotes cellular chemokine synthesis and potentiates mesenchymal stromal cell migration via Rap1 activation." *Mol Med Rep* 14 (2):1283-9. https://doi.org/10.3892/mmr.2016.5398.

Lindersson, E. K., P. Hojrup, W. P. Gai, D. Locker, D. Martin, and P. H. Jensen. 2004. "alpha-Synuclein filaments bind the transcriptional regulator HMGB-1." *Neuroreport* 15 (18):2735-9.

Liu, A. H., Y. T. Wu, and Y. P. Wang. 2017. "MicroRNA-129-5p inhibits the development of autoimmune encephalomyelitis-related epilepsy by targeting HMGB1 through the TLR4/NF-kB signaling pathway." *Brain Res Bull* 132:139-149. https://doi.org/10.1016/j.brainresbull.2017.05.004.

Liu, J., Y. Liu, H. Zhang, G. Chen, K. Wang, and X. Xiao. 2008. "KLF4 promotes the expression, translocation, and releas eof HMGB1 in RAW264.7 macrophages in response to LPS." *Shock* 30 (3):260-6. https://doi.org/10.1097/shk.0b013e318162bef7.

Liu, K., S. Mori, H. K. Takahashi, Y. Tomono, H. Wake, T. Kanke, Y. Sato, N. Hiraga, N. Adachi, T. Yoshino, and M. Nishibori. 2007. "Anti-high mobility group box 1 monoclonal antibody ameliorates brain infarction induced by transient ischemia in rats." *FASEB J* 21 (14):3904-16. https://doi.org/10.1096/fj.07-8770com.

Liu, Y., R. Prasad, W. A. Beard, E. W. Hou, J. K. Horton, C. T. McMurray, and S. H. Wilson. 2009. "Coordination between polymerase beta and FEN1 can modulate CAG repeat expansion." *J Biol Chem* 284 (41):28352-28366. https://doi.org/10.1074/jbc.M109.050286.

Loboda, A., M. Damulewicz, E. Pyza, A. Jozkowicz, and J. Dulak. 2016. "Role of Nrf2/HO-1 system in development, oxidative stress response and diseases: an evolutionarily conserved mechanism." *Cell Mol Life Sci* 73 (17):3221-47. https://doi.org/10.1007/s00018-016-2223-0.

Lo Coco, D., P. Veglianese, E. Allievi, and C. Bendotti. 2007. "Distribution and cellular localization of high mobility group box protein 1 (HMGB1) in the spinal cord of a transgenic mouse model of ALS." *Neurosci Lett* 412 (1):73-7. https://doi.org/10.1016/j.neulet.2006.10.063.

Lu, B., D. J. Antoine, K. Kwan, P. Lundback, H. Wahamaa, H. Schierbeck, M. Robinson, M. A. Van Zoelen, H. Yang, J. Li, H. Erlandsson-Harris, S. S. Chavan, H. Wang, U. Andersson, and K. J. Tracey. 2014. "JAK/STAT1 signaling promotes HMGB1 hyperacetylation and nuclear translocation." *Proc Natl Acad Sci U S A* 111 (8):3068-73. https://doi.org/10.1073/pnas.1316925111.

Luan, G., Q. Gao, F. Zhai, Y. Chen, and T. Li. 2016. "Upregulation of HMGB1, toll-like receptor and RAGE in human Rasmussen's encephalitis." *Epilepsy Res* 123:36-49. https://doi.org/10.1016/j.eplepsyres.2016.03.005.

Lum, H. K., and K. L. Lee. 2001. "The human HMGB1 promoter is modulated by a silencer and an enhancer-containing intron." *Biochim Biophys Acta* 1520 (1):79-84. https://doi.org/10.1016/s0167-4781(01)00243-3.

Luo, L., Y. Jin, I. D. Kim, and J. K. Lee. 2013. "Glycyrrhizin attenuates kainic Acid-induced neuronal cell death in the mouse hippocampus." *Exp Neurobiol* 22 (2):107-15. https://doi.org/10.5607/en.2013.22.2.107.

Luo, L., Y. Jin, I. D. Kim, and J. K. Lee. 2014. "Glycyrrhizin suppresses HMGB1 inductions in the hippocampus and subsequent accumulation in serum of a kainic acid-induced seizure mouse model." *Cell Mol Neurobiol* 34 (7):987-97. https://doi.org/10.1007/s10571-014-0075-4.

Ma, X. L., S. Y. Li, and F. Shang. 2017. "Effect of microRNA-129-5p targeting HMGB1-RAGE signaling pathway on revascularization in a collagenase-induced intracerebral hemorrhage rat model." *Biomed Pharmacother* 93:238-244. https://doi.org/10.1016/j.biopha.2017.06.012.

MacGregor, E. Anne, and Antoinette Maassen van den Brink. 2019. "Transgender and Migraine." In *Gender and Migraine*, edited by Antoinette Maassen van den Brink and E. Anne MacGregor, 113-127. Cham: Springer International Publishing.

Malhotra, S., N. Fissolo, M. Tintore, A. C. Wing, J. Castillo, A. Vidal-Jordana, X. Montalban, and M. Comabella. 2015. "Role of high mobility group box protein 1 (HMGB1) in peripheral blood from patients with multiple sclerosis." *J Neuroinflammation* 12:48. https://doi.org/10.1186/s12974-015-0269-9.

Maroso, M., S. Balosso, T. Ravizza, J. Liu, E. Aronica, A. M. Iyer, C. Rossetti, M. Molteni, M. Casalgrandi, A. A. Manfredi, M. E. Bianchi, and A. Vezzani. 2010. "Toll-like receptor 4 and high-mobility group box-1 are involved in ictogenesis and can be targeted to reduce seizures." *Nat Med* 16 (4):413-9. https://doi.org/10.1038/nm.2127.

Masai, K., K. Kuroda, N. Isooka, R. Kikuoka, S. Murakami, S. Kamimai, D. Wang, K. Liu, I. Miyazaki, M. Nishibori, and M. Asanuma. 2021. "Neuroprotective Effects of Anti-high Mobility Group Box-1 Monoclonal Antibody Against Methamphetamine-Induced Dopaminergic Neurotoxicity." *Neurotox Res* 39 (5):1511-1523. https://doi.org/10.1007/s12640-021-00402-5.

Meyer, M., M. S. Kruse, L. Garay, A. Lima, P. Roig, H. Hunt, J. Belanoff, E. R. de Kloet, M. C. G. Deniselle, and A. F. De Nicola. 2020. "Long-term effects of the glucocorticoid receptor modulator CORT113176 in murine motoneuron degeneration." *Brain Res* 1727:146551. https://doi.org/10.1016/j.brainres.2019.146551.

Min, H. J., E. A. Ko, J. Wu, E. S. Kim, M. K. Kwon, M. S. Kwak, J. E. Choi, J. E. Lee, and J. S. Shin. 2013. "Chaperone-like activity of high-mobility group box 1 protein and its role in reducing the formation of polyglutamine aggregates." *J Immunol* 190 (4):1797-806. https://doi.org/10.4049/jimmunol.1202472.

Misrielal, C., M. Mauthe, F. Reggiori, and B. J. L. Eggen. 2020. "Autophagy in Multiple Sclerosis: Two Sides of the Same Coin." *Front Cell Neurosci* 14:603710. https://doi.org/10.3389/fncel.2020.603710.

Montazeri, M., N. Eskandari, and R. Mansouri. 2021. "Evaluation of the expressed miR-129 and miR-549a in patients with multiple sclerosis." *Adv Biomed Res* 10:48. https://doi.org/10.4103/abr.abr_268_20.

Morales-Sosa, M., S. Orozco-Suarez, A. Vega-Garcia, S. Caballero-Chacon, and I. A. Feria-Romero. 2018. "Immunomodulatory effect of Celecoxib on HMGB1/TLR4 pathway in a recurrent seizures model in immature rats." *Pharmacol Biochem Behav* 170:79-86. https://doi.org/10.1016/j.pbb.2018.05.007.

Mosarrezaii Aghdam, A., S. Rezaei, F. Zarza Nalivan, F. Babaie, M. R. Amiri Nikpour, and S. Torkamandi. 2022. "Downregulation of miR-125a-5p and miR-218-5p in Peripheral Blood Mononuclear Cells of Patients with Relapsing-Remitting Multiple Sclerosis." *Immunol Invest* 51 (5):1149-1161. https://doi.org/10.1080/08820139.2021.1909616.

Muhammad, S., W. Barakat, S. Stoyanov, S. Murikinati, H. Yang, K. J. Tracey, M. Bendszus, G. Rossetti, P. P. Nawroth, A. Bierhaus, and M. Schwaninger. 2008. "The

HMGB1 receptor RAGE mediates ischemic brain damage." *J Neurosci* 28 (46):12023-12031. https://doi.org/10.1523/JNEUROSCI.2435-08.2008.

Murakami, K., M. Koide, T. M. Dumont, S. R. Russell, B. I. Tranmer, and G. C. Wellman. 2011. "Subarachnoid Hemorrhage Induces Gliosis and Increased Expression of the Pro-inflammatory Cytokine High Mobility Group Box 1 Protein." *Transl Stroke Res* 2 (1):72-9. https://doi.org/10.1007/s12975-010-0052-2.

Nakahara, T., R. Tsuruta, T. Kaneko, S. Yamashita, M. Fujita, S. Kasaoka, T. Hashiguchi, M. Suzuki, I. Maruyama, and T. Maekawa. 2009. "High-mobility group box 1 protein in CSF of patients with subarachnoid hemorrhage." *Neurocrit Care* 11 (3):362-8. https://doi.org/10.1007/s12028-009-9276-y.

Nan, K., Y. Han, Q. Fang, C. Huang, L. Yu, W. Ge, F. Xiang, Y. X. Tao, H. Cao, and J. Li. 2019. "HMGB1 gene silencing inhibits neuroinflammation via down-regulation of NF-kappaB signaling in primary hippocampal neurons induced by Abeta25-35." *Int Immunopharmacol* 67:294-301. https://doi.org/10.1016/j.intimp.2018.12.027.

Nasi, M., E. Bianchini, S. De Biasi, L. Gibellini, A. Neroni, M. Mattioli, M. Pinti, A. Iannone, A. V. Mattioli, A. M. Simone, D. Ferraro, F. Vitetta, P. Sola, and A. Cossarizza. 2020. "Increased plasma levels of mitochondrial DNA and pro-inflammatory cytokines in patients with progressive multiple sclerosis." *J Neuroimmunol* 338:577107. https://doi.org/10.1016/j.jneuroim.2019.577107.

Nilson, A. N., K. C. English, J. E. Gerson, T. Barton Whittle, C. Nicolas Crain, J. Xue, U. Sengupta, D. L. Castillo-Carranza, W. Zhang, P. Gupta, and R. Kayed. 2017. "Tau Oligomers Associate with Inflammation in the Brain and Retina of Tauopathy Mice and in Neurodegenerative Diseases." *J Alzheimers Dis* 55 (3):1083-1099. https://doi.org/10.3233/JAD-160912.

O'Connell, K., J. Kinsella, C. McMahon, J. Holian, and S. O'Riordan. 2016. "Posterior reversible encephalopathy syndrome (PRES) associated with liquorice consumption." *Ir J Med Sci* 185 (4):945-947. https://doi.org/10.1007/s11845-014-1214-0.

Ohnishi, M., H. Katsuki, C. Fukutomi, M. Takahashi, M. Motomura, M. Fukunaga, Y. Matsuoka, Y. Isohama, Y. Izumi, T. Kume, A. Inoue, and A. Akaike. 2011. "HMGB1 inhibitor glycyrrhizin attenuates intracerebral hemorrhage-induced injury in rats." *Neuropharmacology* 61 (5-6):975-80. https://doi.org/10.1016/j.neuropharm.2011.06.026.

Orlova, V. V., E. Y. Choi, C. Xie, E. Chavakis, A. Bierhaus, E. Ihanus, C. M. Ballantyne, C. G. Gahmberg, M. E. Bianchi, P. P. Nawroth, and T. Chavakis. 2007. "A novel pathway of HMGB1-mediated inflammatory cell recruitment that requires Mac-1-integrin." *EMBO J* 26 (4):1129-39. https://doi.org/10.1038/sj.emboj.7601552.

Pan, B., H. B. Alam, W. Chong, J. Mobley, B. Liu, Q. Deng, Y. Liang, Y. Wang, E. Chen, T. Wang, M. Tewari, and Y. Li. 2017. "CitH3: a reliable blood biomarker for diagnosis and treatment of endotoxic shock." *Sci Rep* 7 (1):8972. https://doi.org/10.1038/s41598-017-09337-4.

Park, J. S., J. Arcaroli, H. K. Yum, H. Yang, H. Wang, K. Y. Yang, K. H. Choe, D. Strassheim, T. M. Pitts, K. J. Tracey, and E. Abraham. 2003. "Activation of gene expression in human neutrophils by high mobility group box 1 protein." *Am J Physiol Cell Physiol* 284 (4):C870-9. https://doi.org/10.1152/ajpcell.00322.2002.

Paudel, Y. N., S. U. Khan, I. Othman, and M. F. Shaikh. 2021. "Naturally Occurring HMGB1 Inhibitor, Glycyrrhizin, Modulates Chronic Seizures-Induced Memory Dysfunction in Zebrafish Model." *ACS Chem Neurosci* 12 (18):3288-3302. https://doi.org/10.1021/acschemneuro.0c00825.

Paul, S., and E. Candelario-Jalil. 2021. "Emerging neuroprotective strategies for the treatment of ischemic stroke: An overview of clinical and preclinical studies." *Exp Neurol* 335:113518. https://doi.org/10.1016/j.expneurol.2020.113518.

Pauletti, A., G. Terrone, T. Shekh-Ahmad, A. Salamone, T. Ravizza, M. Rizzi, A. Pastore, R. Pascente, L. P. Liang, B. R. Villa, S. Balosso, A. Y. Abramov, E. A. van Vliet, E. Del Giudice, E. Aronica, M. Patel, M. C. Walker, and A. Vezzani. 2019. "Targeting oxidative stress improves disease outcomes in a rat model of acquired epilepsy." *Brain* 142 (7):e39. https://doi.org/10.1093/brain/awz130.

Pedrazzi, M., M. Averna, B. Sparatore, M. Patrone, F. Salamino, M. Marcoli, G. Maura, C. Cervetto, D. Frattaroli, S. Pontremoli, and E. Melloni. 2012. "Potentiation of NMDA receptor-dependent cell responses by extracellular high mobility group box 1 protein." *PLoS One* 7 (8):e44518. https://doi.org/10.1371/journal.pone.0044518.

Pedrazzi, M., M. Patrone, M. Passalacqua, E. Ranzato, D. Colamassaro, B. Sparatore, S. Pontremoli, and E. Melloni. 2007. "Selective proinflammatory activation of astrocytes by high-mobility group box 1 protein signaling." *J Immunol* 179 (12):8525-32. https://doi.org/10.4049/jimmunol.179.12.8525.

Poewe, W., K. Seppi, C. M. Tanner, G. M. Halliday, P. Brundin, J. Volkmann, A. E. Schrag, and A. E. Lang. 2017. "Parkinson disease." *Nat Rev Dis Primers* 3:17013. https://doi.org/10.1038/nrdp.2017.13.

Pulliam, L., M. Liston, B. Sun, and J. Narvid. 2020. "Using neuronal extracellular vesicles and machine learning to predict cognitive deficits in HIV." *J Neurovirol* 26 (6):880-887. https://doi.org/10.1007/s13365-020-00877-6.

Qi, M. L., K. Tagawa, Y. Enokido, N. Yoshimura, Y. Wada, K. Watase, S. Ishiura, I. Kanazawa, J. Botas, M. Saitoe, E. E. Wanker, and H. Okazawa. 2007. "Proteome analysis of soluble nuclear proteins reveals that HMGB1/2 suppress genotoxic stress in polyglutamine diseases." *Nat Cell Biol* 9 (4):402-14. https://doi.org/10.1038/ncb1553.

Qiu, J., M. Nishimura, Y. Wang, J. R. Sims, S. Qiu, S. I. Savitz, S. Salomone, and M. A. Moskowitz. 2008. "Early release of HMGB-1 from neurons after the onset of brain ischemia." *J Cereb Blood Flow Metab* 28 (5):927-38. https://doi.org/10.1038/sj.jcbfm.9600582.

Rendon-Mitchell, B., M. Ochani, J. Li, J. Han, H. Wang, H. Yang, S. Susarla, C. Czura, R. A. Mitchell, G. Chen, A. E. Sama, K. J. Tracey, and H. Wang. 2003. "IFN-gamma induces high mobility group box 1 protein release partly through a TNF-dependent mechanism." *J Immunol* 170 (7):3890-7. https://doi.org/10.4049/jimmunol.170.7.3890.

Robinson, A. P., M. W. Caldis, C. T. Harp, G. E. Goings, and S. D. Miller. 2013. "High-mobility group box 1 protein (HMGB1) neutralization ameliorates experimental autoimmune encephalomyelitis." *J Autoimmun* 43:32-43. https://doi.org/10.1016/j.jaut.2013.02.005.

Sachs, M., H. C. Pape, E. J. Speckmann, and A. Gorji. 2007. "The effect of estrogen and progesterone on spreading depression in rat neocortical tissues." *Neurobiol Dis* 25 (1):27-34. https://doi.org/10.1016/j.nbd.2006.08.013.

Sandweiss, A. J., K. E. Cottier, M. I. McIntosh, G. Dussor, T. P. Davis, T. W. Vanderah, and T. M. Largent-Milnes. 2017. "17-beta-Estradiol induces spreading depression and pain behavior in alert female rats." *Oncotarget* 8 (69):114109-114122. https://doi.org/10.18632/oncotarget.23141.

Santoro, M., W. Maetzler, P. Stathakos, H. L. Martin, M. A. Hobert, T. W. Rattay, T. Gasser, J. V. Forrester, D. Berg, K. J. Tracey, G. Riedel, and P. Teismann. 2016. "In-vivo evidence that high mobility group box 1 exerts deleterious effects in the 1-methyl-4-phenyl-1,2,3,6-tetrahydropyridine model and Parkinson's disease which can be attenuated by glycyrrhizin." *Neurobiol Dis* 91:59-68. https://doi.org/10.1016/j.nbd.2016.02.018.

Sasaki, T., K. Liu, T. Agari, T. Yasuhara, J. Morimoto, M. Okazaki, H. Takeuchi, A. Toyoshima, S. Sasada, A. Shinko, A. Kondo, M. Kameda, I. Miyazaki, M. Asanuma, C. V. Borlongan, M. Nishibori, and I. Date. 2016. "Anti-high mobility group box 1 antibody exerts neuroprotection in a rat model of Parkinson's disease." *Exp Neurol* 275 Pt 1:220-31. https://doi.org/10.1016/j.expneurol.2015.11.003.

Schiraldi, M., A. Raucci, L. M. Munoz, E. Livoti, B. Celona, E. Venereau, T. Apuzzo, F. De Marchis, M. Pedotti, A. Bachi, M. Thelen, L. Varani, M. Mellado, A. Proudfoot, M. E. Bianchi, and M. Uguccioni. 2012. "HMGB1 promotes recruitment of inflammatory cells to damaged tissues by forming a complex with CXCL12 and signaling via CXCR4." *J Exp Med* 209 (3):551-63. https://doi.org/10.1084/jem.20111739.

Schulze, J., D. Zierath, P. Tanzi, K. Cain, D. Shibata, A. Dressel, and K. Becker. 2013. "Severe stroke induces long-lasting alterations of high-mobility group box 1." *Stroke* 44 (1):246-8. https://doi.org/10.1161/STROKEAHA.112.676072.

Seifert, T., J. Bauer, R. Weissert, F. Fazekas, and M. K. Storch. 2005. "Differential expression of sonic hedgehog immunoreactivity during lesion evolution in autoimmune encephalomyelitis." *J Neuropathol Exp Neurol* 64 (5):404-11. https://doi.org/10.1093/jnen/64.5.404.

Shi, Y., L. Zhang, J. Teng, and W. Miao. 2018. "HMGB1 mediates microglia activation via the TLR4/NF-kappaB pathway in coriaria lactone induced epilepsy." *Mol Med Rep* 17 (4):5125-5131. https://doi.org/10.3892/mmr.2018.8485.

Shibata, M., Y. Kayama, T. Shimizu, H. Toriumi, T. Ebine, M. Unekawa, A. Koh, A. Yoshimura, and N. Suzuki. 2017. "High-mobility group box 1 is an important mediator of microglial activation induced by cortical spreading depression." *J Cereb Blood Flow Metab* 37 (3):890-901. https://doi.org/10.1177/0271678X16647398.

Shin, J. H., H. K. Lee, H. B. Lee, Y. Jin, and J. K. Lee. 2014. "Ethyl pyruvate inhibits HMGB1 phosphorylation and secretion in activated microglia and in the postischemic brain." *Neurosci Lett* 558:159-63. https://doi.org/10.1016/j.neulet.2013.11.006.

Sievers, C., M. Meira, F. Hoffmann, P. Fontoura, L. Kappos, and R. L. Lindberg. 2012. "Altered microRNA expression in B lymphocytes in multiple sclerosis: towards a better understanding of treatment effects." *Clin Immunol* 144 (1):70-9. https://doi.org/10.1016/j.clim.2012.04.002.

Singh, J., and G. H. Dixon. 1990. "High mobility group proteins 1 and 2 function as general class II transcription factors." *Biochemistry* 29 (26):6295-302. https://doi.org/10.1021/bi00478a026.

Sokol, B., A. Wozniak, R. Jankowski, S. Jurga, N. Wasik, H. Shahid, and B. Grzeskowiak. 2015. "HMGB1 Level in Cerebrospinal Fluid as a Marker of Treatment Outcome in Patients with Acute Hydrocephalus Following Aneurysmal Subarachnoid Hemorrhage." *J Stroke Cerebrovasc Dis* 24 (8):1897-904. https://doi.org/10.1016/j.jstrokecerebrovasdis.2015.05.002.

Son, S., L. E. Bowie, T. Maiuri, C. L. K. Hung, C. R. Desmond, J. Xia, and R. Truant. 2019. "High-mobility group box 1 links sensing of reactive oxygen species by huntingtin to its nuclear entry." *J Biol Chem* 294 (6):1915-1923. https://doi.org/10.1074/jbc.RA117.001440.

Song, J. X., J. H. Lu, L. F. Liu, L. L. Chen, S. S. Durairajan, Z. Yue, H. Q. Zhang, and M. Li. 2014. "HMGB1 is involved in autophagy inhibition caused by SNCA/alpha-synuclein overexpression: a process modulated by the natural autophagy inducer corynoxine B." *Autophagy* 10 (1):144-54. https://doi.org/10.4161/auto.26751.

Sternberg, Z., C. Kolb, K. Chadha, A. Nir, R. Nir, R. George, J. Johnson, J. Yu, and D. Hojnacki. 2018. "Fingolimod anti-inflammatory and neuroprotective effects modulation of RAGE axis in multiple sclerosis patients." *Neuropharmacology* 130:71-76. https://doi.org/10.1016/j.neuropharm.2017.11.047.

Sternberg, Z., D. Sternberg, T. Chichelli, A. Drake, N. Patel, C. Kolb, K. Chadha, J. Yu, and D. Hojnacki. 2016. "High-mobility group box 1 in multiple sclerosis." *Immunol Res* 64 (2):385-91. https://doi.org/10.1007/s12026-015-8673-x.

Sternberg, Z., B. Weinstock-Guttman, D. Hojnacki, P. Zamboni, R. Zivadinov, K. Chadha, A. Lieberman, L. Kazim, A. Drake, P. Rocco, E. Grazioli, and F. Munschauer. 2008. "Soluble receptor for advanced glycation end products in multiple sclerosis: a potential marker of disease severity." *Mult Scler* 14 (6):759-63. https://doi.org/10.1177/1352458507088105.

Su, M., Y. Ran, Z. He, M. Zhang, G. Hu, W. Tang, D. Zhao, and S. Yu. 2018. "Inhibition of toll-like receptor 4 alleviates hyperalgesia induced by acute dural inflammation in experimental migraine." *Mol Pain* 14:1744806918754612. https://doi.org/10.1177/1744806918754612.

Su, Z., P. Zhang, Y. Yu, H. Lu, Y. Liu, P. Ni, X. Su, D. Wang, Y. Liu, J. Wang, H. Shen, W. Xu, and H. Xu. 2016. "HMGB1 Facilitated Macrophage Reprogramming towards a Proinflammatory M1-like Phenotype in Experimental Autoimmune Myocarditis Development." *Sci Rep* 6:21884. https://doi.org/10.1038/srep21884.

Sun, B., N. Fernandes, and L. Pulliam. 2019. "Profile of neuronal exosomes in HIV cognitive impairment exposes sex differences." *AIDS* 33 (11):1683-1692. https://doi.org/10.1097/QAD.0000000000002272.

Sun, B., P. Dalvi, L. Abadjian, N. Tang, and L. Pulliam. 2017. "Blood neuron-derived exosomes as biomarkers of cognitive impairment in HIV." *AIDS* 31 (14):F9-F17. https://doi.org/10.1097/QAD.0000000000001595.

Sun, Q., W. Wu, Y. C. Hu, H. Li, D. Zhang, S. Li, W. Li, W. D. Li, B. Ma, J. H. Zhu, M. L. Zhou, and C. H. Hang. 2014. "Early release of high-mobility group box 1 (HMGB1) from neurons in experimental subarachnoid hemorrhage in vivo and in vitro." *J Neuroinflammation* 11:106 https://doi.org/10.1186/1742-2094-11-106.

Sun, Y., H. Chen, J. Dai, Z. Wan, P. Xiong, Y. Xu, Z. Han, W. Chai, F. Gong, and F. Zheng. 2018. "Glycyrrhizin Protects Mice Against Experimental Autoimmune Encephalomyelitis by Inhibiting High-Mobility Group Box 1 (HMGB1) Expression and Neuronal HMGB1 Release." *Front Immunol* 9:1518. https://doi.org/10.3389/fimmu.2018.01518.

Tadie, J. M., H. B. Bae, S. Jiang, D. W. Park, C. P. Bell, H. Yang, J. F. Pittet, K. Tracey, V. J. Thannickal, E. Abraham, and J. W. Zmijewski. 2013. "HMGB1 promotes neutrophil extracellular trap formation through interactions with Toll-like receptor 4." *Am J Physiol Lung Cell Mol Physiol* 304 (5):L342-9. https://doi.org/10.1152/ajplung.00151.2012.

Takata, K., Y. Kitamura, D. Tsuchiya, T. Kawasaki, T. Taniguchi, and S. Shimohama. 2004. "High mobility group box protein-1 inhibits microglial Abeta clearance and enhances Abeta neurotoxicity." *J Neurosci Res* 78 (6):880-91. https://doi.org/10.1002/jnr.20340.

Takata, K., Y. Kitamura, J. Kakimura, K. Shibagaki, D. Tsuchiya, T. Taniguchi, M. A. Smith, G. Perry, and S. Shimohama. 2003. "Role of high mobility group protein-1 (HMG1) in amyloid-beta homeostasis." *Biochem Biophys Res Commun* 301 (3):699-703. https://doi.org/10.1016/s0006-291x(03)00024-x.

Takata, K., T. Takada, A. Ito, M. Asai, M. Tawa, Y. Saito, E. Ashihara, H. Tomimoto, Y. Kitamura, and S. Shimohama. 2012. "Microglial Amyloid-beta1-40 Phagocytosis Dysfunction Is Caused by High-Mobility Group Box Protein-1: Implications for the Pathological Progression of Alzheimer's Disease." *Int J Alzheimers Dis* 2012:685739. https://doi.org/10.1155/2012/685739.

Takizawa, T., M. Shibata, Y. Kayama, H. Toriumi, T. Ebine, A. Koh, T. Shimizu, and N. Suzuki. 2016. "Temporal profiles of high-mobility group box 1 expression levels after cortical spreading depression in mice." *Cephalalgia* 36 (1):44-52. https://doi.org/10.1177/0333102415580100.

Tanaka, H., K. Kondo, K. Fujita, H. Homma, K. Tagawa, X. Jin, M. Jin, Y. Yoshioka, S. Takayama, H. Masuda, R. Tokuyama, Y. Nakazaki, T. Saito, T. Saido, S. Murayama, T. Ikura, N. Ito, Y. Yamamori, K. Tomii, M. E. Bianchi, and H. Okazawa. 2021. "HMGB1 signaling phosphorylates Ku70 and impairs DNA damage repair in Alzheimer's disease pathology." *Commun Biol* 4 (1):1175. https://doi.org/10.1038/s42003-021-02671-4.

Tang, D., R. Kang, K. M. Livesey, C. W. Cheh, A. Farkas, P. Loughran, G. Hoppe, M. E. Bianchi, K. J. Tracey, H. J. Zeh, 3rd, and M. T. Lotze. 2010. "Endogenous HMGB1 regulates autophagy." *J Cell Biol* 190 (5):881-92. https://doi.org/10.1083/jcb.200911078.

Thierry, S., J. Gozlan, A. Jaulmes, R. Boniface, N. Nasreddine, F. Strauss, and V. Marechal. 2007. "High-mobility group box 1 protein induces HIV-1 expression from persistently infected cells." *AIDS* 21 (3):283-92. https://doi.org/10.1097/QAD.0b013e3280115b50.

Tian, X., L. Sun, D. Feng, Q. Sun, Y. Dou, C. Liu, F. Zhou, H. Li, H. Shen, Z. Wang, and G. Chen. 2017. "HMGB1 promotes neurovascular remodeling via Rage in the late phase of subarachnoid hemorrhage." *Brain Res* 1670:135-145. https://doi.org/10.1016/j.brainres.2017.06.001.

Tiszlavicz, Z., Z. Gyulai, K. Bencsik, Z. Szolnoki, A. K. Kocsis, F. Somogyvari, L. Vecsei, and Y. Mandi. 2009. "RAGE gene polymorphisms in patients with multiple sclerosis." *J Mol Neurosci* 39 (3):360-5. https://doi.org/10.1007/s12031-009-9291-7.

Tsiskaridze, Alexander, Arne Lindgren, and Adnan Qureshi. 2016. *Treatment-Related Stroke*. Cambridge: Cambridge University Press.

Tsukagawa, T., R. Katsumata, M. Fujita, K. Yasui, C. Akhoon, K. Ono, K. Dohi, and T. Aruga. 2017. "Elevated Serum High-Mobility Group Box-1 Protein Level Is Associated with Poor Functional Outcome in Ischemic Stroke." *J Stroke Cerebrovasc Dis* 26 (10):2404-2411. https://doi.org/10.1016/j.jstrokecerebrovasdis.2017.05.033.

Umahara, T., T. Uchihara, K. Hirao, S. Shimizu, T. Hashimoto, J. Akimoto, M. Kohno, and H. Hanyu. 2020. "Frontotemporal dementia-associated protein "phosphorylated TDP-43" localizes to atherosclerotic lesions of human carotid and main cerebral arteries." *Histol Histopathol* 35 (2):159-167. https://doi.org/10.14670/HH-18-140.

Unnithan, A. K. A., M. Das J, and P. Mehta. 2022. "Hemorrhagic Stroke." In *Stat Pearls*. Treasure Island (FL).

Uramoto, H., H. Izumi, G. Nagatani, H. Ohmori, N. Nagasue, T. Ise, T. Yoshida, K. Yasumoto, and K. Kohno. 2003. "Physical interaction of tumour suppressor p53/p73 with CCAAT-binding transcription factor 2 (CTF2) and differential regulation of human high-mobility group 1 (HMG1) gene expression." *Biochem J* 371 (Pt 2):301-10. https://doi.org/10.1042/BJ20021646.

Uzawa, A., M. Mori, H. Masuda, R. Ohtani, T. Uchida, and S. Kuwabara. 2018. "Recombinant thrombomodulin ameliorates experimental autoimmune encephalomyelitis by suppressing high mobility group box 1 and inflammatory cytokines." *Clin Exp Immunol* 193 (1):47-54. https://doi.org/10.1111/cei.13123.

Uzawa, A., M. Mori, J. Taniguchi, S. Masuda, M. Muto, and S. Kuwabara. 2013. "Anti-high mobility group box 1 monoclonal antibody ameliorates experimental autoimmune encephalomyelitis." *Clin Exp Immunol* 172 (1):37-43. https://doi.org/10.1111/cei.12036.

Uzawa, A., M. Mori, S. Masuda, M. Muto, and S. Kuwabara. 2013. "CSF high-mobility group box 1 is associated with intrathecal inflammation and astrocytic damage in neuromyelitis optica." *J Neurol Neurosurg Psychiatry* 84 (5):517-22. https://doi.org/10.1136/jnnp-2012-304039.

Venereau, E., C. Ceriotti, and M. E. Bianchi. 2015. "DAMPs from Cell Death to New Life." *Front Immunol* 6:422. https://doi.org/10.3389/fimmu.2015.00422.

Volchuk, A., A. Ye, L. Chi, B. E. Steinberg, and N. M. Goldenberg. 2020. "Indirect regulation of HMGB1 release by gasdermin D." *Nat Commun* 11 (1):4561. https://doi.org/10.1038/s41467-020-18443-3.

Volnova, A., V. Tsytsarev, O. Ganina, G. E. Velez-Crespo, J. M. Alves, A. Ignashchenkova, and M. Inyushin. 2022. "The Anti-Epileptic Effects of Carbenoxolone In Vitro and In Vivo." *Int J Mol Sci* 23 (2). https://doi.org/10.3390/ijms23020663.

Walker, L. E., G. J. Sills, A. Jorgensen, T. Alapirtti, J. Peltola, M. J. Brodie, A. G. Marson, A. Vezzani, and M. Pirmohamed. 2022. "High-mobility group box 1 as a predictive biomarker for drug-resistant epilepsy: A proof-of-concept study." *Epilepsia* 63 (1):e1-e6. https://doi.org/10.1111/epi.17116.

Walker, Lauren, Karen Tse, Emanuele Ricci, Thimmasettappa Thippeswamy, Graeme J. Sills, Steve H. White, Daniel J. Antoine, Anthony Marson, and Munir Pirmohamed. 2014. "High mobility group box 1 in the inflammatory pathogenesis of epilepsy: profiling circulating levels after experimental and clinical seizures." *The Lancet* 383:S105. https://doi.org/10.1016/s0140-6736(14)60368-8.

Wang, D., K. Liu, H. Wake, K. Teshigawara, S. Mori, and M. Nishibori. 2017. "Anti-high mobility group box-1 (HMGB1) antibody inhibits hemorrhage-induced brain injury and improved neurological deficits in rats." *Sci Rep* 7:46243. https://doi.org/10.1038/srep46243.

Wang, D., K. Liu, Y. Fukuyasu, K. Teshigawara, L. Fu, H. Wake, A. Ohtsuka, and M. Nishibori. 2020. "HMGB1 Translocation in Neurons after Ischemic Insult: Subcellular Localization in Mitochondria and Peroxisomes." *Cells* 9 (3). https://doi.org/10.3390/cells9030643.

Wang, H., H. Liao, M. Ochani, M. Justiniani, X. Lin, L. Yang, Y. Al-Abed, H. Wang, C. Metz, E. J. Miller, K. J. Tracey, and L. Ulloa. 2004. "Cholinergic agonists inhibit HMGB1 release and improve survival in experimental sepsis." *Nat Med* 10 (11):1216-21. https://doi.org/10.1038/nm1124.

Wang, H., K. Wang, C. Wang, F. Xu, X. Zhong, W. Qiu, and X. Hu. 2013. "Cerebrospinal fluid high-mobility group box protein 1 in neuromyelitis optica and multiple sclerosis." *Neuroimmunomodulation* 20 (2):113-8 https://doi.org/10.1159/000345994.

Wang, J., Y. Jiang, D. Zeng, W. Zhou, and X. Hong. 2020. "Prognostic value of plasma HMGB1 in ischemic stroke patients with cerebral ischemia-reperfusion injury after intravenous thrombolysis." *J Stroke Cerebrovasc Dis* 29 (9):105055. https://doi.org/10.1016/j.jstrokecerebrovasdis.2020.105055.

Wang, K. C., C. P. Tsai, C. L. Lee, S. Y. Chen, L. T. Chin, and S. J. Chen. 2012. "Elevated plasma high-mobility group box 1 protein is a potential marker for neuromyelitis optica." *Neuroscience* 226:510-6. https://doi.org/10.1016/j.neuroscience.2012.08.041.

Wang, K., J. Huang, W. Xie, L. Huang, C. Zhong, and Z. Chen. 2016. "Beclin1 and HMGB1 ameliorate the alpha-synuclein-mediated autophagy inhibition in PC12 cells." *Diagn Pathol* 11:15. https://doi.org/10.1186/s13000-016-0459-5.

Wang, L., Z. Zhang, L. Liang, Y. Wu, J. Zhong, and X. Sun. 2019. "Anti-high mobility group box-1 antibody attenuated vascular smooth muscle cell phenotypic switching and vascular remodelling after subarachnoid haemorrhage in rats." *Neurosci Lett* 708:134338. https://doi.org/10.1016/j.neulet.2019.134338.

Wang, N., H. Liu, B. Ma, T. Zhao, Y. Chen, Y. Yang, P. Zhao, and X. Han. 2021. "CSF high-mobility group box 1 is associated with drug-resistance and symptomatic etiology in adult patients with epilepsy." *Epilepsy Res* 177:106767. https://doi.org/10.1016/j.eplepsyres.2021.106767.

Wang, Q., H. Zhao, Y. Gao, J. Lu, Xie, W. Yu, F. He, W. Liu, I. Hisatome, T. Yamamoto, W. Wang, and J. Cheng. 2021. "Uric acid inhibits HMGB1-TLR4-NF-kappaB signaling to alleviate oxygen-glucose deprivation/reoxygenation injury of microglia."

Biochem Biophys Res Commun 540:22-28. https://doi.org/10.1016/j.bbrc.2020.12.097.

Wang, X. Y., H. R. Zhou, S. Wang, C. Y. Liu, G. C. Qin, Q. Q. Fu, J. Y. Zhou, and L. X. Chen. 2018. "NR2B-Tyr phosphorylation regulates synaptic plasticity in central sensitization in a chronic migraine rat model." *J Headache Pain* 19 (1):102. https://doi.org/10.1186/s10194-018-0935-2.

Wen, L., J. K. Huang, B. H. Johnson, and G. R. Reeck. 1989. "A human placental cDNA clone that encodes nonhistone chromosomal protein HMG-1." *Nucleic Acids Res* 17 (3):1197-214. https://doi.org/10.1093/nar/17.3.1197.

Wolfe, Michael S. 2018. "Chapter 1 - Solving the Puzzle of Neurodegeneration." In *The Molecular and Cellular Basis of Neurodegenerative Diseases*, edited by Michael S. Wolfe, 1-22. Academic Press.

Wolfson, R. K., B. Mapes, and J. G. N. Garcia. 2014. "Excessive mechanical stress increases HMGB1 expression in human lung microvascular endothelial cells via STAT3." *Microvasc Res* 92:50-55. https://doi.org/10.1016/j.mvr.2013.12.005.

Wu, G., J. Liu, S. Li, W. Gao, M. Qiu, C. Yang, Y. Ma, and X. Song. 2020. "Glycyrrhizic acid protects juvenile epileptic rats against hippocampal damage through activation of Sirtuin3." *Brain Res Bull* 164:98-106. https://doi.org/10.1016/j.brainresbull.2020.08.008.

Wu, Y., Y. Zhang, S. Zhu, C. Tian, and Y. Zhang. 2021. "MiRNA-29a serves as a promising diagnostic biomarker in children with temporal lobe epilepsy and regulates seizure-induced cell death and inflammation in hippocampal neurons." *Epileptic Disord* 23 (6):823-832. https://doi.org/10.1684/epd.2021.1331.

Xiao, Y., Y. Sun, W. Liu, F. Zeng, J. Shi, J. Li, H. Chen, C. Tu, Y. Xu, Z. Tan, F. Gong, X. Shu, and F. Zheng. 2021. "HMGB1 Promotes the Release of Sonic Hedgehog From Astrocytes." *Front Immunol* 12:584097. https://doi.org/10.3389/fimmu.2021.584097.

Xiong, L., L. Sun, Y. Zhang, J. Peng, J. Yan, and X. Liu. 2020. "Exosomes from Bone Marrow Mesenchymal Stem Cells Can Alleviate Early Brain Injury After Subarachnoid Hemorrhage Through miRNA129-5p-HMGB1 Pathway." *Stem Cells Dev* 29 (4):212-221. https://doi.org/10.1089/scd.2019.0206.

Xiong, X., L. Gu, Y. Wang, Y. Luo, H. Zhang, J. Lee, S. Krams, S. Zhu, and H. Zhao. 2016. "Glycyrrhizin protects against focal cerebral ischemia via inhibition of T cell activity and HMGB1-mediated mechanisms." *J Neuroinflammation* 13 (1):241. https://doi.org/10.1186/s12974-016-0705-5.

Xu, J., S. M. Firouz, M. Farrokhian, S. Ghoreishizadeh, T. A. Merza Mohamad, A. Rostami, R. Tamjididfar, M. Akbari, N. Shomali, R. E. Zamiri, S. S. Shotorbani, and A. S. Shahi. 2022. "Potential anti-inflammatory effect of anti-HMGB1 in animal models of ICH by downregulating the TLR4 signaling pathway and regulating the inflammatory cytokines along with increasing HO1 and NRF2." *Eur J Pharmacol* 915:174694. https://doi.org/10.1016/j.ejphar.2021.174694.

Xu, J., Y. Jiang, J. Wang, X. Shi, Q. Liu, Z. Liu, Y. Li, M. J. Scott, G. Xiao, S. Li, L. Fan, T. R. Billiar, M. A. Wilson, and J. Fan. 2014. "Macrophage endocytosis of high-mobility group box 1 triggers pyroptosis." *Cell Death Differ* 21 (8):1229-39. https://doi.org/10.1038/cdd.2014.40.

Yang, H., H. Wang, S. S. Chavan, and U. Andersson. 2015. "High Mobility Group Box Protein 1 (HMGB1): The Prototypical Endogenous Danger Molecule." *Mol Med* 21 Suppl 1:S6-S12. https://doi.org/10.2119/molmed.2015.00087.

Yang, H., Q. Zeng, H. A. Silverman, M. Gunasekaran, S. J. George, A. Devarajan, M. E. Addorisio, J. Li, T. Tsaava, V. Shah, T. R. Billiar, H. Wang, M. Brines, U. Andersson, V. A. Pavlov, E. H. Chang, S. S. Chavan, and K. J. Tracey. 2021. "HMGB1 released from nociceptors mediates inflammation." *Proc Natl Acad Sci USA* 118 (33). https://doi.org/10.1073/pnas.2102034118.

Yang, W., J. Li, Y. Shang, L. Zhao, M. Wang, J. Shi, and S. Li. 2017. "HMGB1-TLR4 Axis Plays a Regulatory Role in the Pathogenesis of Mesial Temporal Lobe Epilepsy in Immature Rat Model and Children via the p38MAPK Signaling Pathway." *Neurochem Res* 42 (4):1179-1190. https://doi.org/10.1007/s11064-016-2153-0.

Yang, W. S., N. J. Han, J. J. Kim, M. J. Lee, and S. K. Park. 2016. "TNF-alpha Activates High-Mobility Group Box 1 - Toll-Like Receptor 4 Signaling Pathway in Human Aortic Endothelial Cells." *Cell Physiol Biochem* 38 (6):2139-51. https://doi.org/10.1159/000445570.

Yang, Y., C. Han, L. Guo, and Q. Guan. 2018. "High expression of the HMGB1-TLR4 axis and its downstream signaling factors in patients with Parkinson's disease and the relationship of pathological staging." *Brain Behav* 8 (4):e00948. https://doi.org/10.1002/brb3.948.

Youn, J. H., and J. S. Shin. 2006. "Nucleocytoplasmic shuttling of HMGB1 is regulated by phosphorylation that redirects it toward secretion." *J Immunol* 177 (11):7889-97. https://doi.org/10.4049/jimmunol.177.11.7889.

Yu, S. F., W. Y. Feng, S. Q. Chai, X. B. Meng, Z. X. Dou, and H. Zhu. 2018. "Down-Regulation of miR-218-5p Promotes Apoptosis of Human Umbilical Vein Endothelial Cells Through Regulating High-Mobility Group Box-1 in Henoch-Schonlein Purpura." *Am J Med Sci* 356 (1):64-71. https://doi.org/10.1016/j.amjms.2018.04.001.

Yu, S., H. Zhang, Y. Hei, X. Yi, A. Baskys, W. Liu, and Q. Long. 2019. "High mobility group box-1 (HMGB1) antagonist BoxA suppresses status epilepticus-induced neuroinflammatory responses associated with Toll-like receptor 2/4 down-regulation in rats." *Brain Res* 1717:44-51. https://doi.org/10.1016/j.brainres.2019.04.007.

Zaben, M., N. Haan, F. Sharouf, A. Ahmed, L. E. Sundstrom, and W. P. Gray. 2021. "IL-1beta and HMGB1 are anti-neurogenic to endogenous neural stem cells in the sclerotic epileptic human hippocampus." *J Neuroinflammation* 18 (1):218. https://doi.org/10.1186/s12974-021-02265-1.

Zaniani, N. R., A. Roohbakhsh, A. Moghimi, and S. Mehri. 2022. "Protective effect of Toll-like receptor 4 antagonist on inflammation, EEG, and memory changes following febrile seizure in Wistar rats." *Behav Brain Res* 420:113723. https://doi.org/10.1016/j.bbr.2021.113723.

Zhang, J., H. K. Takahashi, K. Liu, H. Wake, R. Liu, T. Maruo, I. Date, T. Yoshino, A. Ohtsuka, S. Mori, and M. Nishibori. 2011. "Anti-high mobility group box-1 monoclonal antibody protects the blood-brain barrier from ischemia-induced disruption in rats." *Stroke* 42 (5):1420-8. https://doi.org/10.1161/STROKEAHA.110.598334.

Zhang, Y. Y., H. L. Bao, L. X. Dong, Y. Liu, G. W. Zhang, and F. M. An. 2021. "Silenced lncRNA H19 and up-regulated microRNA-129 accelerates viability and restrains apoptosis of PC12 cells induced by Abeta25-35 in a cellular model of Alzheimer's disease." *Cell Cycle* 20 (1):112-125. https://doi.org/10.1080/15384101.2020.1863681.

Zhang, Z., Q. Liu, M. Liu, H. Wang, Y. Dong, T. Ji, X. Liu, Y. Jiang, L. Cai, and Y. Wu. 2018. "Upregulation of HMGB1-TLR4 inflammatory pathway in focal cortical dysplasia type II." *J Neuroinflammation* 15 (1):27. https://doi.org/10.1186/s12974-018-1078-8.

Zhao, J., Y. Wang, C. Xu, K. Liu, Y. Wang, L. Chen, X. Wu, F. Gao, Y. Guo, J. Zhu, S. Wang, M. Nishibori, and Z. Chen. 2017. "Therapeutic potential of an anti-high mobility group box-1 monoclonal antibody in epilepsy." *Brain Behav Immun* 64:308-319. https://doi.org/10.1016/j.bbi.2017.02.002.

Zhao, J., Y. Zheng, K. Liu, J. Chen, N. Lai, F. Fei, J. Shi, C. Xu, S. Wang, M. Nishibori, Y. Wang, and Z. Chen. 2020. "HMGB1 Is a Therapeutic Target and Biomarker in Diazepam-Refractory Status Epilepticus with Wide Time Window." *Neurotherapeutics* 17 (2):710-721. https://doi.org/10.1007/s13311-019-00815-3.

Zhao, P. C., S. N. Xu, Z. S. Huang, G. W. Jiang, P. C. Deng, and Y. M. Zhang. 2021. "Hyperbaric oxygen via mediating SIRT1-induced deacetylation of HMGB1 improved cReperfusion inj/reperfusion injury." *Eur J Neurosci* 54 (9):7318-7331. https://doi.org/10.1111/ejn.15458.

Zhao, X., A. Rouhiainen, Z. Li, S. Guo, and H. Rauvala. 2020. "Regulation of Neurogenesis in Mouse Brain by HMGB1." *Cells* 9 (7). https://doi.org/10.3390/cells9071714.

Zhao, X. D., H. Y. Mao, J. Lv, and X. J. Lu. 2016. "Expression of high-mobility group box-1 (HMGB1) in the basilar artery after experimental subarachnoid hemorrhage." *J Clin Neurosci* 27:161-5. https://doi.org/10.1016/j.jocn.2015.06.034.

Zhao, Y., S. W. Tan, Z. Z. Huang, F. B. Shan, P. Li, Y. L. Ning, S. Y. Ye, Z. A. Zhao, H. Du, R. P. Xiong, N. Yang, Y. Peng, X. Chen, and Y. G. Zhou. 2021. "NLRP3 Inflammasome-Dependent Increases in High Mobility Group Box 1 Involved in the Cognitive Dysfunction Caused by Tau-Overexpression." *Front Aging Neurosci* 13:721474. https://doi.org/10.3389/fnagi.2021.721474.

Zhen, C., X. Feng, Z. Li, Y. Wang, B. Li, L. Li, M. Quan, G. Wang, and L. Guo. 2015. "Suppression of murine experimental autoimmune encephalomyelitis development by 1,25-dihydroxyvitamin D3 with autophagy modulation." *J Neuroimmunol* 280:1-7. https://doi.org/10.1016/j.jneuroim.2015.01.012.

Zhen, C., Y. Wang, D. Li, W. Zhang, H. Zhang, X. Yu, and X. Wang. 2019. "Relationship of High-mobility group box 1 levels and multiple sclerosis: A systematic review and meta-analysis." *Mult Scler Relat Disord* 31:87-92. https://doi.org/10.1016/j.msard.2019.03.030.

Zhou, Y., K. L. Xiong, S. Lin, Q. Zhong, F. L. Lu, H. Liang, J. C. Li, J. Z. Wang, and Q. W. Yang. 2010. "Elevation of high-mobility group protein box-1 in serum correlates with severity of acute intracerebral hemorrhage." *Mediators Inflamm* 2010. https://doi.org/10.1155/2010/142458.

Zhu, H. L., Y. Q. Chen, and Z. F. Zhang. 2020. "Downregulation of lncRNA ZFAS1 and upregulation of microRNA-129 repress endocrine disturbance, increase proliferation and inhibit apoptosis of ovarian granulosa cells in polycystic ovarian syndrome by downregulating HMGB1." *Genomics* 112 (5):3597-3608. https://doi.org/10.1016/j.ygeno.2020.04.011.

Zhu, M., J. Chen, H. Guo, L. Ding, Y. Zhang, and Y. Xu. 2018. "High Mobility Group Protein B1 (HMGB1) and Interleukin-1beta as Prognostic Biomarkers of Epilepsy in Children." *J Child Neurol* 33 (14):909-917. https://doi.org/10.1177/0883073818801654.

Zhu, X. D., J. S. Chen, F. Zhou, Q. C. Liu, G. Chen, and J. M. Zhang. 2012. "Relationship between plasma high mobility group box-1 protein levels and clinical outcomes of aneurysmal subarachnoid hemorrhage." *J Neuroinflammation* 9:194. https://doi.org/10.1186/1742-2094-9-194.

Zurolo, E., A. Iyer, M. Maroso, C. Carbonell, J. J. Anink, T. Ravizza, K. Fluiter, W. G. Spliet, P. C. van Rijen, A. Vezzani, and E. Aronica. 2011. "Activation of Toll-like receptor, RAGE and HMGB1 signalling in malformations of cortical development." *Brain* 134 (Pt 4):1015-32. https://doi.org/10.1093/brain/awr032.

Chapter 6

The Roles of HMGB1 Post-Brain Injury: Implications for Injury and Repair

Malik Zaben[1,*], PhD, Ron Ved[1,2] and Susruta Manivannan[2]
[1]The BRAIN Unit,
Neuroscience and Mental Health Research Institute,
(NMHRI) School of Medicine, Cardiff University, Cardiff, UK
[2]Department of Neurosurgery,
Southampton General Hospital, Southampton, UK

Abstract

Traumatic brain injury (TBI) is amongst the leading causes of mortality and morbidity worldwide. Despite significant medical advances over the previous decades, many survivors continue to suffer with neurocognitive deficits throughout their lives. Widespread recognition of a protracted neuro-inflammatory response following the acute injury has raised the exciting possibility of therapeutic intervention. However, pharmacological strategies in the clinical setting have failed to realise this potential due to: (i) an inadequate understanding of the complex underlying relationship between neuroinflammation and brain repair; and (ii) differences in effects between animal and human models of central nervous system (CNS) injury. Mounting evidence implicates HMGB1 as a unique alternative target following brain injury. It plays a pivotal role as a 'master-switch' of neuro-inflammation following injury and in the regulation of neurogenesis during normal development. Animal models of traumatic, degenerative, and ischaemic brain injury point towards the involvement of HMGB1 signalling in prolonged activation of glial cells

* Corresponding Author's Email: zabenm@cardiff.ac.uk.

In: HMGB1
Editors: Eylem Taskin, Celal Guven and Salih Tunc Kaya
ISBN: 979-8-88697-408-9

and widespread neuronal death. Furthermore, early experimental studies demonstrate positive effects of HMGB1 antagonism on both immunohistochemical and neuro-behavioural parameters following injury. In addition, recent in vitro evidence implicates HMGB1 in mediating both grey and white matter injury, which is thought to cause higher level neuro-cognitive deficits following brain injury. Therefore, modulation of HMGB1 signalling pathways raises the possibility of addressing both the injurious consequences of neuro-inflammation and enhancement of endogenous brain repair. In this chapter, we will outline the role of HMGB1 signalling in post-traumatic neuro-inflammation and explore prospects of therapeutic modulation to enhance recovery.

Keywords: HMGB1, neuroinflammation, traumatic brain injury (TBI), neurogenesis, oligodendrogenesis

Introduction

Traumatic brain injury (TBI) is amongst the leading causes of death and disability worldwide, with a significant impact at individual, societal, and economic levels. Despite improvements in access to healthcare, and rigorous neuro-intensive care approaches to management of TBI, outcomes remain suboptimal (Huebener et al. 2015). There is an increasing awareness that TBI is not characterised by acute traumatic impact alone, but that this is rather the triggering event of a chronic neurodegenerative process with ongoing injury outlasting the initial insult.

Broadly, TBI consists of primary and secondary injury. Primary injury arises from the force of the initial traumatic impact, which causes acceleration-deceleration and rotational injuries. This results in mechanical breakage of axons and triggers a complex cascade of events that constitute the secondary injury (Yang et al. 2015). An explosion of research interest over recent decades have characterised several constituents of secondary injury including glutamate toxicity, mitochondrial dysfunction, pathological calcium influx, and caspase-mediated apoptosis. One driver of these secondary injury processes appears to be release of damage associated molecular patterns (DAMPs), subsequent activation of glial cells, and recruitment of peripheral immune cells, leading to a significant neuroinflammatory response to the traumatic injury (Fink 2007). This has given rise to several candidate biomarkers of injury and potential therapeutic avenues, but pharmacological clinical trials have remained unsuccessful to date. High Mobility Group Box protein 1 (HMGB1), however, is a particularly promising target that has shown encouraging evidence of enhancing brain repair after TBI across *in vitro* and *in vivo* animal models.

HMGB1 Role in Post-TBI Neuroinflammation

High Mobility Group Box 1 (HMGB1) is a 215 amino acid protein with pleiotropic functions. It acts as a non-histone DNA binding protein to modulate gene expression and DNA repair under physiological conditions with a key role in neural development, however, is released passively by necrotic neurons or secreted actively by immune cells as an alarmin under pathological insults (Fink 2007). Two DNA binding domains, known as Box A and Box B, and an acidic C-terminal tail constitute the basic structure of HMGB1 (Ueda and Yoshida 2010) (Figure 1). This reflects its physiological role in co-ordination of gene transcription in its natural physiological milieu within the nucleus.

The dynamics of HMGB1 release following injury have been characterised; and as such it is often depicted as the 'master-switch' of neuroinflammation given its hyper-acute release following injury and stimulation of pro-inflammatory cytokine release.

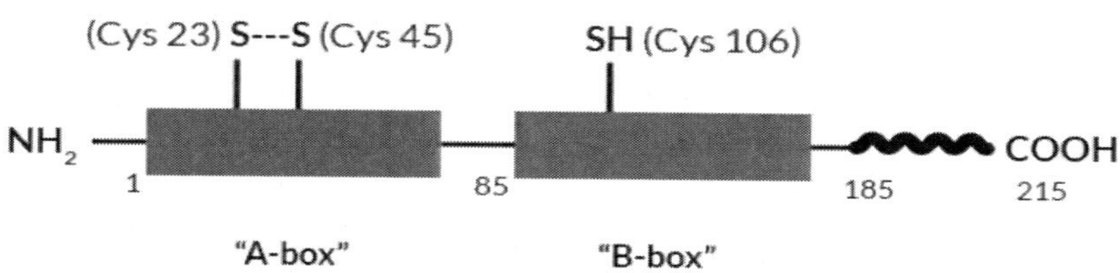

Figure 1. Schematic presentation of HMGB1. The A-Box and B-Box components of HMGB1 determine its binding efficacy with its target receptors (RAGE and TLR4). Modifications to the boxes, such as a disulphide bond within Box A, lead to different isomers of HMGB1, each of which may have disparate roles after TBI. Adapted from He et al. 2018.

HMGB1 as a DAMP

Following brain injury, HMGB1 is secreted passively by dying cells, predominantly neurons, functioning as a DAMP. It is also secreted actively by glial cells including microglia and astrocytes (Manivannan, Wales, and Zaben 2021). Active and passive secretion of HMGB1 may be distinguished based on the acetylation status of lysine residues in the nuclear localisation sequence, with hyper-acetylation indicating active release (Figure 2). The acetylation step prevents HMGB1 from being translocated to the nucleus, and instead promotes its accumulation within secretory vesicles, destined for release into the extracellular environment (Richard et al. 2017). In the extracellular space, HMGB1 triggers the immune system by promoting release of inflammatory cytokines, activating immune cells, and promoting their migration to the site of injury (Bianchi and Manfredi 2007). Consistently, animal studies have identified the accumulation of HMGB1 within the cytosol and extracellular environment locally after models of brain injury (Bohmer et al. 2011; Cohen et al. 2009) and it is also detectable in the cerebrospinal fluid and bloodstream, where its levels may correlate with the severity of the injury (Wang et al. 2012; Au et al. 2012). There is also evidence of an initial peak in HMGB1 release acutely post injury, following by a second wave of HMGB1 release in the 7-14 days post injury (Richard et al. 2017; Au et al. 2012).

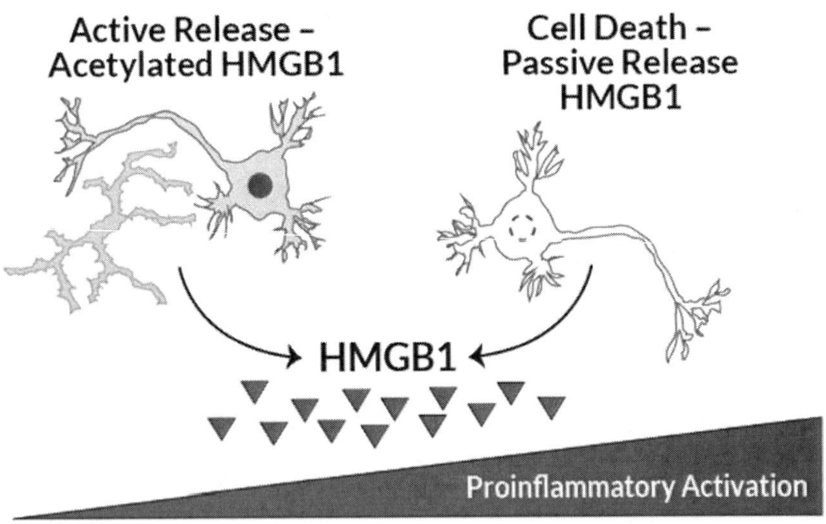

Figure 2. HMGB1 release after neurotrauma. HMGB1 reaches the extracellular milieu following injury via two mechanisms: passive release from damaged cells, and active secretion following acetylation from surviving cells. Both versions may lead to neuro-inflammatory modulation of cellular responses after TBI. Adapted from Manivannan, Wales, and Zaben, 2021.

HMGB1 Downstream Signalling and Subsequent Cytokine Release

Once released after TBI, HMGB1 acts on target cell receptors including Receptor for Advanced Glycation End-Products (RAGE), and Toll-Like Receptor 4 (TLR-4), expressed by glial cells within the CNS, such as astrocytes and microglia (Manivannan, Wales, and Zaben 2021). Through binding of the Box A component (Figure 1), HMGB1 activates RAGE and TLR4 signalling pathways via MyD88-dependent downstream activation of NF-kB and upregulation of pro-inflammatory cytokine gene expression. Figure 3 summarises the various downstream signalling pathways associated with HMGB1 signalling.

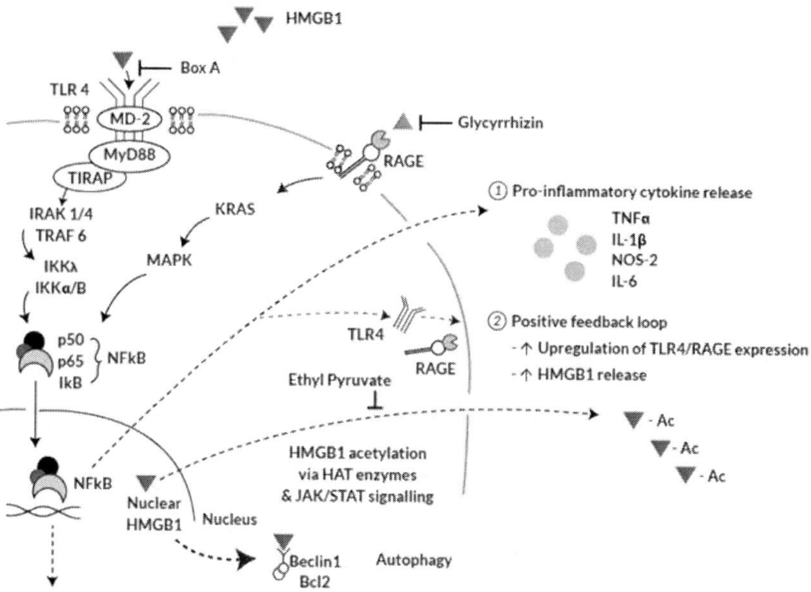

Figure 3. Summary of intracellular signalling following HMGB1 receptor activation. TLR4 and RAGE signalling mechanisms converge on the canonical NFkB pathway to result in up-regulation of the transcription of pro-inflammatory genes such as IL1β, TNF-alpha, and IL-6. These substances go on to modulate tissue damage acute post-injury. There is further positive feedback via up-regulation of TLR4/RAGE expression and stimulation of further, active release of acetylated HMGB1, further promoting neuro-inflammation post-TBI. Adapted from Manivannan, Wales, and Zaben, 2021.

Activation of RAGE and TLR4 by HMGB1 leads to the phosphorylation of several mitogen-activated protein kinases (MAPKs) that activate the canonical pathway of nuclear factor kappa-light-chain-enhancer of activated B cells (NF-κB). This signalling cascade leads to: (i) further release of HMGB1, (ii) upregulation of HMGB1 receptor expression, and (iii) the release of various pro-inflammatory cytokines such as IL1β, TNF-α, IL-6, and interferon gamma (INF-γ) (Parker et al. 2017). These cytokines have pleiotropic effects to regulate tissue damage, such as promotion of apoptosis, blood-brain-barrier (BBB) breakdown, cell proliferation, proteins modulating complement pathways, and microglial activation to a proinflammatory phenotype (Liu et al. 2017). Furthermore, at later time points post injury, neuroinflammation may change its role from deleterious to potentially reparative, although the mechanisms underlying this remain to be fully elucidated (Richard et al. 2017). One emerging theory is that the later waves of HMGB1 release at 7-14 days post injury may be different isomers of HMGB1, which have different roles and signalling pathways to those of the early secretion of HMGB1 (Yang, Lundback, et al. 2012).

HMGB1 Isoforms

The redox status of cysteine and acetylation status of lysine residues within the HMGB1 peptide determine its function and localisation, respectively (Yang, Lundbäck, et al. 2012; Bonaldi et al. 2003). Redox isoforms are determined by the formation of disulphide bonds between three key cysteine residues: Cys23 and Cys45 located in Box A and Cys106 in Box B (Figure 1). Three isoforms of HMGB1 emerge from the redox status of these cysteine residues (Antoine et al. 2014): fr-HMGB1 (fully reduced/thiol HMGB1), ds-HMGB1 (disulphide HMGB1), and ox-HMGB1 (sulphonyl/oxidised HMGB1), in order of increasing oxidation (Figure 4).

The fr-HMGB1 is the main isoform released following cell necrosis and passive HMGB1 release, which forms complexes with CXCL12 and acts on CXCR4 receptors to recruit immune cells to the site of release (Venereau et al. 2012) whilst the ds-HMGB1 is a pro-inflammatory cytokine involved in immune cell activation that interacts with TLR-4 receptors (Schiraldi et al. 2012). The ox-HMGB1, on the other hand, demonstrates neither pro-inflammatory nor chemokine activity (Yang et al. 2013). Acetylation of lysine residues in the nuclear localisation sequence (NLS) of HMGB1 triggers cytoplasmic translocation, which is the first step towards extracellular secretion and subsequent inflammation. Thus, acetylation can distinguish between the two pathways of HMGB1 release: passive release during necrosis and active secretion, the latter necessitating cytoplasmic translocation; this knowledge could be utilised to identify which isoforms of HMGB1 are released via passive or active mechanisms (Bonaldi et al. 2003).

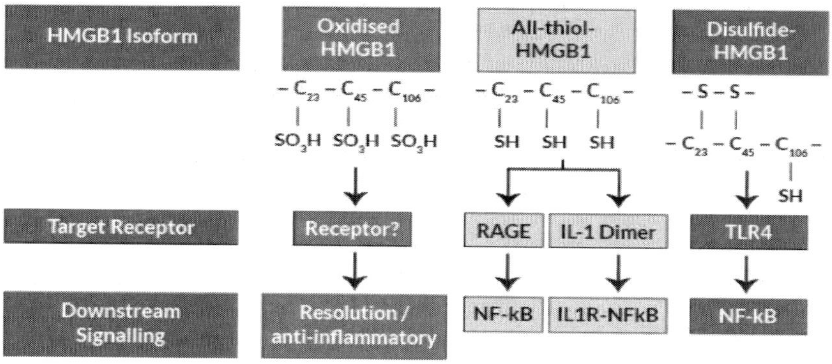

Figure 4. Isoforms of HMGB1. Redox status may determine the pro-inflammatory effects of HMGB1, with disulphide and thiol isomers being associated with pro-inflammatory responses and tissue damage. The role of oxidised HMGB1 in modulation of neuroinflammation at later stages post-TBI remains relatively unexplored. Adapted from Crews and Vetreno, 2016.

In a study which evaluated the effects of different HMGB1 redox isoforms on neuroinflammation, ds- and fr-HMGB1 were administered directly to the cisterna magna of adult rats, and the neuro-inflammatory response within the CNS was evaluated (Frank et al. 2016). Ds-HMGB1, but not fr-HMGB1, stimulated significant upregulation of pro-inflammatory cytokines' gene expressions, including TNF-α, IL-1ß, and IL6 hyper acutely (2 hours) and acutely (24 hours) post-injection. Furthermore, ds-HMGB1 potentiated the inflammatory response to intraperitoneal administration of lipopolysaccharide (LPS) 24 hours post-injection. This translated into altered behavioural performance, with reduced social exploration in mice treated with both LPS and ds-HMGB1 compared with HMGB1 alone. Furthermore, microglia isolated from whole brains did not demonstrate upregulation of pro-inflammatory cytokines including TNF-α, IL-1ß, and IL6. However, subsequent addition of LPS revealed that ds-HMGB1 'primed' microglia and resulted in amplified upregulation of pro-inflammatory gene expression compared to LPS alone. Ds-HMGB1 therefore may act as a switch to activate neuroinflammation following a traumatic, ischaemic, or toxic insult to the CNS.

Furthermore, Aucott et al. examined differences in neuro-inflammatory responses and BBB integrity following intracerebral injection of ds- or fr-HMGB1 in adult rats (Aucott et al. 2018). Injection of both ds- and fr-HMGB1 demonstrated increased BBB permeability radiologically, and via immune-fluorescence studies. There were also increased numbers of IL-1 ß expressing cells in animals treated with both isomers of HMGB1 vs controls. However, a differential effect of isoforms was observed with increased apoptosis on TUNEL assay with ds-HMGB1, but not fr-HMGB1. In keeping with these findings, both ds- and fr-HMGB1 isoforms resulted in significant upregulation of brain TNF-α, but to a greater extent with ds-HMGB1 (Lian et al. 2017). However, a receptor specific mechanism was not identified, given that upregulation was inhibited with the use of both TLR-4 and RAGE antagonists.

Ox-HMGB1 does not appear to have significant chemoattractant or neuroinflammatory effects; for a detailed review please see Lian et al. (2017). It may be that over time, HMGB1 interacts with reactive oxygen species released from dying cells to move towards the oxidised isomer, nullifying its neuroinflammatory properties (Tang, Billiar, and Lotze 2012). Conversely, in the days-weeks post injury, HMGB1 may undergo pre-release modification within microglia to become the oxidised form via an active process, which may then support anti-inflammatory activities at later time points post injury (He et al. 2018; Tang, Billiar, and Lotze 2012).

In summary, ds-HMGB1 and fr-HMGB1 appear to be the strongest candidates as regulators of the pro-inflammatory and chemoattractant properties of HMGB1 after brain injury. Since the isoforms may differ in their affinity to target receptors (He et al. 2018), the differential roles of TLR-4 and RAGE in mediating post-injury responses after exposure to each HMGB1 isoform should be evaluated. Further questions about redox status of HMGB1 include where and how the isoforms are formed, (intracellular vs extracellular; acute vs subacutely post injury) and whether this is a passive process (oxidation of HMGB1 by reactive oxygen species) or actively performed enzymatically. This has translatable consequence, as the latter, for instance, may be amenable to controlled, pharmacological modulation, with a wider therapeutic window than previously trialled medications for TBI (Richard et al. 2017). Finally, the development of a method for the detection of each HMGB1 isoform in clinical specimens is necessary to design experiments to answer these important questions about the isoforms of HMGB1 (Tang, Billiar, and Lotze 2012).

HMGB1 in Grey Matter Injury and Repair

Whilst much animal and human data supports the hypothesis that HMGB1 is released after brain trauma, and stimulates neuroinflammation biochemically, there is also mounting evidence that HMGB1-signalling and its downstream effects of neuroinflammation modulate cellular injury after neurotrauma. These effects may include increasing cell death, altering tissue integrity, and possibly, promoting cellular recovery or replacement at later stages post-injury.

HMGB1 on Neuronal Loss and Glial Proliferation

Models and post-mortem analyses confirm that TBI leads to cell loss and neuronal death at the site of injury, with reduced numbers of cells identified in both *in vitro* and *in vivo* models of TBI (Ved et al. 2021; Taib et al. 2017). The mechanisms of neuronal death include direct mechanical damage, oxidative stress, and likely, secondary death due to neuroinflammatory responses (Jia et al. 2021). Recent *in vitro* experimental data from our laboratory in human tissue corroborates these findings, with marked neuronal death via propidium iodide staining in human cortical cell cultures after weight drop injury (unpublished data). However, some work has identified that HMGB1 may be neuroprotective in certain contexts, such as prolonged ischaemia (Cohen et al. 2009; Fink 2007; Richard et al. 2017). It is possible that the role of HMGB1 upon neuronal injury may switch as time passes from the time of injury. Since this hypothesis may impact upon the utility of HMGB1-signalling modulation as a therapeutic target in TBI, it requires experimental validation.

Proliferation of glial cells is consistently identified in the regions adjacent to traumatic injury (Jia et al. 2021; Mattugini et al. 2018). Whilst the loss of neurons at the site of injury certainly contributes to neurological deficits after TBI, the proliferation of glial cells is potentially more complex. Such astrogliosis has traditionally been thought of as detrimental to recovery, with proliferation of astrocytes and microglia both responding to HMGB1 and other mediators to perpetuate neuroinflammation, connective tissue growth factor expression and tissue oedema (Jia et al. 2021). Furthermore, the presence of a persistent glial scar may impair the ability of surviving neurones to grow and re-form connections with each other post injury. There is evidence that the neuroinflammatory responses mediated via HMGB1 may also stimulate persistent proliferation of glial cells and non-neuronal cells of the neurovascular unit. This includes stimulation pericyte and endothelial cell proliferation (Richard et al. 2017). Furthermore, neuroinflammation may also lead to concomitant impairment of proliferating cell types maturing into functional CNS components, creating a 'block' in the pathway to potential recovery of neurons and axons post injury, which is discussed in later subsections (see *HMGB1 in White Matter Injury and Repair*).

HMGB1 Effect on Brain Oedema and Loss of BBB Function

Cerebral oedema is a key factor causing neurological damage after TBI. HMGB1 signalling via TLR4 was demonstrated to exacerbate brain swelling on magnetic resonance imaging, and increase expression of aquaporin 4, a water channel which effectively increases the permeability of the BBB to water and thus worsens brain swelling after trauma (Laird et al. 2014). HMGB1 signalling also upregulates production of other molecules which may mediate BBB integrity, such as matrix metalloproteases, (MMPs) and through a potential interaction with glutamatergic excitotoxicity, both of which lead to worsening cerebral oedema (Parker et al. 2017; Richard et al. 2017). Furthermore, models wherein antagonists of HMGB1 signalling, such as anti-HMGB1 antibodies, are administered to animals, BBB integrity improves, and aquaporin-4 expression is reduced (Richard et al. 2017; Stein et al. 2004). These models provide further evidence for the impact of HMGB1 signalling on BBB integrity through testing of HMGB1 antagonists, which are discussed in greater detail below (see *Key Potential Drugs*).

HMGB1 Effects on Neurogenesis (in Enhancing Repair or Neuroprotection)

Neurogenesis has been recognised as an important response to brain injury; this is supported by the existence of neural stem progenitor cells (NSPCs) at the site of cortical injury within *in vivo* animal models of TBI (Manivannan et al. 2020). Whilst these cells have the potential to develop into functional neurons and glia, it appears that after injury, these proliferating NSPCs do not mature into functional neurons, but instead fail to survive or remain fixed in an immature, non-functional state (Yi et al. 2013). Identifying the mechanisms that stimulate neurogenesis, and that prevent its completion to develop a pool of newly functional neurones, may highlight ways to optimise the therapeutic strategies of exogenous stem cell transplantation and endogenous neural regeneration to treat TBI patients.

Animal studies have identified that NSPCs proliferate at the site of injury rapidly after injury and for up to 2 weeks thereafter, consisting of a mixture of local cells and migration of progenitors from areas rich in NSPCs, such as the subventricular zone (Braun, Schafer, and Hollt 2002; Dash, Mach, and Moore 2001). When extracted and cultured *in vitro*, these cells can develop into mature neurons (Ahmed et al. 2012). Unfortunately, these cells fail to do so *in vivo*, likely due to a hostile, neuroinflammatory micro-environment, demonstrated by chronic microglial activation and cytokine expression, which are unlikely to support endogenous neurogenesis [33, 36]. The concept of a detrimental effect of neuroinflammation on neurogenesis has been widely studied across other neurological diseases too, such as stroke, epilepsy, and Alzheimer's disease [38, 39]. Whilst this relationship remains relatively unexplored in the context of post-traumatic neurogenesis, pharmacological abolition of HMGB1 signalling (see *Key Potential Drugs*) has been shown to improve local neurogenesis after neurotrauma *in vitro* (Manivannan et al. 2020). Confirming the role of HMGB1 signalling in neurogenesis post-TBI *in vivo,* and in human tissue, are key next steps in this field. Stem cell transplantation remains a promising therapeutic strategy to improve outcomes after TBI, however, manipulating the post-injury microenvironment to support neurogenesis may be necessary to prevent NSPC graft failure (Dash, Mach, and Moore 2001; Yi et al. 2013). Such strategies could target post-traumatic neuroinflammation, for which HMGB1-signalling is a key player.

In summary, HMGB1 may provide a therapeutic target for minimising cellular injury, and optimising neurogenesis after TBI. Several factors including the type of injury, HMGB1 isoform, time since injury, and interaction with other cells must be taken into consideration when deciding which strategies to use to modulate neuroinflammation to optimise neurogenesis post-TBI. Furthermore, most studies analyse the impacts of HMGB1 in the acute phase post injury but there is a need to design experiments to assess the impacts of HMGB1 signalling upon grey matter injury and repair in the days-weeks post-injury. If the role of HMGB1 transitions to a reparative form with time, the timing of potential intervention must be carefully planned. Alternatively, identification of the isomers of HMGB1 active at different time points post injury may provide an alternative paradigm for pharmacological, isomer-specific HMGB1-modulation to improve clinical outcomes after neurotrauma.

HMGB1 in White Matter Injury and Repair

White matter (WM) tracts span neuroanatomically disparate regions of the brain, facilitating connectivity and the formation of large-scale networks. They are composed of axons, enveloped in myelin sheaths provided by oligodendrocytes (OL), astrocytes in close contact at nodes of Ranvier, and blood vessels. Oligodendrocyte precursor cells (OPCs) comprise a sub-population of glia, scattered throughout the brain, which have the potential to differentiate into myelinating OLs in the presence of relevant environmental cues. They thus represent a pool of cells with the potential to support remyelination of damaged axons (Paudel et al. 2020).

The high metabolic rate of WM means that it is particularly vulnerable to the detrimental effects of neuro-inflammation, excitotoxicity, and oxidative stress that follow TBI, and there is an emerging association between immune activation and chronic white matter injury after TBI (Braun et al. 2017). The differential movement of brain tissue during acceleration-deceleration results in widespread, diffuse axonal tearing at grey-white matter junctions (Evans 2006). Furthermore, there is evidence from animal TBI models (Paudel et al. 2020) and neuroimaging experiments in humans (Narayana 2017) that white matter tract degeneration is a progressive process, with objective degradation of axonal tract integrity in the months-years following TBI.

Loss of white matter integrity is strongly associated with long term sensorimotor and neuropsychiatric dysfunction after TBI (Alhilali et al. 2015; Bramlett and Dietrich 2015). Therapies that limit white matter damage/enhance remyelination may thus have the potential to improve long-term clinical outcomes. This provides a larger potential therapeutic window, within which interventions might prevent or even reverse white matter tract injury post-TBI. However, for such a clinical strategy to be pursued, a deeper understanding of the mechanisms of how the initial traumatic event and the manifestation of white matter loss is required (Braun et al. 2017). HMGB1 is released from injured neurones post-TBI (Terrando et al. 2016) and there is mounting evidence that it plays a role in mediating WM responses to trauma.

HMGB1 on Axonal Demyelination/Axotomy and Oligodendrocyte Loss

From a pathophysiological perspective, WM injury from TBI is referred to as traumatic axonal injury (TAI). Long axonal projections that traverse the brain are damaged from forces of torsion in impact-acceleration injuries to the head. Experimental models indicate that TAI does not necessitate death of the corresponding neuron cell body, which improves the chances of subsequent remyelination (Armstrong et al. 2016; Wang et al. 2013). In more severe injuries, TAI occurs in a more widespread pattern, defined as diffuse axonal injury (DAI).

Most of the axon damage in TBI is not due to primary injury, i.e., physical tearing of axons (Wang, Fox, and Povlishock 2013). The axon response to initial mechanical force follows the process of Wallerian degeneration, wherein myelin sheaths collapse and distal axon fragments degenerate, followed by an attempt to remyelinate/re-grow axons if the environment is permissive to do so. This response to axon injury leads to diverse forms of damage, including neuroinflammatory processes, which thus contribute to the mainstay of progressive, axonal injury. If the injury is severe, the presence of myelin debris and release of neuroinflammatory mediators such as HMGB1 has been shown to inhibit the already slow process of Wallerian axonal re-growth, re-myelination, and ultimately the repair of WM pathways (Kotter et al. 2006). This implies the primary injury acts as an 'initiation phase' and the secondary injury, driven by neuroinflammatory mediators, represents a slower but prolonged 'execution phase', providing a therapeutic window within which axonal injury could be mitigated (Armstrong et al. 2016).

The proliferation of these cells and others (astrocytes) after neurotrauma is known to culminate in the development of a 'glial scar' at the site of injury (Adams and Gallo 2018). The compact border of these scars is comprised of proliferated cell populations of astrocytes, microglia and OPCs. The core of these lesions contains a variety of extracellular matrix proteins and cell types, such as further neuroglial cells, macrophages, fibroblasts, and other non-neuronal cells. The glial scar has traditionally been perceived as a barrier to CNS repair. It has been shown that OPCs and microglia proliferate in response to neurotrauma (Ved et al. 2021). However, whilst OPCs can differentiate into myelinating oligodendrocytes, in the glial scar these cells largely fail to mature in this usual manner (Adams and Gallo 2018). Instead, they either remain fixed in their immature state, or are stimulated to differentiate down astrocytic lineage by trauma-related environmental cues, which may include HMGB1 signalling mechanisms (Braun et al. 2017; Foresti, Arisi, and Shapiro 2011; Jia et al. 2021; Xiong, Liu, and Yang 2016). The persistent presence of these immature cell types can prevent axonal repair and remyelination through lack of replenishment of oligodendrocytes, and through active inhibition of repair from neuro and myelinotoxic inflammatory factors secreted by astrocytes and microglia (Adams and Gallo 2018; Paudel et al. 2020; Paudel et al. 2018).

However, there is evidence that complete abolition of glial scar formation does not improve CNS recovery post injury in models of spinal cord injury; removal of glial scar development prevents axonal regrowth in areas adjacent to the site of traumatic injury (Fitch and Silver 2008)]. One explanation for these disparate findings may be that the astrocytes, OPCs and microglial that comprise the boarder of the glial scar, whilst stuck in immature/partially differentiated states, may secrete reparative/anti-inflammatory signals into the peri-lesioned tissue, which support axonal regrowth, repair and remyelination around the lesion. Furthermore, ablation of all astrocytes within the glial scar led to neuronal death and demyelination around the site of CNS injury, which demonstrated increased inflammatory cell infiltration and oedema compared to tissue in which a natural glial scar had formed (Wanner et al. 2013). It is feasible that the glial scar represents a compromised scenario, whereby whilst immature cells are blocked from fully developing into remyelinating or possibly even into neuronal cell types, a fine balance between pro and anti-inflammatory signalling is then modulated by the cells of the scar. This may then facilitate attempts to regenerate CNS tissue in regions adjacent to the site of injury. Studies to better understand of the interplay between these glial scar components, and the substances they secrete, such as HMGB1, will need to consider temporal changes in the behaviour of these cell types as time passes from the initial injury.

There is evolving evidence that HMGB1 and neuroinflammation have roles in initiating and propagating TAI and DAI. Animal models of TBI demonstrate reduced corpus callosum thickness and axon diameter in the weeks following head injury (Sullivan et al. 2013). This process appears to be perturbed in more severe models and cases of neurotrauma (Armstrong et al. 2016). Myelin content within injured WM tracts is markedly reduced in the weight drop animal model of TBI, which also correlated with sensorimotor and cognitive performance for months after the initial injury (Taib et al. 2017). Furthermore, in the days-weeks following injury, WM tracts have been shown to become infiltrated with inflammatory cells, along with higher expression of inflammatory mediators such as HMGB1, IL-1, and IL-6 (Braun et al. 2017; Taib et al. 2017). These factors might affect the local environment of the brain after CNS injury and exacerbate brain damage after TBI (Paudel et al. 2020). An indirect mechanism has been proposed to explain the neuroinflammatory axon and myelin toxicity, wherein HMGB1 and other damage associated substances released from damaged neurones and axons (myelin, ATP) activate and polarize local microglia towards the M1 pro-inflammatory phenotype (Braun et al. 2017). These microglia may then polarise T cells to target WM proteins, culminating in a deleterious, maladaptive immune response against WM, leading to progressive, delayed WM degeneration (Armstrong et al. 2016; Braun et al. 2017). Support for this theory comes from the identification of microglial and macrophage infiltrates within injured brain tissue, induction of OL death in media conditioned by polarised microglia, and improved WM integrity, with improved sensorimotor performance following brain injury in HMGB1 receptor knock-out models (Paudel et al. 2018).

Neuroimaging studies have identified correlations with WM disruption on diffusion tensor MR imaging following neurotrauma in humans and the severity of TBI (Armstrong et al. 2016). In one study, over half of TBI patients demonstrated reduced WM connectivity, which correlated with poorer neurocognitive performance following injury; this reduction of WM integrity progressed with time following injury, doubling in severity between the 1st and 6th month post injury (Jolly et al. 2021). With the ongoing development of high-resolution diffusion weighted MR imaging methodologies, the scope for neuroimaging studies to assess WM integrity in TBI patients is a promising experimental paradigm to further our understanding if WM injury and regeneration post head injury (Wang, Fox, and Povlishock 2013).

In summary, whilst the initial loss of axon integrity may relate to the shearing of axons at the point of head injury, there appears to be an ongoing, myelin-degenerative process that occurs after head injury, which may be at least partly be driven by neuroinflammatory mechanisms.

HMGB1 and Oligodendrocytes and Post-Traumatic Oligodendrogenesis

WM injury that causes TAI will have significant loss of myelin due to loss of axons. However, secondary damage described above may not only impair re-growth and re-myelination of damaged axons, but also cause OL loss, with subsequent demyelination of previously unaffected, viable axons (Armstrong et al. 2016). This oligodendroglial-toxic effect of neurotrauma has led to remyelination being of great interest for promoting repair in WM tracts via rescue of OLs and replenishment of their cell population.

The numbers of myelinating OLs have been shown to be fall due to cell death following fluid percussion and weight-drop injury animal models of TAI (Lotocki et al. 2011). Human post-mortem tissue analysis has also identified neuroinflammatory mechanisms being detrimental to OL number and WM myelin content in human tissue (Johnson et al. 2013). One mechanism underlying these findings might be that OLs have a vulnerability to oxidative stress, and HMGB1 has been shown to induce reactive oxygen species generation via TLR signalling in microglia (Braun et al. 2017; Wang et al. 2013). However, the delayed, degenerative process underlying progressive loss of OLs, and hence reduced WM integrity, may be more influenced by the chronic neuroinflammatory state, also induced via the trauma related HMGB1 release (Armstrong et al. 2016).

The cellular context and temporal pattern of HMGB1-mediated neuroinflammation is also key to unpick, with some evidence that at later stages, and in the context of *in vitro* models of local tissue injury, HMGB1 may then support OL survival (Choi et al. 2017).

Evidence exists for remyelination after injury via OPC recruitment and differentiation into oligodendrocytes. Neural/Glial antigen 2 (NG2) is considered a marker of OPCs in both the developing and mature CNS (Ved et al. 2021). Using an *in vitro* needle scratch injury model, we demonstrated that HMGB1 mediated an anti-proliferative effect on $NG2^+$ cells in rodent cortical neural stem/ progenitor cell cultures via a TLR2/4 receptor dependent mechanism (Ved et al. 2021). This suggests that HMGB1 may impede WM repair by inhibiting the OPC proliferation, but further studies are required to identify the implications for remyelination and neurological recovery in humans. However, is seems that higher exposure to HMGB1 at the time of injury may affect the ability of OPCs to proliferate and therefore impair the potential for later remyelination of injured axons (Kraus et al. 2007). Further analysis of OPC proliferation at further time points post injury may help to identify whether the rate of proliferation of OPCs changes following post-traumatic HMGB1 release (Ved et al. 2021).

Given that pre-existing OL which survive the initial insult may play an important role in remyelination in humans, the effects of HMGB1 on fully differentiated OLs in the later stages post-injury should be clarified. The effects could be via direct mechanisms since there is evidence of TLR2 and RAGE expression by human oligodendrocytes (Taib et al. 2017). Indirect mechanisms should also be explored, as HMGB1 interaction with immune cells, such as microglia as described previously, may influence OL maturation and survival. Overall, animal models of TBI to date suggest that OL survival is reduced following neurotrauma, and when combined with modern neuroimaging data, it seems that more severe head injuries lead to perturbed OL survival.

However, the actions of HMGB1 and other neuroinflammatory mediators are likely related to both concentration, cellular context, and temporal release. This is demonstrated by its promotion of NSPC migration at lower concentrations, but toxic effects at higher concentrations post-injury (Armstrong et al. 2016; Ved et al. 2021). Total inhibition of HMGB1 could therefore negatively impact the prospects of repair. These factors may explain why, in some pre-clinical studies, HMGB1 appears to be supportive of OPC and OL survival (Braun et al. 2017; Choi et al. 2017). Characterising the temporal changes in HMGB1 levels, isoforms and receptor expression may hold the key to unlocking how this molecule and its signalling pathways modulate the responses of oligodendroglial cells, and hence, myelin injury, following head trauma.

HMGB1 Effect on Neurite Growth and the Neurovascular Unit

HMGB1 is established as a mediator of neurite and axonal growth during development of the CNS (Hori et al. 1995). It has emerged in more recent work that it may also mediate neurite/axon repair after degenerative brain injuries (Paudel et al. 2020). However, its effects in these contexts appear to be disparate; during development, active secretion of HMGB1 appears to support neurite growth. By contrast, extracellular HMGB1 released from necrotic cells, acting through a TLR4 dependant pathway, seems to destabilise cytoskeletal components within neurites, an effect which is reversed upon inhibition of HMGB1 signalling (Fujita et al. 2016). This is further evidence that the context in which HMGB1 is generated in the extracellular milieu is a key factor in its overarching effect on cellular processes.

OPCs, as discussed above, are stimulated to proliferate, and migrate to the site of injury after TBI, providing a potential pool of progenitors which may aid in providing myelination of axons which lost their vulnerable OLs due to the injury. However, OPC proliferation may also progress to a detrimental effect, as $NG2^+$ OPCs have been observed to make up a significant proportion of the cells within the 'glial scar', a phenomenon where glial cells proliferate out-of-proportion to their need, acting as a physical blockade for axon re-growth and repair in the vicinity of the injury (Cernak 2005). These $NG2^+$ cells within the glial scar fail to mature into myelinating cell phenotypes, instead remaining $NG2^+$, or evolving to express GFAP and thus developing towards an astrocytic lineage instead (Adams and Gallo 2018). This complicates the interpretation of the impact that HMGB1 has upon OPC after trauma *in vitro*, where it has been shown to modulate NG2+ cellular proliferation (Ved et al. 2021). Is this effect detrimental, insofar as it reduces the number of available OPCs to provide remyelination of axons, or is HMGB1 acting as a handbrake to reduce the potential for over-proliferation and formation of glial scarring? Analysis of HMGB1's impact upon OPCs at later stages post TBI is necessary to answer this question.

The neurovascular unit (NVU) is composed of neurones, glia, endothelial cells, and smooth muscle cells (Dinet, Petry, and Badaut 2019). The interplay between these cells contributes to the blood-brain-barrier (BBB). The impacts that HMGB1 has upon the NVU may also be critical to understanding TBI. HMGB1 has also been shown to increase BBB permeability in response to injury through activation of immune cells to increase levels of MMPs, and reducing expression of proteins that make up tight junctions (Evran et al. 2020) which leads to worsening cerebral oedema (Richard et al. 2017). Furthermore, it may also induce molecular changes to the BBB over time post TBI, in a TLR 4 dependant manner, worsening brain swelling and impairing reparative processes (Venereau et al. 2016). For instance, CNS endothelial cells are stimulated to proliferate upon exposure to HMGB1, which may then contribute to the glial scar that impairs WM recovery (Dinet, Petry, and Badaut 2019). HMGB1 released by mechanically damaged or necrotic cells is indeed capable of binding RAGE and elicit migratory responses in smooth muscle cells towards the injured tissue (Evran et al. 2020) further impeding re-growth of damaged neurones.

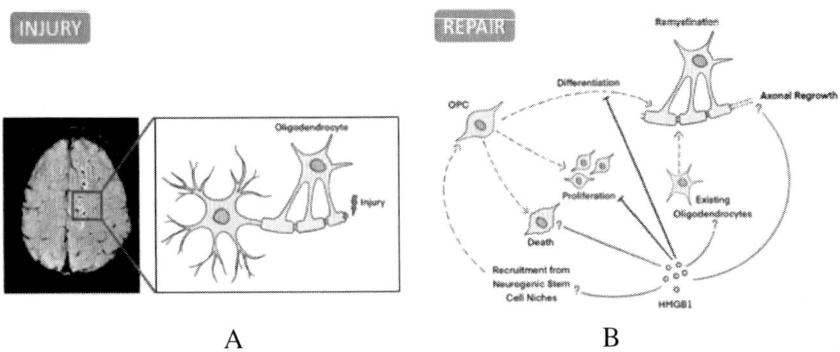

Figure 5. Schematic presentation outlining the response of oligodendrocytes and their progenitors to injury. (A) depicts diffuse axonal injury on susceptibility weighted MRI (left) and injury to the neurone-oligodendrocyte unit (right). (B) summarizes the potential actions HMGB1 may have upon OPCs and mature OLs following neurotrauma. Adapted from Ved et al., 2022.

Summary

Understanding the secondary processes associated with neuroinflammation and focal axon degeneration may provide opportunities for intervention prior to axon fragmentation and disconnection. HMGB1 appears to have marked impacts upon WM injury and potential repair after TBI (Figure 5), and modulation of its signalling pathways may represent an important mechanism for WM protection and functional recovery.

Key Potential Drugs

HMGB1 has emerged as one of the main mediators of neuroinflammation, and thus modulation of its signalling pathways has been investigated. The agents identified thus far can be roughly categorised as: those that inhibit HMGB1 release, neutralising antibodies, and its receptor antagonists.

Inhibition of HMGB1 Release

HMGB1 is primarily a nuclear protein in health, but multiple studies have identified that after in vitro and in vivo modelling of TBI, HMGB1 is mobilised to the cytoplasm, (Paudel et al. 2020; Ved et al. 2021) and plasma and CSF levels of HMGB1 are significantly raised in the hours following TBI (Laird et al. 2014). Furthermore, elevated extracellular HMGB1 appears to correlate with poorer clinical outcomes (Parker et al. 2017). Several existing drugs compounds have been shown to reduce HMGB1 translocation from the nucleus to the cytoplasm and/or extracellular environment in models of septic shock and liver injury, such as metformin, aspirin, and resveratrol (Venereau et al. 2016). However, these are yet to be tested in models of TBI, and the mechanism of HMGB1 release in TBI is likely different to that of septic shock (passive release from injured neurones vs active secretion by immune cells) which may impact on the therapeutic window in which HMGB1 release could pragmatically be modulated in clinical practise after TBI.

Ethyl pyruvate, a compound derived from the endogenous metabolite pyruvic acid, is a safe, established modulator of HMGB1 release acutely post brain trauma. It prevents release of HMGB1 into the extracellular space (Andersson, Yang, and Harris 2018) and reduces pro-inflammatory IL1β and

IL6 expression after cortical impact modelling of TBI, and rodent motor function was significantly improved in head-injured animals treated with ethyl pyruvate compared to placebo treated animals (Su et al. 2011). The acute, parenteral administration of ethyl pyruvate studies may have allowed it to act during the first therapeutic window of HMGB1 release post injury. However, as with many studies, the effects of the treatment were only assessed at short term time-points post-TBI (hours-days). It remains to be determined whether HMGB1 release modulation post-TBI leads to long term improved clinical outcomes.

Global HMGB1- knock-out is lethal at birth due to the roles which HMGB1 plays in neurodevelopment (Paudel et al. 2018). However, inducible HMGB1 knock-out animals are viable when induced in adulthood, albeit with poorer motor and cognitive functioning compared to wild-type animals. These inducible knock-out animals exhibited smaller lesion volumes after in vivo modelling of TBI, but they do not have superior subacute sensorimotor or cognitive performance one-month post-injury compared to sham injured, equivalent genotype animals (Aneja et al. 2019). These disparate findings in the acute and subacute phases of injury led the authors to hypothesise that HMGB1 signalling may induce local tissue damage in the acute stages post injury, but complete loss of its signalling does not impart phenotypic benefits in their animal model of TBI. This might be because HMGB1 related signals have a minimal role in longer-term brain repair post-TBI. However, it may instead be that the benefits of the acute reduction in local lesion volume seen in animals treated with HMGB1 signalling blockers were offset by subsequent loss of the benefits that HMGB1-signalling may have in supporting neuro-regeneration at later (subacute/chronic) time points post injury.

Since HMGB1 may have differing, potentially reparative effects at later time-points post-injury, another strategy requiring investigation is sequestration of released HMGB1 using small molecule binders of HMGB1, or neutralising antibodies in the acute stages after head injury (Tsung et al. 2005).

Neutralising Antibodies and Small Molecule Inhibitors of HMGB1

Monoclonal antibodies (mABs) bind to HMGB1 in the extracellular milieu, preventing its interaction with HMGB1 receptors (Venereau et al. 2016). Neutralising anti-HMGB1 antibodies have reduced systemic neuro-inflammation in models of sepsis, ischaemic injuries and neuropathologies

(Tsung et al. 2005; Venereau et al. 2016). Administration of HMGB1 mABs after *in vivo* models of TBI reduced extracellular levels of HMGB1, attenuated neuronal death, reduced cerebral oedema through protection of BBB integrity and reduced pro-inflammatory cytokine expression, and treated animals also had improved motor outcomes; these benefits persisted into the subacute phase (days-weeks) post injury (Okuma et al. 2019; Paudel et al. 2020; Wang et al. 2017). HMGB1 mAbs have also been shown to reduce proliferation of the neurovascular unit post-injury, potentially reducing the burden of glial scar formation (Chaichana et al. 2010). However, whilst these data have been promising, they are limited to rodent studies thus far, and studies in higher mammals and human tissue are required to ensure adequate BBB penetration and safety of mAbs. This is of particular importance given the history of rare but dangerous side effects in previous human trials of mAbs for other diseases (Hansel et al. 2010). Furthermore, the long-term impacts of acute mAb treatment remains to be established (Okuma et al. 2019).

Box A is a fragment of the HMGB1 molecule, which can bind to its various receptors without activating their associated intracellular cascades (Paudel et al. 2020). Administration of Box A in mouse models of TBI reduced cerebral oedema and improved rodents' neurological outcomes (Parker et al. 2017). Box A has also been shown to ameliorate the anti-proliferative effects of HMGB1 upon glial progenitor cells, which may impact upon the brain's attempts at recover after TBI (Ved et al. 2021). However, there is uncertainty as to the mechanism via which BoxA imparts its anti-HMGB1 signalling effects, and its safety profile in humans remains untested; a better understanding of its chemical interactions with HMGB1 and its receptors is required before it can be considered for clinical trial use (Venereau et al. 2016).

HMGB1 Receptor Antagonists

HMGB1 receptors, (RAGE, TLR4) activate a common signalling pathway that culminates in the activation of NF-κB transcription factors (Andersson, Yang, and Harris 2018; Su et al. 2011). NF-κB activation upregulates gene expression of pro-inflammatory mediators, such as interleukin-1β, tumour necrosis factor-α, and IL-6, leading to tissue damage (Paudel et al. 2018). Therefore, modulating the activity of these receptor pathways may represent a strategy with a pragmatic therapeutic window within which to act to suppress pathological neuroinflammation post-TBI.

RAGE

HMGB1 has been shown to mediate neuronal damage through interactions with RAGE [33]. Data obtained from western blot analysis in an experimental study reported the release profile of HMGB1 and RAGE after TBI, where HMGB1 was released as soon as 30 min after TBI and persists at elevated levels up to 6 h after the injury. However, the expression level of RAGE was elevated later, 6 h after TBI and reached its peak after 24 hours, (Gao et al. 2011) providing a wider therapeutic window within which to test RAGE signalling modulation. RAGE signalling blockade may ameliorate both local neurotoxicity after TBI, and later, systemic complications of head injury, such as neurogenic pulmonary oedema (Weber et al. 2014).

Glycyrrhizin, a compound found in liquorice with a safe side effect profile, binds directly to HMGB1, inhibiting its ability to bind and activate its target receptors such as RAGE (Aneja et al. 2019; Paudel et al. 2020; Paudel et al. 2018). Glycyrrhizin administration has been shown to be neuroprotective in an *in vitro* model of TBI, (Manivannan et al. 2020) and glycyrrhizin treatment to rodents after in vivo TBI ameliorated brain oedema and improved post-injury beam walking distance. It also reduced the up-regulation of inflammatory cytokines identified in the placebo brain injured animals in this study (TNF-α, IL1β, IL-6) (Gu et al. 2014). Furthermore, RAGE knockout mice subjected to TBI do not experience the protective effects from glycyrrhizin following TBI (Okuma et al. 2014).

Early glycyrrhizin treatment also conferred neuroprotection in another study by decreasing brain oedema and reduced the impairment of spatial and motor learning in the days post-injury (Webster et al. 2019). However, in this model, glycyrrhizin did not improve post-traumatic epilepsy in the treated young animals, which may therefore be influenced by alternative pathways such as TLR4, (Paudel et al. 2018) and/or later waves of HMGB1 release. Furthermore, glycyrrhizin had to be administered up to one hour prior to injury in these experiments to have these effects; its protective effects were lost when it was administered post-injury only, potentially limiting its utility as a therapeutic option post-TBI. This is perhaps expected, as HMGB1 is initially released hyper-acutely post-injury, (1-6 hours) (Paudel et al. 2018; Ved et al. 2021) and administration of a HMGB1-binding agent post-injury may thus miss this earliest therapeutic window.

There is a need to investigate the impact of the modulation of downstream HMGB1 receptors directly, for which there may be a longer, more pragmatic, therapeutic window within which to act; these may include TLR4 receptor antagonists (see below) and/or direct RAGE receptor antagonists such as FPS-ZM1, which is in use in pre-clinical and clinical trials for other neurodegenerative diseases (Deane et al. 2012). Furthermore, in studies of spinal cord injury, early RAGE inhibition via injection of anti-RAGE mABs lead to reduced neuronal survival and poor sensorimotor function (Wang et al. 2017). This not only highlights the likely disparate mechanisms of cellular injury in the brain and spinal cord, but also that in certain circumstances, pure inhibition of RAGE signalling may be detrimental to recovery after neurotrauma.

TLR4

TLRs are a family of receptors expressed on cells of the innate immune system that allow for the recognition of invariant molecular motifs of pathogens (Paudel et al. 2020). TLR4 has been linked with TBI, where TLR4 mediates glial phagocytic activity and inflammatory cytokines production (Paudel et al. 2018). Activation of TLR4 has been shown to lead to a cascade of events culminating in marked pro-inflammatory cytokine secretion and worsening cerebral oedema, via higher expression of aquaporin-4 by astrocytes (Richard et al. 2017). *In vivo* studies identified that TLR4 activation leads to release of pro-inflammatory cytokines such as IL1β, IL6 and TNF-α, and subsequent neuronal and glial injury (Hergenroeder et al. 2010; Laird et al. 2014; Ved et al. 2021).

Rats treated with TLR4 modulators post-TBI, such as the established drug simvastatin, had modulated pro-inflammatory cytokine expression, less brain swelling, and improved motor function (Chen et al. 2009; Paudel et al. 2020). Laird et al. used controlled cortical impact *in vivo* model of TBI to demonstrate that both HMGB1 and TLR4 are upregulated after TBI in local microglial cells, with a spatial and temporal overlap with the development of cerebral oedema (Laird et al. 2014). TLR4 knock out models provide further evidence that targeting the HMGB1 receptor TLR4 is one mechanism by which we may be able to ameliorate the impacts of TBI at a cellular level (Jiang et al. 2018). TLR4 knockout mice demonstrate resistance to OL death following neurotoxic insults (Lehnardt et al. 2002). These findings are replicated in more recent works, which demonstrate reduced WM loss and higher myelin-associated proteins following injury (Braun et al. 2017). TLR4 knock-out mice demonstrated reduced cerebral oedema and dampened neuroinflammatory mediator release following TBI, (Ahmad et al. 2013) as did animals pretreated with VGX-1027, a pharmacological TLR4 receptor blocker. VGX-1027 also reduced expression of aquaporin-4 and subsequent brain oedema in brain-injured animals (Laird et al. 2014; Laird et al. 2010).

It is worth noting that HMGB1 knock-down animals still develop cerebral oedema post-TBI; post TBI cerebral oedema may thus be mediated via entirely alternative pathways, or via the TLR4-aquaporin-4 pathway outlined by Laird and colleagues but activated by alternative TLR4 ligands than HMGB1, such as lipopolysacharride and hyaluronan (Aneja et al. 2019; Wietzorrek et al. 2019). P5779 is another small molecule antagonist of TLR4 (Venereau et al. 2016), which has been shown to reduce inflammation in models of liver injury and improved survival in models of sepsis (Parker et al. 2017); its impacts after TBI remain to be studied. Furthermore, disparate reports of the impacts of TLR4 blockade exist (as with RAGE signalling); in spinal cord injury, TLR4 knockout or pharmacological blockade led to worse motor deficits and increased size of glial scarring in a model of spinal cord injury (Braun et al. 2017). TLR4 may thus exhibit protective effects on WM under certain conditions/anatomical locations.

In addition to potentially protective effects of TLR4, activation of TLR2 also was associated with reduced degeneration of central myelinated fibres in the spinal cord, whereas global TLR2−/− mice exhibited greater axonal injury after experimental spinal cord injury (Stirling et al. 2014). Thus, modulation of TLR4 and other TLRs may mediate neurotoxicity, demyelination, or endogenous WM repair, in a context and/or temporal dependant manner. Thus, care must be taken to avoid describing TLR4 signalling as wholly detrimental following neurotrauma; wholesale blockade of HMGB1, and TLR4/RAGE signalling, may not be a desirable strategy to improve patient outcomes after head injuries, particularly in the long term, where data on outcomes after HMGB1 signalling modulation remains sparse.

Conclusion

Although neuroinflammation is reduced by treatment with anti-HMGB1 therapies acutely, the timing, degree, location, and duration of HMGB1 blockade required to confer benefit on motor and behavioural outcomes is complex. In addition to confirming the safety and efficacy of the various HMGB1 modulation strategies outlined above in higher animals and/or humans, there is also a need to better understand the long-term impacts of HMGB1 modulation after TBI, as its effects may be detrimental at initial stages post injury, but beneficial later. The protective effects of acute HMGB1 inhibition have primarily been assessed at acute time points (Gu et al. 2014; Paudel et al. 2018) or do not appear to persist in the few studies assessing effects at chronic time points post injury (Webster et al. 2019). This lends further credence to the need to understand the downstream effects of HMGB1, and how these signalling pathways may be modulated to ameliorate the chronic neurodegeneration or impairment of long-term brain tissue repair following TBI.

Finally, HMGB1 redox state must also be considered (Paudel et al. 2020); current HMGB1 antagonists bind to all the three redox forms of HMGB1 (Andersson, Yang, and Harris 2018) and therefore, future studies should be focused on modulating signalling by specific isoforms of HMGB1 to fully assess the detrimental and reparative inflammatory roles each sub-form of HMGB1 may play following TBI. Notwithstanding these limitations, there is mounting evidence that HMGB1-signalling remains a promising pathway to guide towards the identification of therapeutic targets which may improve patient outcomes following TBI.

References

Adams, K. L., and V. Gallo. 2018. "The diversity and disparity of the glial scar." *Nat Neurosci* 21 (1):9-15. https://doi.org/10.1038/s41593-017-0033-9.

Ahmad, A., R. Crupi, M. Campolo, T. Genovese, E. Esposito, and S. Cuzzocrea. 2013. "Absence of TLR4 reduces neurovascular unit and secondary inflammatory process after traumatic brain injury in mice." *PLoS One* 8 (3):e57208. https://doi.org/10.1371/journal.pone.0057208.

Ahmed, A. I., A. B. Shtaya, M. J. Zaben, E. V. Owens, C. Kiecker, and W. P. Gray. 2012. "Endogenous GFAP-positive neural stem/progenitor cells in the postnatal mouse cortex are activated following traumatic brain injury." *J Neurotrauma* 29 (5):828-42. https://doi.org/10.1089/neu.2011.1923.

Alhilali, L. M., J. A. Delic, S. Gumus, and S. Fakhran. 2015. "Evaluation of White Matter Injury Patterns Underlying Neuropsychiatric Symptoms after Mild Traumatic Brain Injury." *Radiology* 277 (3):793-800. https://doi.org/10.1148/radiol.2015142974.

Andersson, U., H. Yang, and H. Harris. 2018. "Extracellular HMGB1 as a therapeutic target in inflammatory diseases." *Expert Opin Ther Targets* 22 (3):263-277. https://doi.org/10.1080/14728222.2018.1439924.

Aneja, R. K., A. M. Alcamo, J. Cummings, V. Vagni, K. Feldman, Q. Wang, C. E. Dixon, T. R. Billiar, and P. M. Kochanek. 2019. "Lack of Benefit on Brain Edema, Blood-Brain Barrier Permeability, or Cognitive Outcome in Global Inducible High Mobility

Group Box 1 Knockout Mice Despite Tissue Sparing after Experimental Traumatic Brain Injury." *J Neurotrauma* 36 (2):360-369. https://doi.org/10.1089/neu.2018.5664.

Antoine, D. J., H. E. Harris, U. Andersson, K. J. Tracey, and M. E. Bianchi. 2014. "A systematic nomenclature for the redox states of high mobility group box (HMGB) proteins." *Mol Med* 20:135-7. https://doi.org/10.2119/molmed.2014.00022.

Armstrong, R. C., A. J. Mierzwa, C. M. Marion, and G. M. Sullivan. 2016. "White matter involvement after TBI: Clues to axon and myelin repair capacity." *Exp Neurol* 275 Pt 3:328-333. https://doi.org/10.1016/j.expneurol.2015.02.011.

Au, A. K., R. K. Aneja, M. J. Bell, H. Bayir, K. Feldman, P. D. Adelson, E. L. Fink, P. M. Kochanek, and R. S. Clark. 2012. "Cerebrospinal fluid levels of high-mobility group box 1 and cytochrome C predict outcome after pediatric traumatic brain injury." *J Neurotrauma* 29 (11):2013-21. https://doi.org/10.1089/neu.2011.2171.

Aucott, H., J. Lundberg, H. Salo, L. Klevenvall, P. Damberg, L. Ottosson, U. Andersson, S. Holmin, and H. Erlandsson Harris. 2018. "Neuroinflammation in Response to Intracerebral Injections of Different HMGB1 Redox Isoforms." *J Innate Immun* 10 (3):215-227. https://doi.org/10.1159/000487056.

Bianchi, M. E., and A. A. Manfredi. 2007. "High-mobility group box 1 (HMGB1) protein at the crossroads between innate and adaptive immunity." *Immunol Rev* 220:35-46. https://doi.org/10.1111/j.1600-065X.2007.00574.x.

Bohmer, A. E., J. P. Oses, A. P. Schmidt, C. S. Peron, C. L. Krebs, P. P. Oppitz, T. T. D'Avila, D. O. Souza, L. V. Portela, and M. A. Stefani. 2011. "Neuron-specific enolase, S100B, and glial fibrillary acidic protein levels as outcome predictors in patients with severe traumatic brain injury." *Neurosurgery* 68 (6):1624-30; discussion 1630-1. https://doi.org/10.1227/NEU.0b013e318214a81f.

Bonaldi, T., F. Talamo, P. Scaffidi, D. Ferrera, A. Porto, A. Bachi, A. Rubartelli, A. Agresti, and M. E. Bianchi. 2003. "Monocytic cells hyperacetylate chromatin protein HMGB1 to redirect it towards secretion." *EMBO J* 22 (20):5551-60. https://doi.org/10.1093/emboj/cdg516.

Bramlett, H. M., and W. D. Dietrich. 2015. "Long-Term Consequences of Traumatic Brain Injury: Current Status of Potential Mechanisms of Injury and Neurological Outcomes." *J Neurotrauma* 32 (23):1834-48. https://doi.org/10.1089/neu.2014.3352.

Braun, H., K. Schafer, and V. Hollt. 2002. "BetaIII tubulin-expressing neurons reveal enhanced neurogenesis in hippocampal and cortical structures after a contusion trauma in rats." *J Neurotrauma* 19 (8):975-83. https://doi.org/10.1089/089771502320317122.

Braun, M., K. Vaibhav, N. M. Saad, S. Fatima, J. R. Vender, B. Baban, M. N. Hoda, and K. M. Dhandapani. 2017. "White matter damage after traumatic brain injury: A role for damage associated molecular patterns." *Biochim Biophys Acta Mol Basis Dis* 1863 (10 Pt B):2614-2626. https://doi.org/10.1016/j.bbadis.2017.05.020.

Cernak, I. 2005. "Animal models of head trauma." *NeuroRx* 2 (3):410-22. https://doi.org/10.1602/neurorx.2.3.410.

Chaichana, K. L., G. Pradilla, J. Huang, and R. J. Tamargo. 2010. "Role of inflammation (leukocyte-endothelial cell interactions) in vasospasm after subarachnoid hemorrhage." *World Neurosurg* 73 (1):22-41. https://doi.org/10.1016/j.surneu.2009.05.027.

Chen, G., S. Zhang, J. Shi, J. Ai, M. Qi, and C. Hang. 2009. "Simvastatin reduces secondary brain injury caused by cortical contusion in rats: possible involvement of TLR4/NF-kappaB pathway." *Exp Neurol* 216 (2):398-406. https://doi.org/10.1016/j.expneurol.2008.12.019.

Choi, J. Y., Y. Cui, S. T. Chowdhury, and B. G. Kim. 2017. "High-mobility group box-1 as an autocrine trophic factor in white matter stroke." *Proc Natl Acad Sci USA* 114 (25):E4987-E4995. https://doi.org/10.1073/pnas.1702035114.

Cohen, M. J., K. Brohi, C. S. Calfee, P. Rahn, B. B. Chesebro, S. C. Christiaans, M. Carles, M. Howard, and J. F. Pittet. 2009. "Early release of high mobility group box nuclear protein 1 after severe trauma in humans: role of injury severity and tissue hypoperfusion." *Crit Care* 13 (6):R174. https://doi.org/10.1186/cc8152.

Crews, F. T., and R. P. Vetreno. 2016. "Mechanisms of neuroimmune gene induction in alcoholism." *Psychopharmacology (Berl)* 233 (9):1543-57. https://doi.org/10.1007/s00213-015-3906-1.

Dash, P. K., S. A. Mach, and A. N. Moore. 2001. "Enhanced neurogenesis in the rodent hippocampus following traumatic brain injury." *J Neurosci Res* 63 (4):313-9. https://doi.org/10.1002/1097-4547(20010215)63:4<313::AID-JNR1025>3.0.CO;2-4.

Deane, R., I. Singh, A. P. Sagare, R. D. Bell, N. T. Ross, B. LaRue, R. Love, S. Perry, N. Paquette, R. J. Deane, M. Thiyagarajan, T. Zarcone, G. Fritz, A. E. Friedman, B. L. Miller, and B. V. Zlokovic. 2012. "A multimodal RAGE-specific inhibitor reduces amyloid beta-mediated brain disorder in a mouse model of Alzheimer disease." *J Clin Invest* 122 (4):1377-92. https://doi.org/10.1172/JCI58642.

Dinet, V., K. G. Petry, and J. Badaut. 2019. "Brain-Immune Interactions and Neuroinflammation After Traumatic Brain Injury." *Front Neurosci* 13:1178. https://doi.org/10.3389/fnins.2019.01178.

Evans, R. W. 2006. *Neurology and Trauma*: Oxford University Press.

Evran, S., F. Calis, E. Akkaya, O. Baran, S. Cevik, S. Katar, E. G. Gurevin, H. Hanimoglu, M. A. Hatiboglu, E. I. Armutak, E. Karatas, A. Kocyigit, and M. Y. Kaynar. 2020. "The effect of high mobility group box-1 protein on cerebral edema, blood-brain barrier, oxidative stress and apoptosis in an experimental traumatic brain injury model." *Brain Res Bull* 154:68-80. https://doi.org/10.1016/j.brainresbull.2019.10.013.

Fink, M. P. 2007. "Bench-to-bedside review: High-mobility group box 1 and critical illness." *Crit Care* 11 (5):229. https://doi.org/10.1186/cc6088.

Fitch, M. T., and J. Silver. 2008. "CNS injury, glial scars, and inflammation: Inhibitory extracellular matrices and regeneration failure." *Exp Neurol* 209 (2):294-301. https://doi.org/10.1016/j.expneurol.2007.05.014.

Foresti, M. L., G. M. Arisi, and L. A. Shapiro. 2011. "Role of glia in epilepsy-associated neuropathology, neuroinflammation and neurogenesis." *Brain Res Rev* 66 (1-2):115-22. https://doi.org/10.1016/j.brainresrev.2010.09.002.

Frank, M. G., M. D. Weber, L. K. Fonken, S. A. Hershman, L. R. Watkins, and S. F. Maier. 2016. "The redox state of the alarmin HMGB1 is a pivotal factor in neuroinflammatory and microglial priming: A role for the NLRP3 inflammasome." *Brain Behav Immun* 55:215-224. https://doi.org/10.1016/j.bbi.2015.10.009.

Fujita, K., K. Motoki, K. Tagawa, X. Chen, H. Hama, K. Nakajima, H. Homma, T. Tamura, H. Watanabe, M. Katsuno, C. Matsumi, M. Kajikawa, T. Saito, T. Saido, G. Sobue,

A. Miyawaki, and H. Okazawa. 2016. "HMGB1, a pathogenic molecule that induces neurite degeneration via TLR4-MARCKS, is a potential therapeutic target for Alzheimer's disease." *Sci Rep* 6:31895. https://doi.org/10.1038/srep31895.

Gao, H. M., H. Zhou, F. Zhang, B. C. Wilson, W. Kam, and J. S. Hong. 2011. "HMGB1 acts on microglia Mac1 to mediate chronic neuroinflammation that drives progressive neurodegeneration." *J Neurosci* 31 (3):1081-92. https://doi.org/10.1523/JNEUROSCI.3732-10.2011.

Gu, X. J., J. Xu, B. Y. Ma, G. Chen, P. Y. Gu, D. Wei, and W. X. Hu. 2014. "Effect of glycyrrhizin on traumatic brain injury in rats and its mechanism." *Chin J Traumatol* 17 (1):1-7.

Hansel, T. T., H. Kropshofer, T. Singer, J. A. Mitchell, and A. J. George. 2010. "The safety and side effects of monoclonal antibodies." *Nat Rev Drug Discov* 9 (4):325-38. https://doi.org/10.1038/nrd3003.

He, M., M. E. Bianchi, T. R. Coleman, K. J. Tracey, and Y. Al-Abed. 2018. "Exploring the biological functional mechanism of the HMGB1/TLR4/MD-2 complex by surface plasmon resonance." *Mol Med* 24 (1):21. https://doi.org/10.1186/s10020-018-0023-8.

Hergenroeder, G. W., A. N. Moore, J. P. McCoy, Jr., L. Samsel, N. H. Ward, 3rd, G. L. Clifton, and P. K. Dash. 2010. "Serum IL-6: a candidate biomarker for intracranial pressure elevation following isolated traumatic brain injury." *J Neuroinflammation* 7:19. https://doi.org/10.1186/1742-2094-7-19.

Hori, O., Brett, J., Slattery, T., Cao, R., Zhang, J., Chen, J. X., Nagashima, M., Lundh, E. R., Vijay, S., Nitecki, D., Morser, J., Stern, D., & Schmidt, A. M. 1995. "The receptor for advanced glycation end products (RAGE) is a cellular binding site for amphoterin. Mediation of neurite outgrowth and co-expression of rage and amphoterin in the developing nervous system." *J Biol Chem* 270 (43):25752-61. https://doi.org/10.1074/jbc.270.43.25752.

Huebener, P., J. P. Pradere, C. Hernandez, G. Y. Gwak, J. M. Caviglia, X. Mu, J. D. Loike, and R. F. Schwabe. 2015. "The HMGB1/RAGE axis triggers neutrophil-mediated injury amplification following necrosis." *J Clin Invest* 125 (2):539-50. https://doi.org/10.1172/JCI76887.

Jia, Y., G. Wang, Y. Ye, E. Kang, H. Chen, Z. Guo, and X. He. 2021. "Niche Cells Crosstalk In Neuroinflammation After Traumatic Brain Injury." *Int J Biol Sci* 17 (1):368-378. https://doi.org/10.7150/ijbs.52169.

Jiang, H., Y. Wang, X. Liang, X. Xing, X. Xu, and C. Zhou. 2018. "Toll-Like Receptor 4 Knockdown Attenuates Brain Damage and Neuroinflammation After Traumatic Brain Injury via Inhibiting Neuronal Autophagy and Astrocyte Activation." *Cell Mol Neurobiol* 38 (5):1009-1019. https://doi.org/10.1007/s10571-017-0570-5.

Johnson, V. E., J. E. Stewart, F. D. Begbie, J. Q. Trojanowski, D. H. Smith, and W. Stewart. 2013. "Inflammation and white matter degeneration persist for years after a single traumatic brain injury." *Brain* 136 (Pt 1):28-42. https://doi.org/10.1093/brain/aws322.

Jolly, A. E., M. Balaet, A. Azor, D. Friedland, S. Sandrone, N. S. N. Graham, K. Zimmerman, and D. J. Sharp. 2021. "Detecting axonal injury in individual patients after traumatic brain injury." *Brain* 144 (1):92-113. https://doi.org/10.1093/brain/awaa372.

Kotter, M. R., W. W. Li, C. Zhao, and R. J. Franklin. 2006. "Myelin impairs CNS remyelination by inhibiting oligodendrocyte precursor cell differentiation." *J Neurosci* 26 (1):328-32. https://doi.org/10.1523/JNEUROSCI.2615-05.2006.

Kraus, M. F., T. Susmaras, B. P. Caughlin, C. J. Walker, J. A. Sweeney, and D. M. Little. 2007. "White matter integrity and cognition in chronic traumatic brain injury: a diffusion tensor imaging study." *Brain* 130 (Pt 10):2508-19. https://doi.org/10.1093/brain/awm216.

Laird, M. D., J. S. Shields, S. Sukumari-Ramesh, D. E. Kimbler, R. D. Fessler, B. Shakir, P. Youssef, N. Yanasak, J. R. Vender, and K. M. Dhandapani. 2014. "High mobility group box protein-1 promotes cerebral edema after traumatic brain injury via activation of toll-like receptor 4." *Glia* 62 (1):26-38. https://doi.org/10.1002/glia.22581.

Laird, M. D., S. Sukumari-Ramesh, A. E. Swift, S. E. Meiler, J. R. Vender, and K. M. Dhandapani. 2010. "Curcumin attenuates cerebral edema following traumatic brain injury in mice: a possible role for aquaporin-4?" *J Neurochem* 113 (3):637-48. https://doi.org/10.1111/j.1471-4159.2010.06630.x.

Lehnardt, S., C. Lachance, S. Patrizi, S. Lefebvre, P. L. Follett, F. E. Jensen, P. A. Rosenberg, J. J. Volpe, and T. Vartanian. 2002. "The toll-like receptor TLR4 is necessary for lipopolysaccharide-induced oligodendrocyte injury in the CNS." *J Neurosci* 22 (7):2478-86. https://doi.org/20026268.

Lian, Y. J., H. Gong, T. Y. Wu, W. J. Su, Y. Zhang, Y. Y. Yang, W. Peng, T. Zhang, J. R. Zhou, C. L. Jiang, and Y. X. Wang. 2017. "Ds-HMGB1 and fr-HMGB induce depressive behavior through neuroinflammation in contrast to nonoxid-HMGB1." *Brain Behav Immun* 59:322-332. https://doi.org/10.1016/j.bbi.2016.09.017.

Liu, T., L. Zhang, D. Joo, and S. C. Sun. 2017. "NF-kappaB signaling in inflammation." *Signal Transduct Target Ther* 2. https://doi.org/10.1038/sigtrans.2017.23.

Lotocki, G., J. de Rivero Vaccari, O. Alonso, J. S. Molano, R. Nixon, W. D. Dietrich, and H. M. Bramlett. 2011. "Oligodendrocyte Vulnerability Following Traumatic Brain Injury in Rats: Effect of Moderate Hypothermia." *Ther Hypothermia Temp Manag* 1 (1):43-51. https://doi.org/10.1089/ther.2010.0011.

Manivannan, S., E. Wales, and M. Zaben. 2021. "The Role of HMGB1 in Traumatic Brain Injury-Bridging the Gap Between the Laboratory and Clinical Studies." *Curr Neurol Neurosci Rep* 21 (12):75. https://doi.org/10.1007/s11910-021-01158-3.

Manivannan, S., B. Harari, M. Muzaffar, O. Elalfy, S. Hettipathirannahelage, Z. James, F. Sharouf, C. Ormonde, M. Alsaqati, W. Gray, and M. Zaben. 2020. "Glycyrrhizin Blocks the Detrimental Effects of HMGB1 on Cortical Neurogenesis After Traumatic Neuronal Injury." *Brain Sci* 10 (10). https://doi.org/10.3390/brainsci10100760.

Mattugini, N., J. Merl-Pham, E. Petrozziello, L. Schindler, J. Bernhagen, S. M. Hauck, and M. Gotz. 2018. "Influence of white matter injury on gray matter reactive gliosis upon stab wound in the adult murine cerebral cortex." *Glia* 66 (8):1644-1662. https://doi.org/10.1002/glia.23329.

Narayana, P. A. 2017. "White matter changes in patients with mild traumatic brain injury: MRI perspective." *Concussion* 2 (2):CNC35. https://doi.org/10.2217/cnc-2016-0028.

Okuma, Y., K. Liu, H. Wake, R. Liu, Y. Nishimura, Z. Hui, K. Teshigawara, J. Haruma, Y. Yamamoto, H. Yamamoto, I. Date, H. K. Takahashi, S. Mori, and M. Nishibori.

2014. "Glycyrrhizin inhibits traumatic brain injury by reducing HMGB1-RAGE interaction." *Neuropharmacology* 85:18-26. https://doi.org/10.1016/j.neuropharm.2014.05.007.

Okuma, Y., H. Wake, K. Teshigawara, Y. Takahashi, T. Hishikawa, T. Yasuhara, S. Mori, H. K. Takahashi, I. Date, and M. Nishibori. 2019. "Anti-High Mobility Group Box 1 Antibody Therapy May Prevent Cognitive Dysfunction After Traumatic Brain Injury." *World Neurosurg* 122:e864-e871. https://doi.org/10.1016/j.wneu.2018.10.164.

Parker, T. M., A. H. Nguyen, J. R. Rabang, A. A. Patil, and D. K. Agrawal. 2017. "The danger zone: Systematic review of the role of HMGB1 danger signalling in traumatic brain injury." *Brain Inj* 31 (1):2-8. https://doi.org/10.1080/02699052.2016.1217045.

Paudel, Y. N., E. Angelopoulou, C. Piperi, I. Othman, and M. F. Shaikh. 2020. "HMGB1-Mediated Neuroinflammatory Responses in Brain Injuries: Potential Mechanisms and Therapeutic Opportunities." *Int J Mol Sci* 21 (13). https://doi.org/10.3390/ijms21134609.

Paudel, Y. N., M. F. Shaikh, A. Chakraborti, Y. Kumari, A. Aledo-Serrano, K. Aleksovska, M. K. M. Alvim, and I. Othman. 2018. "HMGB1: A Common Biomarker and Potential Target for TBI, Neuroinflammation, Epilepsy, and Cognitive Dysfunction." *Front Neurosci* 12:628. https://doi.org/10.3389/fnins.2018.00628.

Richard, Seidu A., W. Min, Z. Su, and H. Xu. 2017. "High Mobility Group Box 1 and Traumatic Brain Injury " *Journal of Behavioral and Brain Science* 7 (2):50-61. https://doi.org/10.4236/jbbs.2017.72006.

Schiraldi, M., A. Raucci, L. M. Munoz, E. Livoti, B. Celona, E. Venereau, T. Apuzzo, F. De Marchis, M. Pedotti, A. Bachi, M. Thelen, L. Varani, M. Mellado, A. Proudfoot, M. E. Bianchi, and M. Uguccioni. 2012. "HMGB1 promotes recruitment of inflammatory cells to damaged tissues by forming a complex with CXCL12 and signaling via CXCR4." *J Exp Med* 209 (3):551-63. https://doi.org/10.1084/jem.20111739.

Stein, S. C., D. I. Graham, X. H. Chen, and D. H. Smith. 2004. "Association between intravascular microthrombosis and cerebral ischemia in traumatic brain injury." *Neurosurgery* 54 (3):687-91; discussion 691. https://doi.org/10.1227/01.neu.0000108641.98845.88.

Stirling, D. P., K. Cummins, M. Mishra, W. Teo, V. W. Yong, and P. Stys. 2014. "Toll-like receptor 2-mediated alternative activation of microglia is protective after spinal cord injury." *Brain* 137 (Pt 3):707-23. https://doi.org/10.1093/brain/awt341.

Su, X., H. Wang, J. Zhao, H. Pan, and L. Mao. 2011. "Beneficial effects of ethyl pyruvate through inhibiting high-mobility group box 1 expression and TLR4/NF-kappaB pathway after traumatic brain injury in the rat." *Mediators Inflamm* 2011:807142. https://doi.org/10.1155/2011/807142.

Sullivan, G. M., A. J. Mierzwa, N. Kijpaisalratana, H. Tang, Y. Wang, S. K. Song, R. Selwyn, and R. C. Armstrong. 2013. "Oligodendrocyte lineage and subventricular zone response to traumatic axonal injury in the corpus callosum." *J Neuropathol Exp Neurol* 72 (12):1106-25. https://doi.org/10.1097/NEN.0000000000000009.

Taib, T., C. Leconte, J. Van Steenwinckel, A. H. Cho, B. Palmier, E. Torsello, R. Lai Kuen, S. Onyeomah, K. Ecomard, C. Benedetto, B. Coqueran, A. C. Novak, E. Deou, M. Plotkine, P. Gressens, C. Marchand-Leroux, and V. C. Besson. 2017.

"Neuroinflammation, myelin and behavior: Temporal patterns following mild traumatic brain injury in mice." *PLoS One* 12 (9):e0184811. https://doi.org/10.1371/journal.pone.0184811.

Tang, D., T. R. Billiar, and M. T. Lotze. 2012. "A Janus tale of two active high mobility group box 1 (HMGB1) redox states." *Mol Med* 18:1360-2. https://doi.org/10.2119/molmed.2012.00314.

Terrando, N., T. Yang, X. Wang, J. Fang, M. Cao, U. Andersson, H. H. Erlandsson, W. Ouyang, and J. Tong. 2016. "Systemic HMGB1 Neutralization Prevents Postoperative Neurocognitive Dysfunction in Aged Rats." *Front Immunol* 7:441. https://doi.org/10.3389/fimmu.2016.00441.

Tsung, A., R. Sahai, H. Tanaka, A. Nakao, M. P. Fink, M. T. Lotze, H. Yang, J. Li, K. J. Tracey, D. A. Geller, and T. R. Billiar. 2005. "The nuclear factor HMGB1 mediates hepatic injury after murine liver ischemia-reperfusion." *J Exp Med* 201 (7):1135-43. https://doi.org/10.1084/jem.20042614.

Ueda, T., and M. Yoshida. 2010. "HMGB proteins and transcriptional regulation." *Biochim Biophys Acta* 1799 (1-2):114-8. https://doi.org/10.1016/j.bbagrm.2009.11.005.

Ved, R., F. Sharouf, B. Harari, M. Muzaffar, S. Manivannan, C. Ormonde, W. P. Gray, and M. Zaben. 2021. "Disulfide HMGB1 acts via TLR2/4 receptors to reduce the numbers of oligodendrocyte progenitor cells after traumatic injury in vitro." *Sci Rep* 11 (1):6181. https://doi.org/10.1038/s41598-021-84932-0.

Ved, R., S. Manivannan, I. Tasker, and M. Zaben. 2022. "High mobility group box protein 1 and white matter injury following traumatic brain injury: perspectives on mechanisms and therapeutic strategies." *Neural Regen Res* 17 (8):1739-1740. https://doi.org/10.4103/1673-5374.332135.

Venereau, E., F. De Leo, R. Mezzapelle, G. Careccia, G. Musco, and M. E. Bianchi. 2016. "HMGB1 as biomarker and drug target." *Pharmacol Res* 111:534-544. https://doi.org/10.1016/j.phrs.2016.06.031.

Venereau, E., M. Casalgrandi, M. Schiraldi, D. J. Antoine, A. Cattaneo, F. De Marchis, J. Liu, A. Antonelli, A. Preti, L. Raeli, S. S. Shams, H. Yang, L. Varani, U. Andersson, K. J. Tracey, A. Bachi, M. Uguccioni, and M. E. Bianchi. 2012. "Mutually exclusive redox forms of HMGB1 promote cell recruitment or proinflammatory cytokine release." *J Exp Med* 209 (9):1519-28. https://doi.org/10.1084/jem.20120189.

Wang, D., K. Liu, H. Wake, K. Teshigawara, S. Mori, and M. Nishibori. 2017. "Anti-high mobility group box-1 (HMGB1) antibody inhibits hemorrhage-induced brain injury and improved neurological deficits in rats." *Sci Rep* 7:46243. https://doi.org/10.1038/srep46243.

Wang, G., J. Zhang, X. Hu, L. Zhang, L. Mao, X. Jiang, A. K. Liou, R. K. Leak, Y. Gao, and J. Chen. 2013. "Microglia/macrophage polarization dynamics in white matter after traumatic brain injury." *J Cereb Blood Flow Metab* 33 (12):1864-74. https://doi.org/10.1038/jcbfm.2013.146.

Wang, J., M. A. Fox, and J. T. Povlishock. 2013. "Diffuse traumatic axonal injury in the optic nerve does not elicit retinal ganglion cell loss." *J Neuropathol Exp Neurol* 72 (8):768-81. https://doi.org/10.1097/NEN.0b013e31829d8d9d.

Wang, K. Y., G. F. Yu, Z. Y. Zhang, Q. Huang, and X. Q. Dong. 2012. "Plasma high-mobility group box 1 levels and prediction of outcome in patients with traumatic brain injury." *Clin Chim Acta* 413 (21-22):1737-41. https://doi.org/10.1016/j.cca.2012.07.002.

Wanner, I. B., M. A. Anderson, B. Song, J. Levine, A. Fernandez, Z. Gray-Thompson, Y. Ao, and M. V. Sofroniew. 2013. "Glial scar borders are formed by newly proliferated, elongated astrocytes that interact to corral inflammatory and fibrotic cells via STAT3-dependent mechanisms after spinal cord injury." *J Neurosci* 33 (31):12870-86. https://doi.org/10.1523/JNEUROSCI.2121-13.2013.

Weber, D. J., A. S. Gracon, M. S. Ripsch, A. J. Fisher, B. M. Cheon, P. H. Pandya, R. Vittal, M. L. Capitano, Y. Kim, Y. M. Allette, A. A. Riley, B. P. McCarthy, P. R. Territo, G. D. Hutchins, H. E. Broxmeyer, G. E. Sandusky, F. A. White, and D. S. Wilkes. 2014. "The HMGB1-RAGE axis mediates traumatic brain injury-induced pulmonary dysfunction in lung transplantation." *Sci Transl Med* 6 (252):252ra124. https://doi.org/10.1126/scitranslmed.3009443.

Webster, K. M., S. R. Shultz, E. Ozturk, L. K. Dill, M. Sun, P. Casillas-Espinosa, N. C. Jones, P. J. Crack, T. J. O'Brien, and B. D. Semple. 2019. "Targeting high-mobility group box protein 1 (HMGB1) in pediatric traumatic brain injury: Chronic neuroinflammatory, behavioral, and epileptogenic consequences." *Exp Neurol* 320:112979. https://doi.org/10.1016/j.expneurol.2019.112979.

Wietzorrek, G., M. Drexel, M. Trieb, and S. Santos-Sierra. 2019. "Anti-inflammatory activity of small-molecule antagonists of Toll-like receptor 2 (TLR2) in mice." *Immunobiology* 224 (1):1-9. https://doi.org/10.1016/j.imbio.2018.11.004.

Xiong, X. Y., L. Liu, and Q. W. Yang. 2016. "Functions and mechanisms of microglia/macrophages in neuroinflammation and neurogenesis after stroke." *Prog Neurobiol* 142:23-44. https://doi.org/10.1016/j.pneurobio.2016.05.001.

Yang, H., D. J. Antoine, U. Andersson, and K. J. Tracey. 2013. "The many faces of HMGB1: molecular structure-functional activity in inflammation, apoptosis, and chemotaxis." *J Leukoc Biol* 93 (6):865-73. https://doi.org/10.1189/jlb.1212662.

Yang, H., P. Lundback, L. Ottosson, H. Erlandsson-Harris, E. Venereau, M. E. Bianchi, Y. Al-Abed, U. Andersson, K. J. Tracey, and D. J. Antoine. 2012. "Redox modification of cysteine residues regulates the cytokine activity of high mobility group box-1 (HMGB1)." *Mol Med* 18:250-9. https://doi.org/10.2119/molmed.2011.00389.

Yang, H., H. Wang, Z. Ju, A. A. Ragab, P. Lundback, W. Long, S. I. Valdes-Ferrer, M. He, J. P. Pribis, J. Li, B. Lu, D. Gero, C. Szabo, D. J. Antoine, H. E. Harris, D. T. Golenbock, J. Meng, J. Roth, S. S. Chavan, U. Andersson, T. R. Billiar, K. J. Tracey, and Y. Al-Abed. 2015. "MD-2 is required for disulfide HMGB1-dependent TLR4 signaling." *J Exp Med* 212 (1):5-14. https://doi.org/10.1084/jem.20141318.

Yi, X., G. Jin, X. Zhang, W. Mao, H. Li, J. Qin, J. Shi, K. Dai, and F. Zhang. 2013. "Cortical endogenic neural regeneration of adult rat after traumatic brain injury." *PLoS One* 8 (7):e70306. https://doi.org/10.1371/journal.pone.0070306.

Chapter 7

The Functional Role of HMGB1/TLR4/NF-κB Signalling Pathway in Multiple Sclerosis

Firdevs Uluc[1,*], MSc,
Bihter Gokce Bozat[1], PhD,
Seyda Karabork[2], PhD
and Sule Aydın-Turkoglu[3], MD

[1]Department of Interdisciplinary Neuroscience,
Faculty of Medicine,
Bolu Abant Izzet Baysal University, Bolu, Turkey
[2]Department of Medical Microbiology,
Faculty of Medicine,
Bolu Abant Izzet Baysal University, Bolu, Turkey
[3]Department of Neurology,
Faculty of Medicine,
Bolu Abant Izzet Baysal University, Bolu, Turkey

Abstract

Multiple sclerosis (MS) is a chronic autoimmune demyelinating disorder of the central nervous system (CNS) defined by features such as axonal loss, glial cell activation, and immune cell infiltration. To date, the molecular mechanism underlying inflammation and immune response in MS pathogenesis has not been fully understood. Elucidation of the existing pathogenic mechanism is crucial for the development of effective treatment options for this disease. High Mobility Group Box 1 (HMGB1) protein is a proinflammatory-like cytokine protein that has an initiating role in neuroinflammation in triggering MS. In recent years,

* Corresponding Author's Email: frdvserim@gmail.com.

In: HMGB1
Editors: Eylem Taskin, Celal Guven and Salih Tunc Kaya
ISBN: 979-8-88697-408-9

HMGB1 and its related signaling pathways have become a therapeutic target in experimental and clinical MS studies. In particular, HMGB1/TLR4/ NF-κB signaling pathway has an increasing importance. The main goal of this chapter is to explain the effects of HMGB1/TLR4/NF-κB signaling pathway in MS pathophysiology, which enhancing the release of proinflammatory cytokines and causes an inflammatory response. In order to develop effective treatment strategies in MS in the future, this chapter aims to explain the mechanism of HMGB1-induced neuroinflammation.

Keywords: HMGB1, multiple sclerosis, nuclear factor-κappa B, toll-like receptor-4

Introduction

Multiple Sclerosis is an autoimmune demyelinating disorder of CNS that can lead to progressive neuronal-axonal degeneration (McCool et al., 2019). It is classified into 4 different types according to its clinical course:

(a) Relapsing-Remitting MS (RRMS), characterized by periodic relapses with neurological symptoms that can improve with treatment,
(b) Secondary Progressive MS (SPMS) that develops over time after diagnosis of RRMS,
(c) Primary Progressive MS (PPMS), which comprises 10% of patients and progresses gradually with continuous neurological impairment, and
(d) Progressive Relapsing MS (PRMS), similar to PPMS and less than 5% of patients (Ciotti and Cross 2018; Zimmermann et al., 2018).

MS subtypes were revised in 2013 and defined as: (a) Clinically Isolated Syndrome (CIS) is the first clinical form of MS having similar clinical features that suggest MS; (b) RRMS; (c) PPMS and; (d) SPMS, and thus PRMS was excluded from the classification (Lublin et al., 2014). The autoimmune response to myelin components in the CNS has an important role in the pathogenesis of MS (Buc 2013). Pathologically, in the early stages of the disease, infiltration of immune cells such as monocytes, CD4+ and CD8+ T cells and B cells, macrophages, and dendritic cells (DCs) occurs in the CNS.

Leukocyte infiltration causes demyelination, astrocytosis, neuronal and axonal degeneration (Garg and Smith 2015). In addition, another main distinguishing feature of MS pathology is the increased infiltration of T cells specific to the myelin into the CNS. T helper cell 1 (Th1) accomplishes secretion/releasing the proinflammatory cytokines. T helper lymphocytes called T helper 17 are stimulated as a result of releasing proinflammatory cytokines (Wan et al., 2020). Stimulation of Th1 and Th17 cells generate signals initiating inflammation. These signals (such as Th1: IFN-γ, TNF-α, IL-2 and IL-12; Th17: IL-17A, IL-17F, IL-21, IL-22) are known to increase in MS patients (Kaskow and Baecher-Allan 2018; van Langelaar et al., 2018). Th1/Th17 cells play prominent roles in active phase with proinflammatory cytokines compared to Th2/Treg cells. In contrast, Th2/Treg cells and related anti-inflammatory cytokines are observed in the remission period in MS (Trager et al., 2013).

Although MS is defined as an immune system disorder, environmental and genetic factors also have a crucial status in MS onset and clinical course. HLA-DRB1∗15 which is Major Histocompatibility Complex (MHC) Class II allele is considered as MS-related gene involved in immune function (Handunnetthi, Ramagopalan, and Ebers 2010). In this HLA-DRB1 family, HLA-DRB1∗15:01 allele is especially MS-linked, and it has been reported that people carrying HLA-DRB1∗15:01 are approximately three times more probable than non-carriers (Patsopoulos et al., 2013). The effect of varied environmental factors such as viruses (Marrodan et al., 2019), vitamin D deficiency (Pierrot-Deseilligny and Souberbielle 2017), sun exposure (Orton et al., 2011) and dietary habits (Socha et al., 2017) have been reported in the studies. Intestinal microbiota has also been correlated with the functions of the brain by gut-brain axis. It has been claimed that this axis might affect immune functions. It provides antigens and molecules which act as mimicry for self-antigens and stimulate innate immunity, promoting autoreactive T cell activation (Kadowaki et al., 2016; Varol, Zigmond, and Jung 2010). At the present time, studies clarifying the effect of environmental factors in MS/Autoimmune Experimental Encephalomyelitis (EAE), an experimental model of MS, still maintain their importance. Since there is a contribution to the risk of the disease, these factors may also be effective in determining the treatment options in MS.

In recent years, clinical and experimental studies have focused on obtaining effective and targeted results due to the inadequacy of current approaches in the treatment of MS. The fact that MS pathology is still not fully explained is the biggest obstacle to this goal. In the last decade, High Mobility

Group Box 1 (HMGB1) protein has become prominent suspect in the MS pathogenesis studies. HMGB1 is non-histone protein and plays a part in chief role in the releasing of pro-inflammatory cytokines in the pathogenesis of MS and many other autoimmune diseases (Barnie 2019). The main mechanism of HMGB1 is mediated by several receptor proteins and signaling pathways (Yang, Wang, and Andersson 2020). The current studies on the basis of enhancing MS treatment strategy is based on the determining potential HMGB1-related signalling pathways. On behalf of this, there have been many clinical and experimental studies targeting the effective signaling pathways in MS pathophysiology. Based on this concept, the present chapter contains summary information about the role of the HMGB1/TLR4/NF-κB signaling pathway which is thought to be a potential target in MS pathology and related with clinical and experimental studies conducted through this signalling pathway.

High Mobility Group Box 1 (HMGB1): A Potential Therapeutical Target in MS Pathogenesis

High Mobility Group Box 1, also called amphoterin, the endogenous damage-associated molecular pattern (DAMP), is a non-histone DNA-binding protein found in all mammalian cells (Sternberg et al., 2016). The main structure of HMGB1 consists of two homologous HMG-box domains as A box and B box and C-terminal acidic tail (Stros 2010). HMGB1 is acting its role by binding different membrane receptors such as receptors for advanced glycation end products (RAGE) and toll-like receptors (TLRs) for its signaling cascade (Sun et al., 2018). These receptors have been shown in Figure 1 (Janko et al., 2014). RAGE-binding interaction of HMGB1 causes the activation of different signal transduction pathways and results in cellular responses such as cell proliferation and production of cytokines and chemokines. The interaction of TLRs, on the other hand, are related with proinflammatory response of macrophages and neutrophils triggered by HMGB1 (Martinotti, Patrone, and Ranzato 2015). In HMGB1-induced cytokine stimulation, TLR4 signaling pathway is precisely necessary. Since, the releasing of proinflammatory cytokines by macrophages in response to HMGB1 does not take place in the absence of TLR4 (Harris, Andersson, and Pisetsky 2012).

HMGB1 exists in astrocytes, neurons, and microglia in CNS, and in the peripheral nervous system, it is also found in the cytoplasm and extracellular spaces of neurons and Schwann cells (Fang, Schachner, and Shen 2012). It also has higher expressions in brain neurogenesis regions such as dentate gyrus and olfactory bulb (Guazzi et al., 2003). It plays a key role as a DNA chaperone in the nucleus under physiological conditions. It is secreted passively from damaged or dead cells or actively secreted by macrophages, and it may also have a role as a cytokine inducing an early inflammatory response (Harris, Andersson, and Pisetsky 2012; Kwak et al., 2020). Innate immune system does not recognize passively released HMGB1. However, actively released HMGB1 is recognized by the immune system as well as synthesizes and releases proinflammatory cytokines that induce/trigger and potentiate inflammation (Sun et al., 2018). The results from *in vitro* studies show that HMGB1 may affect immune function by inducing T cell activation/proliferation and dendritic cell maturation. It has also shown that naive T cells may play a part in the functional polarization of the proinflammatory Th1 phenotype (Dumitriu et al., 2005; Sundberg et al., 2009).

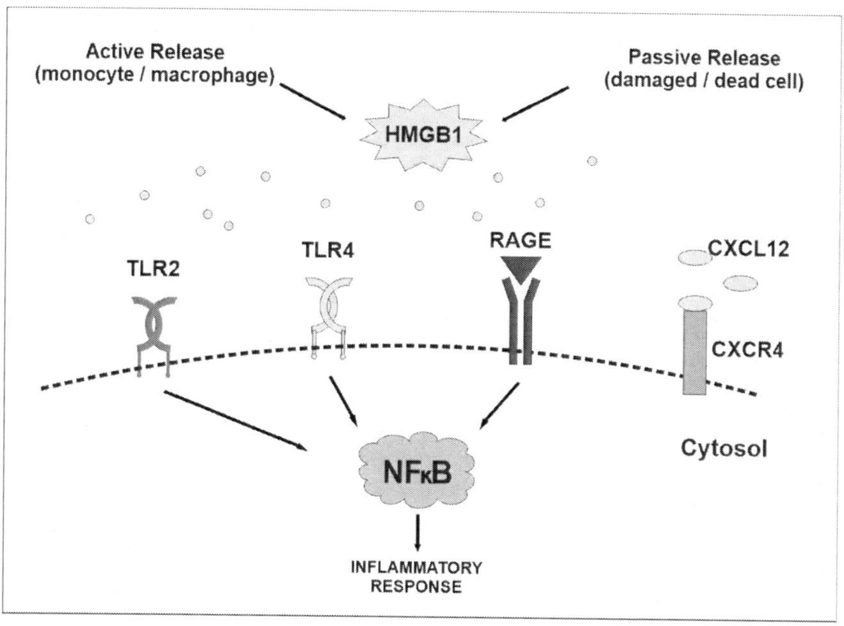

Figure 1. HMGB1-related molecules and its membrane receptors (Janko et al., 2014).

The obvious results have been shown that high serum HMGB1 levels have an association with the disease severity in autoimmune diseases such as rheumatoid arthritis (Chen et al., 2021) and systemic lupus erythematosus (Aragón et al., 2020) and it has also been detected in various neuro-inflammatory diseases (Paudel et al., 2020a) such as epilepsy (Fu et al., 2017), Alzheimer's disease (Fujita et al., 2016), amyotrophic lateral sclerosis (Lo Coco et al., 2007; Paudel et al., 2020b) and Parkinson's disease (Sasaki et al., 2016). A meta-analysis study is stated the apparent relationship between HMGB1 levels in MS patients was revealed and it was emphasized that increased HMGB1 levels in cerebrospinal fluid (CSF), serum and peripheral blood mononuclear cells (PBMC) pose a risk for MS (Zhen et al., 2019). Similarly, PBMC HMGB1 level was detected as higher in MS patients compared to Neuromyelitis Optica patients (Wang et al., 2013) and also, when MS subtypes were compared, an increase in PBMC HMGB1 mRNA level was recorded in patients with RRMS, but no difference was observed in PPMS patients compared to the healthy control (Malhotra et al., 2015).

The obtained results are based on the possibility that releasing of HMGB1 might be take place in damaged cells or stimulated microglia, macrophages and mature DCs during infection and inflammatory signals in the CNS. Consistent with this possibility; in another study, numbers of macrophages and microglial cells with HMGB1 level have been detected as higher in immunohistochemically stained MS white matter compared to healthy subjects (Andersson et al., 2008). When HMGB1 and MS lesions were compared, not only HMGB1 level was correlated with the number of lesions statistically significant but also HMGB1 level was also higher in patients having high Expanded Disability Status Scale (EDSS) Score (Bucova et al., 2020). As a conclusion, results from clinical and experimental studies show that HMGB1 has a potential for being a biomarker for MS disease (Andersson, Yang, and Harris 2018) and also show that it can be distinctive in determining disease severity (Bucova et al., 2020; Malhotra et al., 2015).

Toll-Like Receptors (TLRs): It's Role in the Inflammatory Response

TLRs are sensors called pattern recognition receptors (PRRs) recognizing pathogen-associated molecular patterns (PAMPs), DAMPs and innate immune response (Medzhitov and Janeway 1997; Kawasaki and Kawai 2014).

The existence of TLRs in the innate immune system was defined firstly in Drosophila (Hashimoto, Hudson, and Anderson 1988). After this identification, a new one was identified as a mammalian homologue, which recognizes lipopolysaccharides (LPS) and activates innate and adaptive immune responses (Medzhitov and Janeway 1997; Medzhitov, Preston-Hurlburt, and Janeway 1997). TLRs are synthesized in the endoplasmic reticulum (ER) and are delivered to the cell surface or to an intracellular location. In the literature, at least 10 different TLRs have been described in humans (TLR1–TLR10), and 12 different TLRs (TLR1–TLR9, TLR11–TLR13) in mice, thus far. They are divided into subgroups as cell surface TLRs and intracellular TLRs in terms of their localization. Intracellular localization of TLRs is critical for ligand recognition and prevents TLRs from making a connection with self-nucleic acids for autoimmune response (Lee et al., 2013). While TLR1, TLR2, TLR4, TLR5, TLR6 and TLR10 are localized in the cell surface membrane, TLR3, TLR7, TLR8 and TLR9 are exist in endosome (Zheng et al., 2019). Figure 2 shows these localizations (Goulopoulou, McCarthy, and Webb 2016). TLR1-TLR9 are preserved in both humans and mice. However, TLR10 in mice and TLR11, TLR12 and TLR13 in humans are not functional (Celhar, Magalhaes, and Fairhurst 2012; Kawai and Akira 2010).

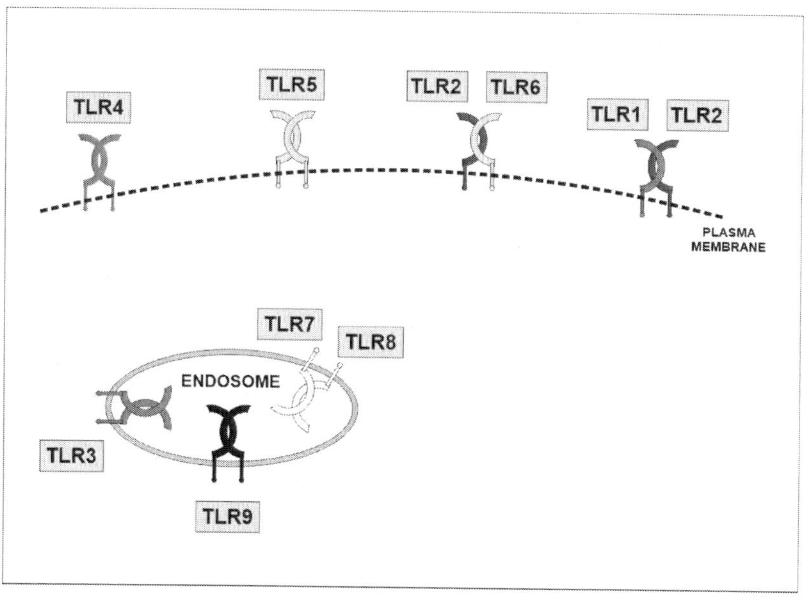

Figure 2. Toll-like receptor localization (Goulopoulou, McCarthy, and Webb 2016).

TLR4 is a cell surface receptor that recognizes bacterial lipopolysaccharides (LPS) (Kawasaki and Kawai 2014) and is also known as HMGB1 receptor for cytokine production. It is expressed in oligodendrocytes, astrocytes, microglia, and neurons in brain (Bsibsi et al., 2002; Bowman et al., 2003; Olson and Miller 2004; Wadachi and Hargreaves 2006). TLR4 plays a key role in deconstructing a connection between innate and adaptive immunity (Andersson, Yang, and Harris 2018; Sun et al., 2020).

HMGB1/TLR4 signaling is crucial in pathologies associated with inflammation (Frank et al., 2015; Liu, Wu, and Wang 2017), mainly by modulating/regulating inflammatory responses (Ding et al., 2017). The significant increase in the expression of TLR2 and TLR4 in innate immunity has been stated in MS/EAE studies (Andersson et al., 2008; Zheng et al., 2019). HMGB1 shows its biological activity by binding its TLR4 receptor, but there is limited evidence to support the key role of HMGB1-induced TLR4 activation in MS pathogenesis. The interaction of HMGB1 and TLR4 triggers the upregulation of nuclear factor-kappa B (NF-κB) and causes the releasing of proinflammatory chemokines, cytokines, and enzymes (Trotta et al., 2014; Li et al., 2018). In MS patients, increased TLRs expression has been reported in lesions and especially raising of TLR1-TLR8 expression in glial cells has been shown. Additionally, active lesions have been linked with the high expression of TLR3 and TLR4 on astrocytes and microglia, however, no TLRs expression in healthy white matter has been detected (Bsibsi et al., 2002). On the other hand, there is no precise statement about the certain role of TLR3 and TLR4 activity in these lesions.

Although relationship with the inflammation process of MS/EAE and TLR4 activation has been known, there is no accurate definition about how TLRs may alter the course of disease. Some opposite opinions have been stated about the role of TLR4 in MS/EAE (Marta et al., 2008). In TLR4-knockdown mice, less severity of EAE symptoms has been reported (Kerfoot et al., 2004). However, in opposite, increasing the severity of EAE in lacking TLR4 has been stated (Marta et al., 2008). Furthermore, in TLR4-knockdown mice, higher levels of serum IL-17 and Th17 cells have been shown. According to these results, disputable role of TLR4 in MS/EAE might be probable (Marta, Meier, and Lobell 2009). There are some hypotheses for the explanation of this disputable role. In the first one, TLR3 and TLR4 activation has been held responsible in transforming microglia to antigen presenting cells and facilitating T cell infiltration (Olson, Girvin, and Miller 2001). The second one is the releasing of proinflammatory cytokines may originate from TLRs, and thus cellular damage occurs. LPS-induced microglial TLR4 activation causes damage and death to oligodendrocytes and neurons (Lehnardt et al., 2002; Lehnardt et al., 2003). In consistent with, in the neonatal rat brain, LPS-induced TLR4 activation which resulted in oligodendrocyte and myelin loss mediated by proinflammatory cytokine IL-1β has been reported (Pang, Cai, and Rhodes 2003). All the obtained results support the disputable role of TLR4 in MS/EAE.

Nuclear Factor Kappa B (NF-κB): A View from Multiple Sclerosis Perspective

NF-κB is one of the main transcription factors regulating the expression of genes comprising cell death and survival, inflammation, proliferation and cell differentiation (Oeckinghaus, Hayden, and Ghosh 2011). In NF-κB family, five transcription factors such as p50, p52, P65, c-Rel and Rel B are defined, and they are detectable in mammalian cells. These regulate the expression of various genes differently (Hoesel and Schmid 2013; Zhang, Lenardo, and Baltimore 2017). In resting condition, NF-κB is termed as an inactive cytoplasmic protein complex bound to inhibitor κB (IκB) proteins. When it is stimulated, IκB kinase (IKK) phosphorylates IκB and ubiquitination and proteasomal degradation are facilitated. This allows the entrance of NF-κB into the nucleus and provides transcriptional activity of the nucleus (Zhou et al., 2020).

There are two basic NF-κB activation pathways: canonical (classic) and non-canonical (alternative) pathways. These are shown in Figure 3 (Verstrepen and Beyaert 2014). The canonical pathway which is the most studied pathway is stimulated by signals from cytokine receptors for IL-1, TNF and TLRs causing in the activation of IKK complex. This complex provides phosphorylation of IκB proteins and results in the release of p50 and p65 which are the cytoplasmic NF-κB dimers from the degraded IκB. The canonical pathway takes part in a wide variety of biological functions (Sun 2017). The non-canonical pathway is triggered by B cell activating factor (BAF), CD40 ligand, lymphotoxin β (LTβ) receptor and receptor activator of NF-κB (RANK). Activation of NF-κB inducing kinase (NIK) occurs by these stimuli. NIK activates and phosphorylates IKK1. Thus, the phosphorylation of p100 takes place. This causes its ubiquitination and degradation to p52. Then, p52 translocates to the nucleus by the way of its dimerization with RelB.

This pathway is substantially slower and associated with the limited number of cellular responses (Cildir, Low, and Tergaonkar 2016; Sun 2017).

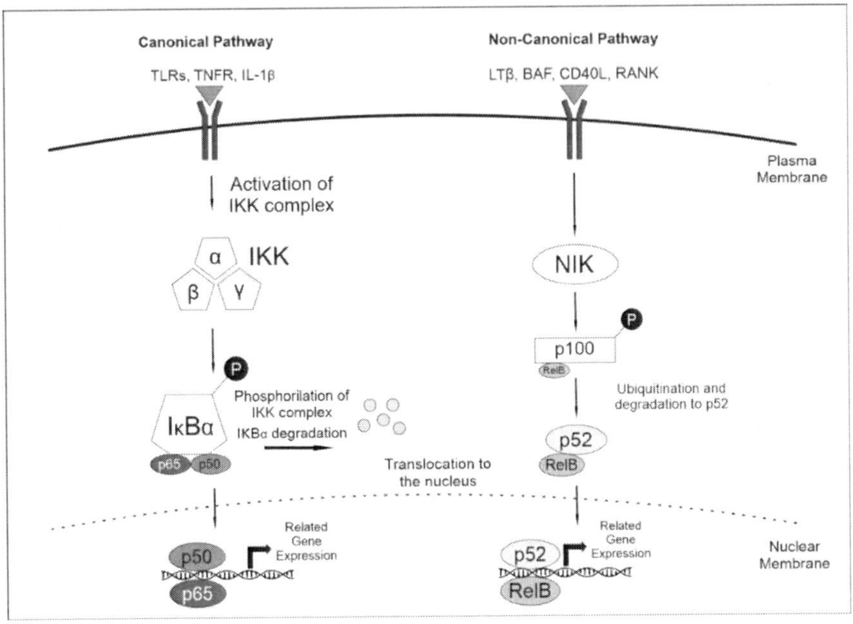

Figure 3. Canonical and non-canonical pathways (Verstrepen and Beyaert 2014).

NF-κB plays a part in as a regulator in the nervous system and controls physiological functions in CNS. It also takes part in the maintenance of neuronal plasticity by affecting memory, learning and behavior (Kaltschmidt and Kaltschmidt 2009). NF-κB regulates the differentiation of Th1 and Th17 cells and the activity of inflammatory cells (Mc Guire et al., 2013). Dysregulation of NF-κB has been related with diseases such as multiple sclerosis, psoriasis, viral infection, and cancer (Hoesel and Schmid 2013; Mc Guire et al., 2013). In recent years, with the advances in genetic technologies, numerous MS susceptibility variants associated with genes regulating NF-κB signaling pathway have been defined (Beecham et al., 2013; Sawcer et al., 2011). Rs7665090-g, known as the MS risk allele, has been reported as associated in upregulation of NF-κB signaling and this allele-related target genes in astrocytes related with increased MS lesion size and lymphocyte infiltration (Ponath et al., 2018). Also, increased NF-κB-related genes in T cells has been associated with relapses of MS (Lindsey, Agarwal, and Tan 2011; Satoh et al., 2008).

The presence of NF-κB activation is known in MS lesions localized to microglia, astrocytes and oligodendrocytes (Zhou et al., 2020). Regulation of the viability of cells in inflammatory lesions by NF-κB is known (Park and Hong 2016). TLR4/NF-κB signaling pathway has a prominent role in microglia-mediated inflammation. The increasing NF-κB levels in CD3+/CD4+ T cells and monocytes in PBMCs isolated from MS patients has been reported (Yan et al., 2018). Upregulation of NF-κB results in increased expression and release of tumor necrosis factor-alpha (TNF-α) and interleukin-1β (IL-1β). In experimental studies, it has been observed that these inflammatory markers gradually increase in the later stages of EAE and NF-κB inhibition suppresses the development of EAE (Sun et al., 2018; Yi et al., 2014). All these data suggest that inhibiting HMGB1/TLR4/NF-κB signaling has a potential to suppress MS/EAE, apart from the current known treatment possibilities (Paudel et al., 2019).

Current Treatment Strategies and Importance of This Pathway

One of the goals, trying to achieve in the treatment of MS, is the suppression of pathogenic autoimmune response without causing a loss in adaptive immune system response. Although various drugs such as Ocrelizumab, Fingolimod, Dimethyl Fumarate, Mitoxantrone and Natalizumab are used to prevent MS relapses (English and Aloi 2015) treatment options for acute MS relapses are still limited (Sun et al., 2018). Currently, disease-modifying therapy (DMT) options have limitation due to high cost (Ciotti and Cross 2018; Zimmermann et al., 2018). Research of specific biomarkers and development of DMT strategies for MS treatment is needed.

In recent years, HMGB1 protein has raised as a potential candidate for DMT due to its effect on the pathogenesis of MS/EAE via the HMGB1/TLR4/NF-κB signaling pathway (Paudel et al., 2019). Considering the role of HMGB1 in different pathologies, the goal of suppressing the autoimmune response with the identification of HMGB1 inhibitors has been achieved by experimental and clinical studies. For this purpose, anti-HMGB1 antibodies have been used in MS/EAE to demonstrate the effectiveness of autoimmune suppression in the course of the disease. It has been reported that administration of anti-HMGB1 monoclonal antibodies inhibit T cell infiltration and activation (Robinson et al., 2013). Similarly, ability of Glycyrrhizin extracted from licorice root, to inhibit the functions of HMGB1 in MS/EAE has been stated. In addition, it has been shown that Glycyrrhizin provides a significant decrease in the severity of EAE as well as inflammatory cell response and demyelination, decreases IL-17A, IL-6, TNF-α, IFN-γ and TGF-B1 in both serum and CSF, and decreases IL-4 level (Sun et al., 2018). Amelioration of EAE and significant reduction in IL-17 upregulation by HMGB1 monoclonal antibodies have been detected (Uzawa et al., 2013). As a result of these data, HMGB1/TLR4/NF-κB signaling pathway and agents with a strong neuroprotective effect targeting this signaling pathway may be considered as having an effective potential in the treatment of MS in future.

Conclusion

When considering treatment strategies, lots of treatment alternatives available for MS patients, especially those in the active phase. However, due to their short-term and limited efficacy, their side effects, and their inability to prevent relapses, advanced treatment options that are specific to the disease are needed. The pathogenesis-related signaling pathways appear as important key points in experimental and clinical studies for the development of effective treatment options. Based on the literature data, it is thought that HMGB1/TLR4/NF-κB signaling pathway might has a potential being a key target for new and promising therapeutic agents in experimental and clinical trials.

References

Andersson, Åsa, Covacu Ruxandra, Sunnemark Dan, Danilov I. Alexandre, Bianco Dal Assunta, Khademi Mohsen, Wallström Erik, Lobell Anna, Brundin Lou, Lassmann Hans, Harris A. Robert. 2008. "Pivotal advance: HMGB1 expression in active lesions of human and experimental multiple sclerosis." *J Leukoc Biol* 84 (5):1248-1255.

Andersson, U., H. Yang, and H. Harris. 2018. "Extracellular HMGB1 as a therapeutic target in inflammatory diseases." *Expert Opin. Ther. Targets* 22 (3):263-277. https://doi.org/10.1080/14728222.2018.1439924.

Aragón, Cristian C, AlfredoLlanos Julio, Osorio IvánPosso, Aristizábal IvanaNieto, Cardona NataliaPerea, FelipeOrtiz Andrés, Tobón Gabriel J. 2020. "The role of high mobility proteins (HMGB1) in systemic lupus erythematosus." *Colombian Journal of Rheumatology* 27 (1):37-45.

Barnie, Prince Amoah. 2019. "HMGB1 as A Therapeutic Target in Autoimmune Diseases: The Journey So Far." *Canadian Journal of Applied Science and Technology* 7 (3).

Beecham, Ashley H., Patsopoulos Nikolaos A., Xifara Dionysia K., Davis Mary F., Kemppinen Anu, Cotsapas Chris, Shah Tejas S., Spencer Chris, Booth David, Goris An, Oturai Annette, Saarela Janna, Fontaine Bertrand, Hemmer Bernhard, Martin Claes, Zipp Frauke, D'Alfonso Sandra, Bonesschi Filippo M., Taylor Bruce, Harbo Hanne F., Kockum Ingrid, Hillert Jan, Olsson Tomas, Ban Maria, Oksenberg Jorge R., Hintzen Rogier, Barcellos Lisa F. 2013. "Analysis of immune-related loci identifies 48 new susceptibility variants for multiple sclerosis." *Nat. Genet.* 45 (11):1353-60. https://doi.org/10.1038/ng.2770.

Bowman, Christal C, Rasley Amy, Tranguch Susanne L, Marriott Ian. 2003. "Cultured astrocytes express toll-like receptors for bacterial products." *Glia* 43 (3):281-291.

Bsibsi, Malika, Ravid Rivka, Gveric Djordje, van Noort Johannes M. 2002. "Broad expression of Toll-like receptors in the human central nervous system." *J Neuropathol Exp Neurol* 61 (11):1013-1021.

Buc, Milan J Mediators of inflammation. 2013. "Role of regulatory T cells in pathogenesis and biological therapy of multiple sclerosis." *Mediators Inflamm.* 2013.

Bucova, Maria, Majernikova Beata, Durmanova Vladimira, Cudrakova Daniela, Gmitterova Karin, Lisa Iveta, Klimova Eleonora, Kluckova Kristina, Buc Milan. 2020. "HMGB1 as a potential new marker of disease activity in patients with multiple sclerosis." *Neurol Sci.* 41 (3):599-604.

Celhar, T, R Magalhaes, and A-M J Immunologic research Fairhurst. 2012. "TLR7 and TLR9 in SLE: when sensing self goes wrong." *Immunol Res.* 53 (1-3):58-77.

Chen, Mingkun, Zhu Li, Xue Miao, Zhu Rongrong, Jing Liling, Wang Huaizhou, Qin Yanghua. 2021. "HMGB1, anti-HMGB1 antibodies, and ratio of HMGB1/anti-HMGB1 antibodies as diagnosis indicator in fever of unknown origin." *Sci Rep.* 11 (1):1-7.

Cildir, G., K. C. Low, and V. Tergaonkar. 2016. "Noncanonical NF-kappaB Signaling in Health and Disease." *Trends Mol. Med.* 22 (5):414-429. https://doi.org/10.1016/j.molmed. 2016.03.002.

Ciotti, J. R., and A. H. Cross. 2018. "Disease-Modifying Treatment in Progressive Multiple Sclerosis." *Curr. Treat. Options Neurol.* 20 (5):12. https://doi.org/10.1007/s11940-018-0496-3.

Ding, Junying, Cui, Xuran, Liu QingQuan. 2017. "Emerging role of HMGB1 in lung diseases: friend or foe." *J Cell Mol Med.* 21 (6):1046-1057.

Dumitriu, Ingrid E, Baruah Paramita, Valentinis Barbara, Voll Reinhard E., Herrmann Martin, Nawroth Peter P., Arnold Bernd, Bianchi Marco E., Manfredi Angelo A., Rovere-Querini Patrizia. 2005. "Release of high mobility group box 1 by dendritic cells controls T cell activation via the receptor for advanced glycation end products." *The Journal of Immunology* 174 (12):7506-7515.

English, C., and J. J. Aloi. 2015. "New FDA-Approved Disease-Modifying Therapies for Multiple Sclerosis." *Clin. Ther.* 37 (4):691-715. https://doi.org/10.1016/j.clinthera.2015.03.001.

Fang, P., M. Schachner, and Y. Q. Shen. 2012. "HMGB1 in development and diseases of the central nervous system." *Mol. Neurobiol.* 45 (3):499-506. https://doi.org/10.1007/s12035-012-8264-y.

Frank, Matthew G, Weber Michael D, Watkins linda R, Maier Steven F. 2015. "Stress sounds the alarmin: the role of the danger-associated molecular pattern HMGB1 in stress-induced neuroinflammatory priming." *Brain Behav Immun.* 48:1-7.

Fu, Li, Liu Keyue, Wake Hidenoi, Tesshigawara Kiyoshi, Yoshino Tadashi, Takahashi Hideo, Mori Shuji, Nishibori Masahiro. 2017. "Therapeutic effects of anti-HMGB1 monoclonal antibody on pilocarpine-induced status epilepticus in mice." *Sci Rep.* 7 (1):1-13.

Fujita, Kyota, Motoki Kazumi, Tagawa Kazuhiko, Chen Xiqui, Hama Hiroshi, Kazuyuki Nakajima, Homma Hidenori, Tamura Takuya, Watanabe Hirohisa, Masahisa Katsuno, Matsumi Chiemi, Kajikawa Masunori, Saito Takashi, Saido Takaomi, Sobue gen, Miyawaki Atsushi, Okazawa Hitoshi. 2016. "HMGB1, a pathogenic molecule that induces neurite degeneration via TLR4-MARCKS, is a potential therapeutic target for Alzheimer's disease." *Sci Rep.* 6 (1):1-15.

Garg, N., and T. W. Smith. 2015. "An update on immunopathogenesis, diagnosis, and treatment of multiple sclerosis." *Brain Behav* 5 (9):e00362. https://doi.org/10.1002/brb3.362.

Goulopoulou, S., C. G. McCarthy, and R. C. Webb. 2016. "Toll-like Receptors in the Vascular System: Sensing the Dangers Within." *Pharmacol. Rev.* 68 (1):142-67. https://doi.org/10.1124/pr.114.010090.

Guazzi, Stefania, Strangio Antonella, Franzi Adriano T, Bianchi Marco E. 2003. "HMGB1, an architectural chromatin protein and extracellular signalling factor, has a spatially and temporally restricted expression pattern in mouse brain." *Gene Expression Patterns* 3 (1):29-33.

Handunnetthi, L., S. V. Ramagopalan, and G. C. Ebers. 2010. "Multiple sclerosis, vitamin D, and HLA-DRB1*15." *Neurology* 74 (23):1905-10. https://doi.org/10.1212/WNL.0b013e3181e24124.

Harris, H. E., U. Andersson, and D. S. Pisetsky. 2012. "HMGB1: a multifunctional alarmin driving autoimmune and inflammatory disease." *Nat. Rev. Rheumatol.* 8 (4):195-202. https://doi.org/10.1038/nrrheum.2011.222.

Hashimoto, C., K. L. Hudson, and K. V. Anderson. 1988. "The Toll gene of Drosophila, required for dorsal-ventral embryonic polarity, appears to encode a transmembrane protein." *Cell* 52 (2):269-79. https://doi.org/10.1016/0092-8674(88)90516-8.

Hoesel, B., and J. A. Schmid. 2013. "The complexity of NF-kappaB signaling in inflammation and cancer." *Mol. Cancer* 12:86. https://doi.org/10.1186/1476-4598-12-86.

Janko, Christina, Filipović Milos, Munoz Luis E, Schorn Christine, Schett Georg, Ivanović-Burmazović Ivana, Herrmann Martin. 2014. "Redox modulation of HMGB1-related signaling." *Antioxid. Redox. Signal.* 20 (7):1075-85. https://doi.org/10.1089/ars.2013.5179.

Kadowaki, Atsushi, Miyake Sachiko, Saga Ryoko, Chiba Asako, Mochizuki Hideki, Yamamura Takashi. 2016. "Gut environment-induced intraepithelial autoreactive CD4(+) T cells suppress central nervous system autoimmunity via LAG-3." *Nat. Commun.* 7:11639. https://doi.org/10.1038/ncomms11639.

Kaltschmidt, B., and C. Kaltschmidt. 2009. "NF-kappaB in the nervous system." *Cold Spring Harb. Perspect. Biol.* 1 (3):a001271. https://doi.org/10.1101/cshperspect.a001271.

Kaskow, B. J., and C. Baecher-Allan. 2018. "Effector T Cells in Multiple Sclerosis." *Cold Spring Harb. Perspect. Med* 8 (4). https://doi.org/10.1101/cshperspect.a029025.

Kawai, T., and S. Akira. 2010. "The role of pattern-recognition receptors in innate immunity: update on Toll-like receptors." *Nat. Immunol.* 11 (5):373-84. https://doi.org/10.1038/ni.1863.

Kawasaki, T., and T. Kawai. 2014. "Toll-like receptor signaling pathways." *Front. Immunol.* 5:461. https://doi.org/10.3389/fimmu.2014.00461.

Kerfoot, Steven M., Long Elizabeth M, Hickey Michael J, Andonequi Graciela, Lapointe Benoit M, Zanardo Renata C O, Bonder Claudine, Jamer Will G, Robbins Stephen M, Kubes Paul. 2004. "TLR4 contributes to disease-inducing mechanisms resulting in central nervous system autoimmune disease." *J. Immunol.* 173 (11):7070-7. https://doi.org/10.4049/jimmunol.173.11.7070.

Kwak, Man S., Kim hee S, Lee Bin, Kim Young H, Son Myoungsun, Shin Jeon-Soo. 2020. "Immunological Significance of HMGB1 Post-Translational Modification and Redox Biology." *Front. Immunol.* 11:1189. https://doi.org/10.3389/fimmu.2020.01189.

Lee, Bettina L., moon Joanne E, Shu Jeffrey H, Yuan Lin, Newman Zachary R, Schekman Randy, Barton Gregory M. 2013. "UNC93B1 mediates differential trafficking of endosomal TLRs." *Elife* 2:e00291. https://doi.org/10.7554/eLife.00291.

Lehnardt, Seija, Lachance Christian, Patrizi Silvia, Lefebvre Sharon, Follett Pamela L, Jensen Frances E, Rosenberg Paul A, Volpe Joseph J, Vartania Timothy. 2002. "The toll-like receptor TLR4 is necessary for lipopolysaccharide-induced oligodendrocyte injury in the CNS." *J. Neurosci.* 22 (7):2478-86. https://doi.org/20026268.

Lehnardt, Seija, Massillon Leon, Follett Pamela, Jensen Frances E, Ratan Rajiv, Rosenberg Paul A, Volpe Joseph J, Vartanian Timothy. 2003. "Activation of innate immunity in the CNS triggers neurodegeneration through a Toll-like receptor 4-dependent pathway." *Proc. Natl. Acad. Sci. U S A* 100 (14):8514-9. https://doi.org/10.1073/pnas.1432609100.

Li, Jun, Shi Junyu, Sun Yan, Zheng Fang. 2018. "Glycyrrhizin, a Potential Drug for Autoimmune Encephalomyelitis by Inhibiting High-Mobility Group Box 1." *DNA Cell Biol.* 37 (12):941-946. https://doi.org/10.1089/dna.2018.4444.

Lindsey, J. W., S. K. Agarwal, and F. K. Tan. 2011. "Gene expression changes in multiple sclerosis relapse suggest activation of T and non-T cells." *Mol. Med.* 17 (1-2):95-102. https://doi.org/10.2119/molmed.2010.00071.

Liu, A. H., Y. T. Wu, and Y. P. Wang. 2017. "MicroRNA-129-5p inhibits the development of autoimmune encephalomyelitis-related epilepsy by targeting HMGB1 through the TLR4/NF-kB signaling pathway." *Brain Res Bull.* 132:139-149. https://doi.org/10.1016/j.brainresbull.2017.05.004.

Lo Coco, Daniele, Veglianese, Pietro, Allievi Elisa, Bendotti Caterina. 2007. "Distribution and cellular localization of high mobility group box protein 1 (HMGB1) in the spinal cord of a transgenic mouse model of ALS." *Neurosci. Lett.* 412 (1):73-7. https://doi.org/10.1016/j.neulet.2006.10.063.

Lublin, Fred D., reingold Stephen C, Cohen Jeffrey A, Cutter Gary R, Sorensen Per Soelberg, Thompson Alan J, Wolinsky Jerry S, Balcer Laura J,Banwell Brenda, Barkhof Frederik, Bebo Jr Bruce, Calabresi Peter A, Clanet Michel, Comi Giancarlo, Fox Robert J, Freedman Mark S, Goodman Andrew D, Inglese Matilde, Kappos Ludwig, Kieseier Bernd C, Lincoln John A, Lubetzki Catherine, Miller Aaron E, Montalban Xavier, O'Connor Paul W, Petkau John, Pozzilli Carlo, Rudick Richard A, Sormani Maria Pia, Stüve Olaf, Waubant Emmanuelle, Polman Chris H. 2014. "Defining the clinical course of multiple sclerosis: the 2013 revisions." *Neurology* 83 (3):278-86. https://doi.org/10.1212/wnl.0000000000000560.

Malhotra, Sunny, Fissolo Nicolas, Tintoré Mar, Wing Ana Cristina, Castilló Joaquin, Vidal-Jordana Angela, Montalban Xavier, Comabella Manuel. 2015. "Role of high mobility group box protein 1 (HMGB1) in peripheral blood from patients with multiple sclerosis." *J. Neuroinflammation* 12:48. https://doi.org/10.1186/s12974-015-0269-9.

Marrodan, Mariano, Alessandro Lucas, Farez Mauricio F, Correale Jorge. 2019. "The role of infections in multiple sclerosis." *Mult. Scler.* 25 (7):891-901. https://doi.org/10.1177/1352458518823940.

Marta, M., Andersson Asa, Isaksson Magnus, Kämpe Olle, Lobell Anna. 2008. "Unexpected regulatory roles of TLR4 and TLR9 in experimental autoimmune encephalomyelitis." *Eur. J. Immunol.* 38 (2):565-75. https://doi.org/10.1002/eji.2007 37187.

Marta, M., U. C. Meier, and A. Lobell. 2009. "Regulation of autoimmune encephalomyelitis by toll-like receptors." *Autoimmun. Rev.* 8 (6):506-9. https://doi.org/10.1 016/j.autrev.2009.01.006.

Martinotti, S., M. Patrone, and E. Ranzato. 2015. "Emerging roles for HMGB1 protein in immunity, inflammation, and cancer." *Immunotargets Ther.* 4:101-9. https://doi.org/10.2147/itt.S58064.

Mc Guire, Conor, Prinz Marco, Beyaert Rudi, van Loo Geert. 2013. "Nuclear factor kappa B (NF-kappaB) in multiple sclerosis pathology." *Trends Mol. Med.* 19 (10):604-13. https://doi.org/10.1016/j.molmed.2013.08.001.

McCool, Rachael, Wilson Katy, Arber Mick, Fleetwood Kelly, Toupin Sydney, Thom Howard, Bennett Iain, Edwards Susan. 2019. "Systematic review and network meta-analysis comparing ocrelizumab with other treatments for relapsing multiple sclerosis." *Mult. Scler. Relat. Disord.* 29:55-61. https://doi.org/10.1016/j.msard.2018.12.040.

Medzhitov, R., and C. A. Janeway, Jr. 1997. "Innate immunity: the virtues of a nonclonal system of recognition." *Cell* 91 (3):295-8. https://doi.org/10.1016/s0092-8674(00) 80412-2.

Medzhitov, R., P. Preston-Hurlburt, and C. A. Janeway, Jr. 1997. "A human homologue of the Drosophila Toll protein signals activation of adaptive immunity." *Nature* 388 (6640):394-7. https://doi.org/10.1038/41131.

Oeckinghaus, A., M. S. Hayden, and S. Ghosh. 2011. "Crosstalk in NF-κB signaling pathways." *Nat. Immunol.* 12 (8):695-708. https://doi.org/10.1038/ni.2065.

Olson, J. K., A. M. Girvin, and S. D. Miller. 2001. "Direct activation of innate and antigen-presenting functions of microglia following infection with Theiler's virus." *J. Virol.* 75 (20):9780-9. https://doi.org/10.1128/jvi.75.20.9780-9789.2001.

Olson, J. K., and S. D. Miller. 2004. "Microglia initiate central nervous system innate and adaptive immune responses through multiple TLRs." *J. Immunol.* 173 (6):3916-24. https://doi.org/10.4049/jimmunol.173.6.3916.

Orton, S. M., Wald L, Confavreux C, Vukusic S, Krohn J P, Ramagopalan S V, Herrera B M, Sadovnick A D, Ebers G C. 2011. "Association of UV radiation with multiple sclerosis prevalence and sex ratio in France." *Neurology* 76 (5):425-31. https://doi.org/10.1212/WNL.0b013e31820a0a9f.

Pang, Y., Z. Cai, and P. G. Rhodes. 2003. "Disturbance of oligodendrocyte development, hypomyelination and white matter injury in the neonatal rat brain after intracerebral injection of lipopolysaccharide." *Brain Res. Dev. Brain Res.* 140 (2):205-14. https://doi.org/10.1016/s0165-3806(02)00606-5.

Park, M. H., and J. T. Hong. 2016. "Roles of NF-κB in Cancer and Inflammatory Diseases and Their Therapeutic Approaches." *Cells* 5 (2). https://doi.org/10.3390/cells5020015.

Patsopoulos, Nikolaos A., Barcellos Lisa F, Hintzen Rogier Q, Schaefer Catherine, van Duijn Cornelia M, Noble Janella A, Oksenberg Jorge, Olsson Tomas, Taylor Bruce V, Sawcer Stephen, Hafler David A, Carrington Mary, De Jager Philip L, de Bakker Paul I W. 2013. "Fine-mapping the genetic association of the major histocompatibility complex in multiple sclerosis: HLA and non-HLA effects." *PLoS Genet* 9 (11):e1003926. https://doi.org/10.1371/journal.pgen.1003926.

Paudel, Yam. N., Angelopoulou Efthalia, K C Bhuvan, Piperi Christina, Othman lekhsan. 2019. "High mobility group box 1 (HMGB1) protein in Multiple Sclerosis (MS): Mechanisms and therapeutic potential." *Life Sci.* 238:116924. https://doi.org/10.1016/j.lfs.2019.116924.

Paudel, Yam N., Angelopoulou Efthalia, Piperi Christina, Othman lekhsan, Shaikh Mohd F. 2020a. "HMGB1-Mediated Neuroinflammatory Responses in Brain Injuries: Potential Mechanisms and Therapeutic Opportunities." *Int. J. Mol. Sci.* 21 (13). https://doi.org/10.3390/ijms21134609.

Paudel, Yam N., Angelopoulou Efthalia, Piperi Christina, Othman lekhsan, Shaikh Mohd F. 2020b. "Implication of HMGB1 signaling pathways in Amyotrophic lateral sclerosis (ALS): From molecular mechanisms to pre-clinical results." *Pharmacol. Res.* 156:104792. https://doi.org/10.1016/j.phrs.2020.104792.

Pierrot-Deseilligny, C., and J. C. Souberbielle. 2017. "Vitamin D and multiple sclerosis: An update." *Mult. Scler. Relat. Disord.* 14:35-45. https://doi.org/10.1016/j.msard.2017.03.014.

Ponath, Gerald, Lincoln Matthew R, Levine-Ritterman Maya, Park Calvin, Dahlawi Somiah, Mubarak Mayyan, Sumida Tomokazu, Airas Laura, zhang Shun, Isitan Cigdem, Nguyen Tnanh D, Raine Cedric s, Hafler David a, Pitt David. 2018. "Enhanced astrocyte responses are driven by a genetic risk allele associated with multiple sclerosis." *Nat. Commun.* 9 (1):5337. https://doi.org/10.1038/s41467-018-07785-8.

Robinson, Andrew P., Caldis Matthew W, Harp Christopher T, Goings Gwendolyn E, Miller Stephen D. 2013. "High-mobility group box 1 protein (HMGB1) neutralization ameliorates experimental autoimmune encephalomyelitis." *J. Autoimmun.* 43:32-43. https://doi.org/10.1016/j.jaut.2013.02.005.

Sasaki, Tatsuya, Liu Keyue, Agari Takashi, Yasuhara Takao, Morimoto Jun, Okazaki Mihoko, Takeuchi Hayato, Toyoshima Atsuhiko, Sasada Susumu, Shinko Aiko, Kondo Akihiko, Kameda Masahiro, Miyazaki Ikuko, Asanuma Masato, Borlongan Cesario V, Nishibori Masahiro, Date Isao. 2016. "Anti-high mobility group box 1 antibody exerts neuroprotection in a rat model of Parkinson's disease." *Exp. Neurol.* 275 Pt 1:220-31. https://doi.org/10.1016/j.expneurol.2015.11.003.

Satoh, Jun-ichi., Misawa Tamako, Tabunoki Hiroko, Yamamura Takashi. 2008. "Molecular network analysis of T-cell transcriptome suggests aberrant regulation of gene expression by NF-kappaB as a biomarker for relapse of multiple sclerosis." *Dis. Markers* 25 (1):27-35. https://doi.org/10.1155/2008/824640.

Sawcer S, Hellenthal G, Pirinen M, Spencer CC. 2011. "Genetic risk and a primary role for cell-mediated immune mechanisms in multiple sclerosis." *Nature* 476 (7359):214-9. https://doi.org/10.1038/nature10251.

Socha, Katarzyna, Karpińska Elżbieta, Kochanowicz Jan, Soroczyńska Jolanta, Jakoniuk Marta, Wilkiel Marianna, Mariak Zenon D, Borawska Maria H. 2017. "Dietary habits; concentration of copper, zinc, and Cu-to-Zn ratio in serum and ability status of patients with relapsing-remitting multiple sclerosis." *Nutrition* 39-40:76-81. https://doi.org/10.1016/j.nut.2017.03.009.

Sternberg, Zohara, Sternberg Daniel, Chichelli Trevor, Drake Allison, Patel Neel, Kolb Chana, Chadha Kailash, Yu Jinhee, Hojnacki David. 2016. "High-mobility group box 1 in multiple sclerosis." *Immunol. Res.* 64 (2):385-91. https://doi.org/10.1007/s12026-015-8673-x.

Stros, M. 2010. "HMGB proteins: interactions with DNA and chromatin." *Biochim. Biophys. Acta* 1799 (1-2):101-13. doi: 10.1016/j.bbagrm.2009.09.008.

Sun, Shao-Cong. 2017. "The non-canonical NF-κB pathway in immunity and inflammation." *Nature Reviews Immunology* 17 (9):545-558. https://doi.org/10.1038/nri.2017.52.

Sun, Yan, Chen Huoying, Dai Jiapei, Wan Zhongjun, Xiong Ping, Xu Yong, Han Zhengrong, Chai Weitai, Gong Feili, Zheng Fang. 2018. "Glycyrrhizin Protects Mice Against Experimental Autoimmune Encephalomyelitis by Inhibiting High-Mobility Group Box 1 (HMGB1) Expression and Neuronal HMGB1 Release." *Front. Immunol.* 9:1518. https://doi.org/10.3389/fimmu.2018.01518.

Sun, Zhezhe, Nyanzu Mark, Yang Su, Zhu Xiaohong, Wang Kankai, Ru Junnan, Yu Enxing, Zhang Hengli, Wang Zhenzzhong, Shen Jie, Zhuge Qichuan, Huang Lijie. 2020. "VX765 Attenuates Pyroptosis and HMGB1/TLR4/NF-kappaB Pathways to Improve Functional Outcomes in TBI Mice." *Oxid Med. Cell Longev.* 2020:7879629. https://doi.org/10.1155/2020/7879629.

Sundberg, Eric, Fasth Andreas E r, Palmblad Karin, Harris Erlandsson Helena, Andersson Ulf. 2009. "High mobility group box chromosomal protein 1 acts as a proliferation signal for activated T lymphocytes." *Immunobiology* 214 (4):303-9. https://doi.org/10.1016/j.imbio.2008.09.006.

Trager, Nicole, Butler Jonathan T, Hague Azizul, Ray Swapan K, Beeson Craig, Banik Naren L. 2013. "The Involvement of Calpain in CD4(+) T Helper Cell Bias in Multple Sclerosis." *J. Clin. Cell Immunol.* 4 (4):1000153. https://doi.org/10.4172/2155-9899.1000153.

Trotta, Teresa, Porro Chiara, Calvello Rosa, Panaro Maria A. 2014. "Biological role of Toll-like receptor-4 in the brain." *J. Neuroimmunol.* 268 (1-2):1-12. https://doi.org/10.1016/j.jneuroim.2014.01.014.

Uzawa, A., Mori M, Taniguchi J, Masuda S, Muto M, Kuwabara S. 2013. "Anti-high mobility group box 1 monoclonal antibody ameliorates experimental autoimmune encephalomyelitis." *Clin. Exp. Immunol.* 172 (1):37-43. https://doi.org/10.1111/cei.12036.

van Langelaar, Jamie, van der Vuurst de Vries Roos M, Janssen Malou, Wierenga-Wolf Annet F, Spilt Isis M, Siepman Theodora A, Dankers Wendy, Verjans Georges M G M, de Vries Helga E, Lubberts Erik, Hintzen Rogier Q, van Luijn Marvin M. 2018. "T helper 17.1 cells associate with multiple sclerosis disease activity: perspectives for early intervention." *Brain* 141 (5):1334-1349. https://doi.org/10.1093/brain/awy069.

Varol, C., E. Zigmond, and S. Jung. 2010. "Securing the immune tightrope: mononuclear phagocytes in the intestinal lamina propria." *Nat. Rev. Immunol.* 10 (6):415-26. https://doi.org/10.1038/nri2778.

Verstrepen, L., and R. Beyaert. 2014. "Receptor proximal kinases in NF-κB signaling as potential therapeutic targets in cancer and inflammation." *Biochem. Pharmacol.* 92 (4):519-29. https://doi.org/10.1016/j.bcp.2014.10.017.

Wadachi, R., and K. M. Hargreaves. 2006. "Trigeminal nociceptors express TLR-4 and CD14: a mechanism for pain due to infection." *J. Dent. Res.* 85 (1):49-53. https://doi.org/10.1177/154405910608500108.

Wan, Zhikai, Zhou Zhifeng, Liu Yao, Lai Yuhan, Luo Yuan, Peng Xiaoping, Zou Wei. 2020. "Regulatory T cells and T helper 17 cells in viral infection." *Scand. J. Immunol.* 91 (5):e12873. https://doi.org/10.1111/sji.12873.

Wang, Honghao, Wang Kai, Wang Conghui, Xu Fangcheng, Zhong Xiaonan, Qiu Wei, Hu Xueqiang. 2013. "Cerebrospinal fluid high-mobility group box protein 1 in neuromyelitis optica and multiple sclerosis." *Neuroimmunomodulation* 20 (2):113-8. https://doi.org/10.1159/000345994.

Yan, Jun, Winterford Clay M, Catts Vibeke S, Pat Betty K, Pender Michael P, McCombe Pamela A, Greer Judith M. 2018. "Increased constitutive activation of NF-κB p65 (RelA) in peripheral blood cells of patients with progressive multiple sclerosis." *J. Neuroimmunol.* 320:111-116. https://doi.org/10.1016/j.jneuroim.2018.04.002.

Yang, H., H. Wang, and U. Andersson. 2020. "Targeting Inflammation Driven by HMGB1." *Front. Immunol.* 11:484. https://doi.org/10.3389/fimmu.2020.00484.

Yi, Hongwei, Bai Ying, Zhu Xianjian, Lin Lin, Zhao Lei, Wu Xiaodong, Buch Shilpa, Wang Longxin, Chao Jie, Yao Honghong. 2014. "IL-17A induces MIP-1α expression in primary astrocytes via Src/MAPK/PI3K/NF-kB pathways: implications for multiple sclerosis." *J. Neuroimmune Pharmacol.* 9 (5):629-41. https://doi.org/10.1007/s11481-014-9553-1.

Zhang, Q., M. J. Lenardo, and D. Baltimore. 2017. "30 Years of NF-kappaB: A Blossoming of Relevance to Human Pathobiology." *Cell* 168 (1-2):37-57. https://doi.org/10.1016/j.cell.2016.12.012.

Zhen, C., Zhen, Chao, Wang Yabo, Li Dongmei, Zhang Weijie, Zhang Hanzhe, Yu Xinjuan, Wang Xin.2019. "Relationship of High-mobility group box 1 levels and multiple sclerosis: A systematic review and meta-analysis." *Mult Scler Relat Disord* 31:87-92. https://doi.org/10.1016/j.msard.2019.03.030.

Zheng, Chao, Chen Jingtao, Chu Fengna, Zhu Jie, Jin Tao. 2019. "Inflammatory Role of TLR-MyD88 Signaling in Multiple Sclerosis." *Front Mol Neurosci* 12:314. https://doi.org/10.3389/fnmol.2019.00314.

Zhou, Yifan, Cui Chunping, Ma Xiaoyu, Luo Wenjing, Zheng Song G, Qiu Wei. 2020. "Nuclear Factor κB (NF-κB)-Mediated Inflammation in Multiple Sclerosis." *Front Immunol* 11:391. https://doi.org/10.3389/fimmu.2020.00391.

Zimmermann, Marita, Brouwer Elizabeth, Tice Jeffrey A, Seidner Matt, Loos Anne M, Liu Shanshan, Chapman Richard H, Kumar Varun, Carlson Josh J. 2018. "Disease-Modifying Therapies for Relapsing-Remitting and Primary Progressive Multiple Sclerosis: A Cost-Utility Analysis." *CNS Drugs* 32 (12):1145-1157. https://doi.org/10.1007/s40263-018-0566-9.

Chapter 8

HMGB1 Related Pathways in Epilepsy

Hayriye Soyturk[1], PhD, Umit Kilic[2], PhD, Cansu Onal[3], PhD and Aysegül Yildiz[4], MSc

[1]Bolu Abant Izzet Baysal University,
Department of Interdisciplinary Neuroscience, Bolu, Turkey
[2]Duzce University,
Vocational School of Health Services, Duzce, Turkey
[3]Bolu Abant Izzet Baysal University,
Department of Biology, Bolu, Turkey
[4]Bolu Abant Izzet Baysal University,
Department of Physiology, Bolu, Turkey

Abstract

Epilepsy is one of the most common neurological illnesses, affecting an estimated 70 million individuals worldwide. A mechanistic approach to discovering crucial pathogenic brain abnormalities driving seizure onset, recurrence, and advancement is critical for generating novel and rational treatment methods that may modify or prevent disease progression. In this context, an increasing body of evidence suggests that neuroinflammation plays a role in drug-resistant epilepsies. The high mobility group box 1 (HMGB1) protein, a key regulator of neuroinflammation, has been associated with several neurological disorders, including epilepsy. Increased levels of HMGB1, one of the most prominent pro-inflammatory cytokines, have been linked to seizure length, recurrence, and epilepsy development. Therefore, HMGB1 has attracted a lot of attention and has been developed quickly. According to clinical and experimental data, HMGB1 isoforms may serve as neurobiological biomarkers for epileptogenesis and drug-resistant epilepsy. These novel findings suggest that the HMGB1 system could be targeted to prevent seizure generation and provide therapeutically

In: HMGB1
Editors: Eylem Taskin, Celal Guven and Salih Tunc Kaya
ISBN: 979-8-88697-408-9

relevant prognostic markers that can also predict a patient's response to therapy.

This section will recap epileptogenesis theory and examine the relationships between epilepsy and neuroinflammatory pathways using experimental and clinical evidence.

Keywords: epilepsy, HMGB1, neuroinflammation

Introduction

Epilepsy is one of the most common major neurological illnesses, affecting around 70 million people worldwide (Wang et al. 2021). Although antiepileptic drugs, special diets, vagal nerve stimulation, and surgical methods are used in the treatment, epilepsy still cannot be controlled in approximately 30% of the patients. Despite all the advances, the mechanism of epileptogenesis is still unclear.

The epileptogenesis process includes a variety of molecular and cellular changes that can be used as potential targets for the treatment and prevention of epilepsy. Neuroinflammation is considered the primary factor in epileptogenesis. It is known that reactive oxygen species (ROS) play a very important role as mediators in the process of neuronal damage. It is thought that ROS may play a very important epileptic role in epilepsy, including proinflammatory cytokine production and microglial activation.

Extensive studies on epilepsy in recent years have revealed the complexity of epileptogenesis in which immunological processes, epigenetic modifications, and structural changes in neuronal tissues have been identified as playing a crucial role. High mobility group box-1 (HMGB1) has recently emerged as a potential biomarker of epilepsy. This section has been prepared to provide a broad perspective on neuroinflammation and epilepsy and to discuss the latest developments in the High mobility group box-1 (HMGB1), one of the biomarkers of epilepsy.

Epilepsy Definition and Classifications

Epilepsy is characterized by a lengthy history of spontaneous epileptic seizures and is associated with a variety of neurological, cognitive, and psychosocial repercussions (Fisher et al. 2014). The causes of epilepsy are unknown to many people. Furthermore, epilepsy is frequently associated with neurological comorbidities such as cognitive deficits, anxiety, and depression, as well as mental abnormalities such as autism spectrum disorders, all of which have a severe influence on the quality of life of patients. The prevalence and severity of these comorbidities differ according to causes of epilepsy and age at which the illness reveals itself (Devinsky et al. 2018).

It is most frequent in children and the second most common in adults. Epilepsy is classified into four kinds based on its etiology: idiopathic, cryptogenic, symptomatic, and induced. Symptomatic epilepsy is described as epilepsy in which neuroanatomical and neuropathological reasons such as trauma, infection, and space-occupying lesions have been established as the etiological cause. Cryptogenic epilepsy affects 20% of all epilepsy patients (Shorvon 2011).

Several types of brain trauma, including brain trauma, which is one of the most common causes of idiopathic epilepsy, can result in acquired epilepsy (Hunt, Boychuk, and Smith 2013). Epilepsy is defined as two unprovoked seizures occurring more than 24 hours apart (Fisher et al. 2014). Epilepsy is considered resolved in those who had an age-dependent disease but have passed the required age and are seizure-free, or in people who have been seizure-free for the preceding 10 years with no medication for the previous 5 years in other cases of epilepsy (Fisher et al. 2014).

People are frequently diagnosed with epilepsy, even though the diagnosis should be as specific and clear as possible. Seizures, epilepsy, and syndromes are classified into three types. Causes and comorbidities should be examined at each stage because they may have treatment consequences. There are six types of causes: genetic, structural, metabolic, infectious, immunological, and unknown (Scheffer et al. 2017).

Seizures are classified initially based on their onset, which might be focal, generalized, or indeterminate. The degree of consciousness differs between focused seizures with intact awareness and those with impaired awareness. The earliest and most obvious motor or non-motor manifestation of focal seizures is utilized to further characterize them (Fisher et al. 2017). Seizures are classed as either motor or non-motor (absence). Seizures with an unknown onset may have features that can still be diagnosed. A frequent occurrence is someone presenting with convulsions but no clinical indicators of a concentrated or generalized onset. These seizures are classified as tonic-clonic seizures with unknown onset (Fisher et al. 2017).

On the other hand, epilepsy, defined by spontaneous and recurrent seizures, is a common brain condition associated with an elevated death rate, reduced social engagement, and worse quality of life (Fukata and Fukata 2017; Henshall et al. 2016; Krishnamurthy 2016). Epilepsy etiology is linked to large changes in the expression of genes that govern neurotransmitter signaling, ion channels, synaptic structure, neuronal death, gliosis, and inflammation (Henshall et al. 2016). Inflammation caused by a stroke, infection, or neurotrauma, for example, appears to be linked to an increase in the occurrence of epilepsy. Furthermore, several inflammatory mediators, such as cytokines, complement factors, and prostaglandins, have been discovered to have roles in epileptogenesis (de Vries et al. 2016; Vezzani et al. 2011; Vezzani, Lang, and Aronica 2015; Walker and Sills 2012).

High Mobility Group Box 1 (HMGB1)

High mobility group box 1 (HMGB1) is highly conserved throughout evolution (Muller, Ronfani, and Bianchi 2004), and while it is produced by nearly all cell types that stabilize nucleosome formation and facilitate transcription (van Beijnum, Buurman, and Griffioen 2008), its nuclear localization can change with development and age (Muller, Ronfani, and Bianchi 2004).

The high mobility group protein (HMG) is a structural protein found in the eukaryotic nucleus, which alters, bends, and changes the structure of DNA (Bianchi 2004; Stros 2010; Tang et al. 2011). The HMGB1 protein ordinarily translocates from the cytosol to the nucleus, where it interacts with DNA and regulates transcription (Bianchi 2004). At least two nuclear localization signals (NLS) govern HMGB1 nuclear translocation: NLS1 (amino acids 28–44) and NLS2 (amino acids 180–185) (Bonaldi et al. 2003).

The HMG superfamily is made up of three subfamilies: HMGA, HMGB, and HMGN (Bustin 2001). There are three proteins in the HMGB protein family: HMGB1, HMGB2, and HMGB3 (previously HMG4 or HMG2B). Previous studies have revealed that the pro-inflammatory cytokine HMGB1 exists in two forms: intracellular HMGB1, which is primarily involved in gene transcription and regulation, and extracellular HMGB1, which functions as an inflammatory factor and promotes tumor metastasis and inflammatory response (Venereau et al. 2016). HMGB1 can connect to a variety of receptors, including the receptor for advanced glycation end products (RAGE), toll-like receptor (TLR) 2, and TLR4 (Xie et al. 2017). By binding to TLR2, TLR4, or RAGE, HMGB1 can enhance cytokine activation, altering downstream inflammatory variables that play a significant role in the control of neuronal excitability and the development of epilepsy (Paudel et al. 2018).

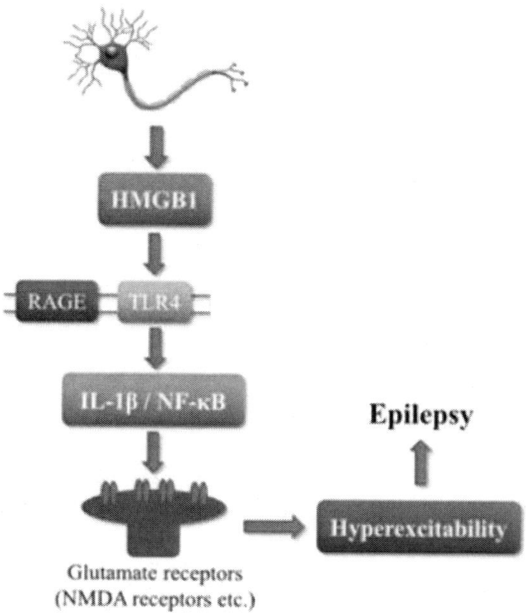

Figure 1. Views on the role of HMGB1 in epilepsy (Dai et al. 2021).

HMGB1- deficient cell lines grow normally, but activation of gene expression by numerous stimuli, including glucocorticoid receptors, is hampered (Calogero et al. 1999). HMGB1 acts as a "necrotic signal" for the immune system when it is present in the extracellular environment. A new study found that damaged or necrotic cells can release HMGB1 into the extracellular environment, resulting in inflammatory reactions. In contrast to necrosis, cells that undergo programmed cell death or apoptosis cause very little inflammation in the surrounding tissue (Magna and Pisetsky 2014; Savill et al. 2002). This is due in part to the retention of HMGB1 within apoptotic cells (Scaffidi, Misteli, and Bianchi 2002). Indeed, cells release HMGB1 into the extracellular environment in two ways. The first way includes "passive release" of HMGB1 from wounded or necrotic cells. HMGB1 is an intracellular marker used by the innate immune system to detect tissue damage and activate reparative responses in this context (Dumitriu et al. 2005; Scaffidi, Misteli, and Bianchi 2002).

Extracellular HMGB1 functions as an immune-stimulatory signal that shows the level of tissue injury (Scaffidi, Misteli, and Bianchi 2002), encourages the recruitment of mononuclear cells to remove cellular debris, and protects against infection, which frequently follows trauma (Bianchi 2004). The second mechanism is an "active release" of HMGB1 from immune cells to operate as a pro-inflammatory cytokine following an immunological challenge (Bonaldi et al. 2003; Gardella et al. 2002). Extracellular HMGB1 can orchestrate defensive inflammatory responses to ischemia, injury, infection, or sepsis and drive tissue regeneration. HMGB1 is a suitable immunological marker employed by the innate immune system to communicate tissue injury from an evolutionary standpoint (Scaffidi, Misteli, and Bianchi 2002). HMGB1 can also be "actively released" by immune cells to encourage tissue repair via a necrotic-like process (Park et al. 2003).

HMGB1 is acetylated on lysine residues inside nuclear localization signals when macrophages are activated by an immunological insult. Acetylation of these domains is thought to inhibit nuclear HMGB1 translocation, resulting in hyper-acetylated HMGB1 accumulating in the cytosol and packing into specialized secretory lysosomes (Gardella et al. 2002).

Although the physiological effects of the HMGB1 release mechanism are unknown, acetylation disrupts HMGB1's interactions with other nuclear proteins, potentially affecting its biological activity and inflammatory potential (Ugrinova et al. 2001). The discovery of HMGB1 isoforms and their contributions to viral and inflammatory illnesses (Yang et al. 2004) is of great interest. Which may also be exposed to ADP ribose, glycosylation, phosphorylation, and methylation (Ugrinova et al. 2001), each of which may contribute to the biological and inflammatory activity of HMGB in varying degrees. The autocrine effect of HMGB1 on the cells from which it is secreted causes a delay in the synthesis of proinflammatory cytokines in neutrophils (Andersson et al. 2000).

HMGB1 is a ligand for the RAGE (Huttunen et al. 2002). RAGE is a member of the immunoglobulin superfamily whose gene is found in humans and mice near the major histocompatibility complex (MHC) (Schmidt and Stern 2001). RAGE has been found on neurons (Huttunen et al. 2000), endothelium (Kuniyasu, Chihara, and Takahashi 2003), macrophages (Kokkola et al. 2005), and T and immature dendritic cells (Dumitriu et al. 2005). Even though RAGE is expressed on T cells, it is unknown if RAGE or other receptors play a role in HMGB1-induced T cell survival. Dendritic cell maturation is promoted by HMGB1 via RAGE, but anti-RAGE neutralizing antibodies significantly reduce dendritic cell maturation in response to Lipopolysaccharide (LPS) stimulation (Dumitriu et al. 2005). RAGE appears to be significant in inflammatory responses, but other receptors are also likely to play a role in HMGB1-mediated adaptive immune responses.

HMGB1 stimulates p38 Mitogen-Activated Protein Kinase (MAPK) in murine macrophages and human neutrophils (Li et al. 2003). It also promotes transient phosphorylation of p38MAPK, p44/p42 MAPK, and stress-activated MAP kinase (SAPKs)/ c-Jun N-terminal kinase (JNK) SAPK/JNK in rat macrophages. These kinases were acting independently of RAGE, as RAGE knockout mice showed no alterations. RAGE signaling, on the other hand, leads to HMGB1-induced pro-inflammatory cytokine production, which was reduced by 70% in RAGE knockout animals. Because this effect was not full, it suggests that other receptors are involved in HMGB1 signaling (Kokkola et al. 2005).

In rat macrophages (Kokkola et al. 2005), human dendritic cells (Messmer et al. 2004), and human neutrophils (Park et al. 2003), HMGB1 induces nuclear translocation of NF-kB. In human myeloid dendritic cells, recombinant B-box stimulates nuclear translocation and activation of NF-kB family members p65 and p50 (Messmer et al., 2004), but only p65 in RAW264.7 cells (Li et al. 2003). HMGB1 activates p38 MAPK in RAW264.7 cells (Li et al. 2003) and increases CD83 and interleukin-6 (IL-6) production in human dendritic cells through a p38 MAPK dependent mechanism (Kokkola et al. 2005).

Endogenous HMGB1 produced by LPS-stimulated dendritic cells activates ERK1/2, NF-kB, and, to a lesser extent, p38 MAPK (Dumitriu et al. 2005). Dendritic cell survival and development are influenced by many signaling systems. Survival is governed by the ERK pathway, whereas dendritic cell maturation is governed by NF-kB (Rescigno et al. 1998).

By activating both routes, HMGB1 promotes dendritic cell development and survival. TLR2 and TLR4 are also implicated in HMGB1-induced NF-kB activation, according to transfection experiments using dominant-negative toll-like receptors in murine macrophage-like cell cultures. The ability of necrotic cells to optimally boost NF-kB activity in macrophages is dependent, at least in part, on a TLR-mediated signaling pathway, and both TLR-2 and TLR-4 have been implicated in macrophage activation by HMGB1 (Park et al. 2004).

Experiments with TLR2-knockout mice and C3H/HeJ mice, both of which lack a functional TLR4 receptor, revealed that neither receptor is required for HMGB1-induced NF-kB activation in murine neutrophils (Park et al. 2004). TNF-alpha production from HMGB1-stimulated murine macrophages is unchanged in TLR2-deficient mice (Kokkola et al. 2005). TLR2, TLR4, TLR9, and Myeloid differentiation primary response 88 (Myd88) were found to be inactive in murine dendritic cells during recombinant B-box-induced cytokine production.

In a comprehensive study by (Das et al. 2016) they showed that the C terminal tail of HMGB1 binding to TLR5 initiates activation of the NF-kB signaling pathway in a MyD88-dependent manner, resulting in pro-inflammatory cytokine production and increased pain *in vivo*.

Inflammation and Neuroinflammation in the Nervous System

Inflammation is a vital biological process that occurs when cells or tissues are injured, unwell, or traumatized. Invading microorganisms are removed by an effective inflammatory response mechanism, which also initiates wound healing and angiogenesis (Carson, Thrash, and Walter 2006).

Activated microglial cells have a key role in immunological and inflammatory responses in the central nervous system, as well as in neurodegenerative diseases. Many pro-apoptotic pathways are mediated by signaling molecules produced during neuroinflammation (Shabab et al. 2017).

NF-kB, a transcription factor, initiates and regulates the development of several inflammatory processes in glial cells during inflammation, which has been connected to the pathophysiology of several neurodegenerative diseases.

Neuroinflammation is a response that affects all central nervous system (CNS) cells, including neurons and microglia. Some factors, such as the initial insult, genetic background, environmental circumstances, age, or previous experiences, all contribute to microglia activation and the complicated neuroinflammatory cascade (Bradl and Hohlfeld 2003; Carson, Thrash, and Walter 2006). Many neurodegenerative diseases, including multiple sclerosis (MS), Alzheimer's disease (AD), Parkinson's disease (PD), and autism, are characterized by neuroinflammation (Carson, Thrash, and Walter 2006).

LPS, an endotoxin found in Gram-negative bacteria's outer membrane, causes systemic inflammatory response syndrome via TLR activation (Lyman et al. 2014). LPS binding to TLR4 on the surface of microglia activates multiple signaling pathways, including PI3K/AKT, mitogen-activated protein kinase (MAPK), and mammalian target of rapamycin (mTOR), which all activate NF-KB. Following that, activation of NF-KB induces the production of pro-inflammatory cytokines, chemokines, and inducible enzymes such as inducible nitric oxide synthase (iNOS) and cyclooxygenase-2 (COX-2), all of which contribute to neuroinflammation (Glass et al. 2010; Park et al. 2011).

HMGB1 is released from necrotic neurons during traumatic brain damage via an N-methyl-D-aspartate (NMDA) receptor subunit 2 B-mediated mechanism. HMGB1 production is associated with post-traumatic coagulation difficulties, complement activation, and significant systemic inflammatory responses after severe injury and tissue hypoperfusion (Richard et al. 2017). HMGB1 is also released by macrophages, according to other research (Lamkanfi et al. 2010) findings indicate that HMGB1 is produced by neurons, glial cells like astrocytes and microglia, and immunological cells like macrophages (Lamkanfi et al. 2010). Furthermore, immune cells like macrophages and monocytes can actively release HMGB1 when stimulated by cytokines like interferon, TNF, and IL-1, or pathogen-derived chemicals like lipopolysaccharide (Willingham et al. 2009). Additionally, macrophages express HMGB1 when triggered by NOD-like receptors-3 (NLRP3) or NLRC4 (inflammasomes, leucine-rich repeats, and nucleotide-binding oligomerization domains containing caspase recruitment domain-containing-4) (Lamkanfi et al. 2010).

Inflammasomes can direct the release of cytoplasmic HMGB1 from activated immune cells. Further study has shown that hyperexcitability can activate the Janus kinase/signal transducer and activator of transcription (JAK/STAT1) pathway, resulting in HMGB1 translocation from the nucleus to the cytoplasm and cytoplasmic HMGB1 release (Lu et al. 2014).

Frank et al. demonstrated that the redox state of HMGB1 is an important molecular feature since the reduced form of HMGB1 is chemotactic, but the disulfide form of HMGB1 (dsHMGB1) is pro-inflammatory. The NLRP3 inflammasome may be implicated in dsHMGB1 priming effects (Frank et al. 2016).

These findings imply that cytokines, pathogen-derived chemicals, and inflammasomes are the primary causes of HMGB1 upregulation and that ds-HMGB1 is essential for hyperexcitability. Some research has concentrated on the signaling pathways linked to extracellular HMGB1. Before being endocytosed by RAGE into the endolysosomal system, extracellular HMGB1 binds to other inflammatory chemicals. Internalized HMGB1 causes lysosomal membranes to collapse, allowing HMGB1 partner molecules to enter the cytoplasm and boost inflammatory factor transcription or regulation. Extracellular HMGB1 inhibition has been shown to improve therapeutic outcomes in a variety of sterile and viral inflammatory conditions. This therapeutic potential has already been demonstrated in experimental sepsis models, where extracellular HMGB1 can successfully target TLR4 and RAGE even at later periods (Andersson, Yang, and Harris 2018). However, it is unknown how intracellular HMGB1 is linked to epileptic seizures. Furthermore, investigations have revealed that the degradation of the blood-brain barrier (BBB) may be a significant mediator in the deleterious effect of HMGB1 on epileptogenesis. HMGB1 can cause BBB degradation which can worsen epilepsy (Nishibori et al. 2020). As a result, inhibiting HMGB1 expression and release may be a promising therapeutic option for epilepsy treatment. It is uncertain, however, how HMGB1 causes BBB deterioration. Changes in gene expression, it has been argued, cell. Finally, HMGB1 can work on downstream receptors like TLR4 and RAGE, stimulating their downstream IL-1 and NF-B, which then act with NMDA receptors to worsen hyperexcitability and epilepsy.

The Role of HMGB1 in Epilepsy

Some studies have revealed possible cell types that contribute to the release of HMGB1. Devinsky et al. discovered that HMGB1 pathways mediated by astrocyte and microglia activation play an important role in the development of epilepsy (Devinsky et al. 2013).

Epileptogenesis is defined as complex anatomical changes in the brain that transform a normal brain into a brain crippled by recurrent seizure activity (Sloviter and Bumanglag 2013). Neurodegeneration (Pitkanen and Lukasiuk 2009) disruption of BBB (Bar-Klein et al. 2017), the glutamatergic system (Aroniadou-Anderjaska et al. 2008), oxidative stress (Ashrafi et al. 2007), and epigenetic modification of DNA (Hauser, Annegers, and Kurland 1993) are all involved in epileptogenesis and aggravates the process.

Epilepsy and neuroinflammation are closely related. In epileptic people and various animal models of epilepsy, high mobility group protein B1 (HMGB1), a critical inflammatory factor, was discovered to have increased expression and a larger proportion of translocation from the nucleus to the cytoplasm. HMGB1 can activate TLR4 and (NF-B), which in turn activate glutamate receptors such as NMDA receptors, causing hyperexcitability and epilepsy. Hyperexcitability can then cause HMGB1 expression and translocation.

Antiepileptic medication treatment may involve blocking HMGB1 and its downstream signaling pathways (Dai et al. 2021). Due to the absence of disease-modifying impact in standard AEDs, a thorough study of disease pathophysiology and the development of innovative treatment approaches for epilepsy-related hyperexcitability is a current necessity. In the search for molecular mediators of epileptogenesis in animal models, substantial emphasis has been dedicated to the possible pathogenic function of HMGB1 in the production and recurrence of seizures (Kaneko et al. 2017; Yang et al. 2017).

Despite major research developments, the biology of epilepsy remains complex; however, brain inflammation is likely to have a role (Riazi, Galic, and Pittman 2010). Emerging data suggest that neuroinflammatory processes play a role in the pathophysiology of seizures/epilepsy, and HMGB1 has been shown to operate as an inflammatory cytokine in response to epileptogenic stimuli (Kaneko et al. 2017).

Glial cell activation has been linked to the onset of epilepsy, and HMGB1 has been shown to enhance microglial activation during seizures via the TLR4/NF-B signaling pathway (Shi et al. 2018). HMGB1 enhances IL-1R/TLR signaling in neurons and is involved in seizure initiation and recurrence via NMDA-NR2B receptor activation catalyzed by rapid sarcoma family kinases (Vezzani et al. 2012). HMGB1 also plays an important role in epileptogenesis via microglial activation and TLR4-NF-B signaling pathway activation (Shi et al. 2018). Several studies have revealed the function of HMGB1 in seizure creation, however, the particular mechanism by which HMGB1 causes seizure development is not fully characterized.

Based on the anti-convulsant activity of TLR4 inhibitors and Box A, a competitor of endogenous HMGB1, the role of HMGB1 released from glia and neurons, as well as its signaling with TLR4, in generating and perpetuating seizures has been proposed, but the study lacks a detailed mechanism for how the HMGB1/TLR4 axis leads to seizure generation (Maroso et al. 2010). Although HMGB1 stimulates both TLR4 and RAGE, RAGE's role in seizures is less clear than TLR4's (Iori et al. 2013).

Mesial temporal lobe epilepsy (MTLE) is one of the most common refractory focal epilepsy syndromes (Palleria et al. 2015), and the role of HMGB1 in its pathogenesis is unknown. An experimental MTLE study discovered a significant increase in HMGB1 and TLR4 gene expression in rat hippocampi, and this overexpression of HMGB1 and TLR4 was connected to the development of MTLE in young rats (Yang et al. 2017). Pharmacological and genetic studies on animal and clinical brain specimens demonstrated that HMGB1 translocation and release occur in pathological epileptogenic foci of different forms of epilepsy (Iori et al. 2013; Maroso et al. 2010).

It is worth noting that the HMGB1/TLR4 axis not only reduced seizure frequency and duration but also accelerated seizure onset, which usually occurs within minutes in kainite and bicuculline-induced seizure models, implying that HMGB1 plays an important role in the precipitation of the first seizure following pro-convulsant administration (Maroso et al. 2010). Extracellular HMGB1 levels increased as seizures began, establishing HMGB1 as a critical pathophysiological contributor to the onset of epilepsy-related hyperexcitability (Kaneko et al. 2017).

TLR4 activation in neurons and astrocytes by HMGB1 proteins is a critical mechanism of seizure development and blocking TLR4 signaling with an antagonist may also reduce the severity of epilepsy (Iori et al. 2013). In the brains of epilepsy patients, elevated levels of HMGB1, TLR4, RAGE, NF-B, p65, and inducible nitric oxide synthase (iNOS) were identified, as well as increased levels of IL-1, IL-6, TNF-, TGF-, and IL-10 (Shi et al. 2018).

In recent years, there has been a surge in interest in HMGB1/TLR4/RAGE signaling as a treatment for epilepsy. MicroRNA-129-5p reduced HMGB1 expression and the TLR4/NF-B signaling pathway, which inhibited the development of autoimmune encephalomyelitis-related epilepsy (Liu, Wu, and Wang 2017).

In C57BL/6 mice, HMGB1 and TLR4 antagonists slowed seizure onset and prevented acute and chronic seizure recurrence, and increased expression of HMGB1 and TLR4 in human epileptogenic tissue, which is similar to a mouse model of chronic seizures and suggests a role for the HMGB1- TLR4 axis in human epilepsy (Maroso et al. 2010). Furthermore, HMGB1/ RAGE/TLR4 signaling may have a role in the initiation and maintenance of seizures in humans and can be successfully targeted to achieve anti-convulsant effects in drug-resistant epilepsies. After 24 hours in a KA-induced model of SE (Kainic Acid-Induced Post-Status Epilepticus Models of Temporal Lobe Epilepsy), the expression level of HMGB1 was drastically elevated in the hippocampus and cortex, showing that the HMGB1 protein plays an important role in epilepsy (Walker et al. 2014).

HMGB1 enhances hypothermia-induced seizures and contributes to the pathophysiology of febrile seizures. However, the precise mechanism of HMGB1 in febrile seizures is uncertain(Ito et al. 2017). There is limited data available on the therapeutic efficacy of HMGB1 inhibitors in animal models of epilepsy. Glycyrrhizin, on the other hand, shows neuroprotection against lithium/pilocarpine-induced SE in rats, as well as ameliorating pilocarpine-induced oxidative stress and inflammatory responses via suppression of IL-1 and TNF-α (Gonzalez-Reyes et al. 2016), but no anti-epileptic action.

Dynamic changes in HMGB1 expression in the mouse brain's hippocampus were observed after KA administration, and glycyrrhizin exerts neuroprotective but not anti-epileptic effects by suppressing both acute and delayed HMGB1 inductions in the hippocampal cornu ammonis (CA)1 and CA3 regions, as well as serum HMGB1 accumulation (Luo et al. 2014).

Anti-HMGB1 mAb has an anti-seizure effect, as evidenced by the lack of disturbance in the physical EEG rhythm and fundamental physical functioning since it blocks the translocation of HMGB1 from nuclei during a seizure. This anti-seizure effect was not detected in TLR4 knockout animals (Zhao et al. 2017).

Furthermore, anti-HMGB1 mAb revealed a disease-modifying anti-epileptogenic impact on epileptogenesis following SE, as shown by decreased seizure frequency and better cognitive performance (Zhao et al. 2017). Seizure frequency and duration can be reduced by HMGB1 inhibitors, which are thought to function by targeting the HMGB1/RAGE/TLR4 axis and delaying seizure precipitation by inhibiting HMGB1 translocation and protecting BBB integrity. Anti-HMGB1 mAb, on the other hand, inhibited BBB leakage and pilocarpine induced HMGB1 translocation. In addition to preventing BBB permeability and reducing HMGB1 translocation (Fu et al. 2017). Zhao et al. (2017) examined the anti-epileptic effect of anti-HMGB1 mAb on human brain slices from drug-resistant epilepsy patients, showing that anti-HMGB1 mAb binds to HMGB1 and has long-lasting anti-epileptic properties, which are similar to the previously estimated long half-time elimination in the brain (Zhang et al. 2011).

An extensive study has revealed that HMGB1 may have a role in seizure formation, as well as that it is expressed at a greater level in epileptic brains (Chen et al. 2015). The inhibition of seizure precipitation by HMGB1 inhibitors (Chen et al. 2015). shows that HMGB1 is involved in all phases of seizure generation and seizure retardation, making HMGB1 an excellent option for a reliable biomarker for epileptogenesis. Furthermore, the current prognostic indicators for seizure recurrence and seizure remission in epilepsy patients are exclusively dependent on supplemental characteristics such as age, seizure type, EEG, and MRI, but their accuracy is restricted (Ravizza et al. 2018). An earlier study advocated HMGB1 isoforms as mechanistic biomarkers for epileptogenesis, in which they investigated the utility of blood HMGB1 in predicting epilepsy development as well as separating epileptogenic from non-epileptogenic animals following SE (Walker et al. 2017).

As an epileptogenesis biomarker, HMGB1 will eventually provide a deeper understanding of normal biological mechanisms, pathogenic processes, or responses to an exposure or intervention, such as therapeutic interventions with broad clinical applications capable of arresting disease progression or trying to improve its clinical course.

Furthermore, detailed knowledge of the molecular mechanism through which HMGB1 induces seizures via inflammatory signaling will be critical in devising novel medicines that target inflammatory pathways to reduce seizures. However, the link between HMGB1 and seizures has to be investigated further. Overall, the data imply that inhibiting the HMGB1/TLR4/RAGE regulation axis might be a unique strategy for treating epilepsy.

Recent research has given further insight into the numerous functions of HMGB1 in a variety of disorders such as brain damage, epilepsy, neuroinflammation, and cognitive decline. Treatments based on HMGB1 antagonists that target extracellular HMGB1 have produced hopeful results in a variety of experimental models of the aforementioned HMGB1-mediated diseases, while clinical studies have yet to be published.

Alteration in HMGB1 in Epileptic Patients

Inflammation occurs when HMGB1 is produced as a result of an injury or illness. HMGB1 can activate inflammatory responses and play a crucial role in epilepsy via TLR pathways, according to recent research. HMGB1 has also been studied for its potential application in the treatment of epilepsy (Wang et al. 2021). Zurolo et al. Discovered the increased expression of HMGB1 and its downstream receptors, such as TLR2, TLR4, and RAGE, in pathological brain tissues from patients with focal cortical dysplasia (FCD) for the first time. They discovered that HMGB1 nuclear transfer stimulated the production of IL-1 in glial cells, which then mediated downstream inflammatory signaling pathways (Zurolo et al. 2011). In pathological tissues of FCD type II, Zhang et al. discovered significantly higher levels of TLR4, cytoplasm HMGB1, and inflammation factors such as IL1 and tumor necrosis factor- (TNF-) compared to peri-FCD controls, as well as increased HMGB1 translocation from the nucleus to the cytoplasm (Zhang et al. 2018).

Traditional epilepsy therapy typically targets ion channels to decrease cell excitability, ultimately limiting the creation and progression of seizures. Long-term use of these medications, however, may impair patients' normal physiological processes. As an important inflammatory factor, HMGB1 interacts with downstream receptors including TLR4 and RAGE, initiating neuroinflammatory responses via signaling pathways like NMDA receptors. In this situation, neurons may get overexcited, causing the brain network to change and lowering the seizure attack threshold (Zurolo et al. 2011). Given the significant link between HMGB1 activation and seizures, it is unclear whether HMGB1 serum levels may be used as a biomarker of epilepsy for severity, prognosis, or therapeutic response. Furthermore, the sensitivity of serum HMGB1 levels in predicting seizure severity in various epilepsy types needs to be explored further. The HMGB1 translocation ratio has been proposed as a potential predictor of epilepsy risk (Yang et al. 2017).

The precise role of HMGB1 in epileptogenesis is only partially understood. The critical molecules in the HMGB1-related pathway, as well as whether HMGB1 overexpression underpins the activation of several brain regions or the switch-on of specific brain circuits, are crucial concerns to be addressed. Anticonvulsant activities of anti-HMGB1 mAb have been suggested in previous studies. The therapeutic impact must have been demonstrated in a more specific context, i.e., during various phases of epilepsy and for epilepsy of diverse etiologies (Zhang et al. 2018).

Clinical studies are also necessary to determine if HMGB1 inhibition has an anticonvulsant or antiepileptogenic effect. Furthermore, whether specific inhibition of neuroglia, such as astrocytes or microglia, to downregulate the HMGB1 level might provide a more focused anti-seizure effect warrants further investigation (Dey et al. 2016). TLRs are proteins that play an important role in immune system activation and the inflammatory response. TLR4 is a lipopolysaccharide-sensing receptor that activates gene transcription to cause inflammation. It belongs to the TLRs family, which has received a lot of interest in a variety of diseases (Dey et al. 2016).

Activation of TLR 2, 3, 4, and 9 increases the expression of cytokines and co-stimulatory molecules. TLR signaling pathways can recognize molecules released from damaged tissues, such as injury-associated molecular patterns/proteins (DAMPs). Of these molecules, HMGB1 is secreted passively by necrotic cells, not by programmed apoptotic cells, and it is actively released by cells exposed to extreme stress and is defined as DAMP with this feature (Venereau et al. 2016).

Damaged tissues emit high-mobility group box-1 (HMGB1), an endogenous TLR4 ligand and one of the damage-associated molecular patterns (DAMPs) (Liang et al. 2014; Mazarati et al. 2011). Both HMGB1 and TLR4 have been linked to the prevalence and durability of seizures in studies

(Huang et al. 2015). HMGB1, for example, is responsible for the unfavorable inflammatory effects found in epilepsy. TLR4 knockout mice were found to be resistant to kainite-induced seizures (Chiavegato et al. 2014; Maroso et al. 2010). However, the majority of prior investigations on the function of HMGB1 and TLR4 in epilepsy were conducted in experimental animal models or brain tissues; few clinical studies on the functions of HMGB1 and TLR4 in epileptic patients have been conducted (Iori et al. 2013). Also, few studies have looked at the role of HMGB1 and TLR4 in medication resistance in epileptic patients (Yang et al. 2017).

A major clinical need for epilepsy is the development of novel medications for seizure management in persons with pharmacoresistant epilepsy. Furthermore, the existing treatments are symptomatic; hence, disease-modifying medicines for avoiding illness start or progression are lacking. Although several signaling pathways have lately been reported to have a pathogenic role in illness onset and progression, the essential molecular processes underpinning disease onset and progression remain unknown (Pitkanen and Lukasiuk 2011).

In epileptogenic foci, experimental and clinical evidence from animal models of epilepsy and human brain tissues from diverse drug-resistant kinds of epilepsy show activation of innate and adaptive immune pathways, as well as the development of associated inflammatory processes (Aronica and Crino 2011; Vezzani et al. 2011).

Potential Role of HMGB1 in the Diagnosis and Treatment of Epilepsy

According to recent research, the brain system governs innate immune responses and regulates macrophage HMGB1 release. These discoveries are particularly interesting since glucocorticoids and other neuroimmune mediators reflect physiological processes that can be used to treat a variety of inflammatory illnesses (Wang et al. 2004). According to a recent study, the vagus nerve can prevent reticuloendothelial system tissue macrophages from producing pro-inflammatory cytokines. The principal neurotransmitter of the vagus nerve, acetylcholine, was discovered to be the first physiological inhibitor of HMGB1 release from human macrophages (Wang et al. 2004). Through a mechanism involving the a7-nicotinic acetylcholine receptor (nAChR), nicotine, a more selective agonist, decreases NF-kB pathway activation and inhibits HMGB1 synthesis (Wang et al. 2004). The discovery of this "nicotinic anti-inflammatory pathway" gives researchers a leg up in creating novel pharmaceutical strategies to restrict HMGB1's pathogenic potential. However, due to its nonspecific and possibly hazardous effects, nicotine's therapeutic applicability as an anti-inflammatory medication is restricted. Recent research is focusing on developing selective a7nAChR agonists to modulate HMGB1 release and overcome nicotine's therapeutic limitations. The fact that the mechanism of HMGB1 production from immune cells is unclear is a key constraint in developing pharmaceutical therapies against it.

There are various pharmacological techniques for inhibiting HMGB1 secretion converge in the suppression of the NF-kB pathway, implying a connection between these two processes. Recent research suggests that ethyl pyruvate can block NF-kB directly by covalently altering a cysteine residue (Cys 38) on the DNA-binding surface of p65relA (Han et al. 2005). This link between the NF-kB system and macrophage extracellular secretion of HMGB1 poses a difficult paradox because NF-kB generally influences pro-inflammatory cytokine production at the transcriptional level. HMGB1 continual protein production, on the other hand, is necessary for survival, and its extracellular secretion appears to be controlled at the post-transcriptional level. Nicotine and lysophosphatidylcholine, on the other hand, can decrease HMGB1 secretion from macrophages by reducing its translocation from the nucleus to the cytoplasm, which is required for extracellular secretion (Xu 2002). Regardless of the underlying cause, these viral and inflammatory illnesses appear to follow a similar clinical trajectory, with high circulating HMGB1 levels (Wang et al. 2004). Closing comments Extracellular HMGB1 appears to be a sufficient mediator of systemic inflammation, acting as a potent chemotactic and activating signal for immune cells that mediate inflammation, such as neutrophils, monocytes, and macrophages (Lotze and Tracey 2005).

HMGB1's multiple roles in several illnesses, including brain injury, epilepsy, neuroinflammation, and cognitive decline, have been shown in the research. Treatments based on extracellular HMGB1 antagonists have shown promise in a range of animal models of the aforementioned HMGB1-mediated illnesses, while clinical trials have yet to be reported. However, because the precise mechanism by which HMGB1 mediates various neurological disorders is unknown, the relationship between HMGB1 and brain injury, epilepsy, neuroinflammation-mediated illnesses, and cognitive decline is believed to require further exploration. Treatments based on HMGB1 antagonists that target extracellular HMGB1 have generated encouraging results in a range of animal models of the aforementioned HMGB1-mediated illnesses, but clinical studies have yet to be reported. However, the complex biology of HMGB1 is not fully understood, and there is a belief that the association between HMGB1 and brain injury, epilepsy, neuroinflammation-mediated pathologies, and cognitive decline requires further investigation, as the precise mechanism by which HMGB1 mediates these neurological conditions is unknown.

Neuroinflammation has been connected to several neurological disorders, including epilepsy and memory loss. HMGB1, a neuroinflammation mediator, has been proposed to play a significant role in, neuroinflammatory diseases, and epileptogenesis via an unknown mechanism. HMGB1 levels in serum and CSF have been observed to be elevated in various neuroinflammation-mediated disorders. This confirms our notion that HMGB1 is a common biomarker in neuroinflammation, epilepsy, and cognitive decline. Biomarker development, in conjunction with research into innovative pharmaceutical choices, would make substantial advances in the treatment of epilepsy, memory impairment, and neuroinflammation by focusing on the mechanistic approach rather than symptomatic therapy.

Conclusion

The current review aimed to connect the dots between HMGB1 and its possible role in a variety of brain injuries, neuroinflammation-mediated disorders, epilepsy, and cognitive decline by using experimental and clinical evidence. There are some fundamental limitations to the topic that should not be neglected, including practicality and viability concerns in making HMGB1 a common functional biomarker for the neuroinflammation-mediated illnesses discussed here. Is it possible that inhibiting the HMGB1/RAGE/TLR4 axis is a common epilepsy target? Can the therapy effects derived from experimental data on HMGB1 involvement in epilepsy be easily applied to clinical settings?

Despite accumulating scientific evidence that HMGB1 plays a role in neuroinflammation-mediated illnesses, there has been no attempt to portray HMGB1 as a common biomarker and target for various HMGB1-mediated neurological disorders. HMGB1 proteins, we believe, are a promising non-invasive, common biomarker of neuroinflammation, epilepsy, and cognitive impairment since they meet multiple criteria for being a common biomarker. It is relatively stable in blood and can be readily and cheaply measured in blood.

References

Andersson, U., H. Yang, and H. Harris. 2018. "Extracellular HMGB1 as a therapeutic target in inflammatory diseases." *Expert Opin Ther Targets* 22 (3):263-277. https://doi.org/10.1080/14728222.2018.1439924.

Andersson, U., Wang, H., Aveberger, A. C., Bloom, O., Erlandsson-Harris, H., Janson, A., Kokkola, R., Zhang, M., Yang, H., Tracey, K. J. 2000. "High mobility group 1 protein (HMG-1) stimulates proinflammatory cytokine synthesis in human monocytes." *J Exp Med* 192 (4):565-70. https://doi.org/10.1084/jem.192.4.565.

Andersson, U., Yang, H., & Harris, H. 2008. "Pathology and pathophysiology of the amygdala in epileptogenesis and epilepsy." *Epilepsy Res* 78 (2-3):102-16. https://doi.org/10.1016/j.eplepsyres.2007.11.011.

Aronica, E., and P. B. Crino. 2011. "Inflammation in epilepsy: clinical observations." *Epilepsia* 52 Suppl 3:26-32. https://doi.org/10.1111/j.1528-1167.2011.03033.x.

Ashrafi, M. R., Shams, S., Nouri, M., Mohseni, M., Shabanian, R., Yekaninejad, M. S., Chegini, N., Khodadad, A., Safaralizadeh, R. 2007. "A probable causative factor for an old problem: selenium and glutathione peroxidase appear to play important roles in epilepsy pathogenesis." *Epilepsia* 48 (9):1750-1755. https://doi.org/10.1111/j.1528-1167.2007.01143.x.

Bar-Klein, G., Lublinsky, L., Kamintsky, L., Noyman, I., Veksler, R., Dalipaj, H., Senatorov, V. V., Swissa, E., Rosenbach, D., Elazary, N., Milikovsky, D. Z., Milk, N., Kassirer, M., Rosman, Y., Serlin, Y., Eisenkraft, A., Chassidim, Y., Parmet, Y., Kaufer, D., Friedman, A. 2017. "Imaging blood-brain barrier dysfunction as a biomarker for epileptogenesis." *Brain* 140 (6):1692-1705. https://doi.org/10.1093/brain/awx073.

Bianchi, M. E. 2004. "Significant (re)location: how to use chromatin and/or abundant proteins as messages of life and death." *Trends Cell Biol* 14 (6):287-93. https://doi.org/10.1016/j.tcb.2004.04.004.

Bonaldi, T., Talamo, F., Scaffidi, P., Ferrera, D., Porto, A., Bachi, A., Rubartelli, A., Agresti, A., Bianchi, M. E. 2003. "Monocytic cells hyperacetylate chromatin protein HMGB1 to redirect it towards secretion." *EMBO J* 22 (20):5551-60. https://doi.org/10.1093/emboj/cdg516.

Bradl, M., and R. Hohlfeld. 2003. "Molecular pathogenesis of neuroinflammation." *J Neurol Neurosurg Psychiatry* 74 (10):1364-70. https://doi.org/10.1136/jnnp.74.10.1364.

Bustin, M. 2001. "Revised nomenclature for high mobility group (HMG) chromosomal proteins." *Trends Biochem Sci* 26 (3):152-3. https://doi.org/10.1016/s0968-0004(00)01777-1.

Calogero, S., Calogero, S., Grassi, F., Aguzzi, A., Voigtländer, T., Ferrier, P., Ferrari, S., & Bianchi, M. E. 1999. "The lack of chromosomal protein Hmg1 does not disrupt cell growth but causes lethal hypoglycaemia in newborn mice." *Nat Genet* 22 (3):276-80. https://doi.org/10.1038/10338.

Carson, M. J., J. C. Thrash, and B. Walter. 2006. "The cellular response in neuroinflammation: The role of leukocytes, microglia and astrocytes in neuronal death and survival." *Clin Neurosci Res* 6 (5):237-245. https://doi.org/10.1016/j.cnr.2006.09.004.

Chen, Y., Xian-Jing Huang, X. J., Yu, N., Xie, Y., Zhang, K., Wen, F., Liu, H., Di, Q. 2015. "HMGB1 Contributes to the Expression of P-Glycoprotein in Mouse Epileptic Brain through Toll-Like Receptor 4 and Receptor for Advanced Glycation End Products." *PLoS One* 10 (10):e0140918. https://doi.org/10.1371/journal.pone.0140918.

Chiavegato, A., Song, L., Zhang, X., Zhang, J., Fang, P. 2014. "The inflammatory molecules IL-1beta and HMGB1 can rapidly enhance focal seizure generation in a brain slice model of temporal lobe epilepsy." *Front Cell Neurosci* 8:155. https://doi.org/10.3389/fncel.2014.00155.

Dai, S., Zheng, Y., Wang, Y., Chen, Z. 2021. "HMGB1, neuronal excitability and epilepsy." *Acta Epileptologica* 3 (1):13. https://doi.org/10.1186/s42494-021-00048-y.

Das, N., Dewan, V., Grace, P. M., Gunn, R. J., Tamura, R., Tzarum, N., Watkins, L. R., Wilson, I. A., Yin, H. 2016. "HMGB1 Activates Proinflammatory Signaling via TLR5 Leading to Allodynia." *Cell Rep* 17 (4):1128-1140. https://doi.org/10.1016/j.celrep.2016.09.076.

de Vries, E. E., an den Munckhof, B., Braun, K. P. J., van Royen-Kerkhof, A., de Jager, W., Jansen, F. E. 2016. "Inflammatory mediators in human epilepsy: A systematic review and meta-analysis." *Neurosci Biobehav Rev* 63:177-90. https://doi.org/10.1016/j.neubiorev.2016.02.007.

Devinsky, O., Vezzani, A., Najjar, S., de Lanerolle., N. C., Rogawski, M. A. 2013. "Glia and epilepsy: excitability and inflammation." *Trends Neurosci* 36 (3):174-84. https://doi.org/10.1016/j.tins.2012.11.008.

Devinsky, O., Vezzani A., O'Brien, T. J., Jette, N., Scheffer I., de Curtis, M., Perucca, P. 2018. "Epilepsy." *Nat Rev Dis Primers* 4 (1):18024. https://doi.org/10.1038/nrdp.2018.24.

Dey, A., Kang, X., Qiu, J., Du, Y., Jiang, J. 2016. "Anti-Inflammatory Small Molecules To Treat Seizures and Epilepsy: From Bench to Bedside." *Trends Pharmacol Sci* 37 (6):463-484. https://doi.org/10.1016/j.tips.2016.03.001.

Dumitriu, I. E., Baruah, P., Manfredi, A. A., Bianchi, M. E., Rovere-Querini, P. 2005. "HMGB1: guiding immunity from within." *Trends Immunol* 26 (7):381-7. https://doi.org/10.1016/j.it.2005.04.009.

Fisher, R. S., Acevedo, C., Arzimanoglou, A., Bogacz, A., Cross, J. H., Elger, C. E., Engel, J., Forsgren, L., French, J. A., Glynn, M., Hesdorffer, D. C., Lee, B. I., Mathern, G. W., Moshe, S. L., Perucca, E., Scheffer, I. E., Tomson, T., Watanabe, M., Wiebe, S. 2014. "ILAE official report: a practical clinical definition of epilepsy." *Epilepsia* 55 (4):475-82. https://doi.org/10.1111/epi.12550.

Fisher, R. S., Cross, J. H., French, J. A., Higurashi, N., Hirsch, E., Jansen, F. E., Lagae, L., Moshé, S. L., Peltola, J., Roulet Perez, E., Scheffer, I. E., & Zuberi, S. M. 2017. "Operational classification of seizure types by the International League Against Epilepsy: Position Paper of the ILAE Commission for Classification and Terminology." *Epilepsia* 58 (4):522-530. https://doi.org/10.1111/epi.13670.

Frank, M. G., Weber, M. D., Fonken, L. K., Hershman, S. A., Watkins, L. R., & Maier, S. F. 2016. "The redox state of the alarmin HMGB1 is a pivotal factor in neuroinflammatory and microglial priming: A role for the NLRP3 inflammasome." *Brain Behav Immun* 55:215-224. https://doi.org/10.1016/j.bbi.2015.10.009.

Fu, L., Liu, K., Wake, H., Teshigawara, K., Yoshino, T., Takahashi, H., Mori, S., & Nishibori, M. 2017. "Therapeutic effects of anti-HMGB1 monoclonal antibody on pilocarpine-induced status epilepticus in mice." *Sci Rep* 7 (1):1179. https://doi.org/10.1038/s41598-017-01325-y.

Fukata, Y., and M. Fukata. 2017. "Epilepsy and synaptic proteins." *Curr Opin Neurobiol* 45:1-8. https://doi.org/10.1016/j.conb.2017.02.001.

Gardella, S., Andrei, C., Ferrera, D., Lotti, L. V., Torrisi, M. R., Bianchi, M. E., & Rubartelli, A. 2002. "The nuclear protein HMGB1 is secreted by monocytes via a non-classical, vesicle-mediated secretory pathway." *EMBO Rep* 3 (10):995-1001. https://doi.org/10.1093/embo-reports/kvf198.

Glass, C. K., Saijo, K., Winner, B., Marchetto, M. C., & Gage, F. H. 2010. "Mechanisms underlying inflammation in neurodegeneration." *Cell* 140 (6):918-34. https://doi.org/10.1016/j.cell.2010.02.016.

González-Reyes, S., Santillán-Cigales, J. J., Jiménez-Osorio, A. S., Pedraza-Chaverri, J., & Guevara-Guzmán, R. 2016. "Glycyrrhizin ameliorates oxidative stress and inflammation in hippocampus and olfactory bulb in lithium/pilocarpine-induced status epilepticus in rats." *Epilepsy Res* 126:126-33. https://doi.org/10.1016/j.eplepsyres.2016.07.007.

Han, Y., Englert, J. A., Yang, R., Delude, R. L., & Fink, M. P. 2005. "Ethyl pyruvate inhibits nuclear factor-kappaB-dependent signaling by directly targeting p65." *J Pharmacol Exp Ther* 312 (3):1097-105. https://doi.org/10.1124/jpet.104.079707.

Hauser, W. A., J. F. Annegers, and L. T. Kurland. 1993. "Incidence of epilepsy and unprovoked seizures in Rochester, Minnesota: 1935-1984." *Epilepsia* 34 (3):453-68. https://doi.org/10.1111/j.1528-1157.1993.tb02586.x.

Henshall, D. C., Hamer, H. M., Pasterkamp, R. J., Goldstein, D. B., Kjems, J., Prehn, J. H. M., Schorge, S., Lamottke, K., & Rosenow, F. 2016. "MicroRNAs in epilepsy: pathophysiology and clinical utility." *Lancet Neurol* 15 (13):1368-1376. https://doi.org/10.1016/S1474-4422(16)30246-0.

Hunt, R. F., J. A. Boychuk, and B. N. Smith. 2013. "Neural circuit mechanisms of post-traumatic epilepsy." *Front Cell Neurosci* 7:89. https://doi.org/10.3389/fncel.2013.00089.

Huttunen, H. J., Fages, C., Kuga-Panula, J., Ridley A. J., Ruavala, H. 2002. "Receptor for advanced glycation end products-binding COOH-terminal motif of amphoterin inhibits invasive migration and metastasis." *Cancer Res* 62 (16):4805-11.

Huttunen, H. J., Kuja-Panula, J., Sorci, G., Agneletti, A. L., Donato, R., & Rauvala, H. 2000. "Coregulation of neurite outgrowth and cell survival by amphoterin and S100 proteins through receptor for advanced glycation end products (RAGE) activation." *J Biol Chem* 275 (51):40096-105. https://doi.org/10.1074/jbc.M006993200.

Iori, V., Maroso, M., Rizzi, M., Iyer, A. M., Vertemara, R., Carli, M., Agresti, A., Antonelli, A., Bianchi, M. E., Aronica, E., Ravizza, T., & Vezzani, A. 2013. "Receptor for Advanced Glycation Endproducts is upregulated in temporal lobe epilepsy and contributes to experimental seizures." *Neurobiol Dis* 58:102-14. https://doi.org/10.1016/j.nbd.2013.03.006.

Ito, M., Takahashi, H., Yano, H., Shimizu, Y. I., Yano, Y., Ishizaki, Y., Tanaka, J., Ishii, E., & Fukuda, M. 2017. "High mobility group box 1 enhances hyperthermia-induced seizures and secondary epilepsy associated with prolonged hyperthermia-induced seizures in developing rats." *Metab Brain Dis* 32 (6):2095-2104. https://doi.org/10.1007/s11011-017-0103-4.

Kaneko, Y., Pappas, C., Malapira, T., Vale, F. Ĺ., Tajiri, N., & Borlongan, C. V. 2017. "Extracellular HMGB1 Modulates Glutamate Metabolism Associated with Kainic Acid-Induced Epilepsy-Like Hyperactivity in Primary Rat Neural Cells." *Cell Physiol Biochem* 41 (3):947-959. https://doi.org/10.1159/000460513.

Kokkola, R., Andersson, A., Mullins, G., Ostberg, T., Treutiger, C.-J., Arnold, B., Nawroth, P., Andersson, U., Harris, R. A., & Harris, H. E. 2005. "RAGE is the major receptor for the proinflammatory activity of HMGB1 in rodent macrophages." *Scand J Immunol* 61 (1):1-9. https://doi.org/10.1111/j.0300-9475.2005.01534.x.

Krishnamurthy, K. B. 2016. "Epilepsy." *Ann Intern Med* 164 (3):ITC17-32. https://doi.org/10.7326/AITC201602020.

Kuniyasu, H., Y. Chihara, and T. Takahashi. 2003. "Co-expression of receptor for advanced glycation end products and the ligand amphoterin associates closely with metastasis of colorectal cancer." *Oncol Rep* 10 (2):445-8.

Lamkanfi, M., Sarkar, A., Vande Walle, L., Vitari, A. C., Amer, A. O., Wewers, M. D., Tracey, K. J., Kanneganti, T.-D., & Dixit, V. M. 2010. "Inflammasome-dependent release of the alarmin HMGB1 in endotoxemia." *J Immunol* 185 (7):4385-92. https://doi.org/10.4049/jimmunol.1000803.

Li, J., Kokkola, R., Tabibzadeh, S., Yang, R., Ochani, M., Qiang, X., Harris, H. E., Czura, C. J., Wang, H., Ulloa, L., Wang, H., Warren, H. S., Moldawer, L. L., Fink, M. P., Andersson, U., Tracey, K. J., & Yang, H. 2003. "Structural basis for the proinflammatory cytokine activity of high mobility group box 1." *Mol Med* 9 (1-2):37-45.

Liang, Y., Lei, Z., Zhang, H., Xu, Z., Cui, Q., & Xu, Z. C. 2014. "Toll-like receptor 4 is associated with seizures following ischemia with hyperglycemia." *Brain Res* 1590:75-84. https://doi.org/10.1016/j.brainres.2014. 09.020.

Liu, A. H., Y. T. Wu, and Y. P. Wang. 2017. "MicroRNA-129-5p inhibits the development of autoimmune encephalomyelitis-related epilepsy by targeting HMGB1 through the TLR4/NF-kB signaling pathway." *Brain Res Bull* 132:139-149. https://doi.org/10.1016/j.brainresbull.2017.05.004.

Lotze, M. T., and K. J. Tracey. 2005. "High-mobility group box 1 protein (HMGB1): nuclear weapon in the immune arsenal." *Nat Rev Immunol* 5 (4):331-42. https://doi.org/10.1038/nri1594.

Lu, B., Antoine, D. J., Kwan, K., Lundbäck, P., Wähämaa, H., Schierbeck, H., Robinson, M., Van Zoelen, M. A. D., Yang, H., Li, J., Erlandsson-Harris, H., Chavan, S. S., Wang, H., Andersson, U., & Tracey, K. J. 2014. "JAK/STAT1 signaling promotes HMGB1 hyperacetylation and nuclear translocation." *Proc Natl Acad Sci U S A* 111 (8):3068-73. https://doi.org/10.1073/pnas.1316925111.

Luo, L., Jin, Y., Kim, I. D., & Lee, J. K. 2014. "Glycyrrhizin suppresses HMGB1 inductions in the hippocampus and subsequent accumulation in serum of a kainic acid-induced seizure mouse model." *Cell Mol Neurobiol* 34 (7):987-97. doi: 10.1007/s10571-014-0075-4.

Lyman, M., Lloyd, D. G., Ji, X., Vizcaychipi, M. P., & Ma, D. 2014. "Neuroinflammation: the role and consequences." *Neurosci Res* 79:1-12. https://doi.org/10.1016/j.neures.2013.10.004.

Magna, M., and D. S. Pisetsky. 2014. "The role of HMGB1 in the pathogenesis of inflammatory and autoimmune diseases." *Mol Med* 20:138-46. https://doi.org/10.2119/molmed.2013.00164.

Maroso, M., Balosso, S., Ravizza, T., Liu, J., Aronica, E., Iyer, A. M., Rossetti, C., Molteni, M., Casalgrandi, M., Manfredi, A. A., Bianchi, M. E., & Vezzani, A. 2010. "Toll-like receptor 4 and high-mobility group box-1 are involved in ictogenesis and can be targeted to reduce seizures." *Nat Med* 16 (4):413-9. https://doi.org/10.1038/nm.2127.

Mazarati, A., Maroso, M., Iori, V., Vezzani, A., & Carli, M. 2011. "High-mobility group box-1 impairs memory in mice through both toll-like receptor 4 and Receptor for Advanced Glycation End Products." *Exp Neurol* 232 (2):143-8. https://doi.org/10.1016/j.expneurol.2011.08.012.

Messmer, D., Yang, H., Telusma, G., Knoll, F., Li, J., Messmer, B., Tracey, K. J., & Chiorazzi, N. 2004. "High mobility group box protein 1: an endogenous signal for dendritic cell maturation and Th1 polarization." *J Immunol* 173 (1):307-13. https://doi.org/10.4049/jimmunol.173.1.307.

Muller, S., L. Ronfani, and M. E. Bianchi. 2004. "Regulated expression and subcellular localization of HMGB1, a chromatin protein with a cytokine function." *J Intern Med* 255 (3):332-43. https://doi.org/10.1111/j.1365-2796.2003.01296.x.

Nishibori, M., Wang, D., Ousaka, D., & Wake, H. 2020. "High Mobility Group Box-1 and Blood-Brain Barrier Disruption." *Cells* 9 (12). https://doi.org/10.3390/cells9122650.

Palleria, C., Coppola, A., Citraro, R., Del Gaudio, L., Striano, S., De Sarro, G., & Russo, E. 2015. "Perspectives on treatment options for mesial temporal lobe epilepsy with hippocampal sclerosis." *Expert Opin Pharmacother* 16 (15):2355-71. https://doi.org/10.1517/14656566.2015.1084504.

Park, J. S., Arcaroli, J., Yum, H. K., Yang, H., Wang, H., Yang, K. Y., Choe, K. H., Strassheim, D., Pitts, T. M., Tracey, K. J., & Abraham, E. 2003. "Activation of gene expression in human neutrophils by high mobility group box 1 protein." *Am J Physiol Cell Physiol* 284 (4):C870-9. https://doi.org/10.1152/ajpcell.00322.2002.

Park, J. S., Svetkauskaite, D., He, Q., Kim, J.-Y., Strassheim, D., Ishizaka, A., & Abraham, E. 2004. "Involvement of toll-like receptors 2 and 4 in cellular activation by high mobility group box 1 protein." *J Biol Chem* 279 (9):7370-7. https://doi.org/10.1074/jbc.M306793200.

Park, S., Sapkota, K., Kim, S., Kim, H., & Kim, S. 2011. "Kaempferol acts through mitogen-activated protein kinases and protein kinase B/AKT to elicit protection in a model of neuroinflammation in BV2 microglial cells." *Br J Pharmacol* 164 (3):1008-25. https://doi.org/10.1111/j.1476-5381.2011.01389.x.

Paudel, Y. N., Shaikh, Mohd. F., Shah, S., Kumari, Y., & Othman, I. 2018. "Role of inflammation in epilepsy and neurobehavioral comorbidities: Implication for therapy." *Eur J Pharmacol* 837:145-155. https://doi.org/10.1016/j.ejphar.2018.08.020.

Pitkanen, A., and K. Lukasiuk. 2009. "Molecular and cellular basis of epileptogenesis in symptomatic epilepsy." *Epilepsy Behav* 14 Suppl 1:16-25. https://doi.org/10.1016/j.yebeh.2008.09.023.

Pitkanen, A., and K. Lukasiuk. 2011. "Mechanisms of epileptogenesis and potential treatment targets." *Lancet Neurol* 10 (2):173-86. https://doi.org/10.1016/S1474-4422(10)70310-0.

Ravizza, T., Terrone, G., Salamone, A., Frigerio, F., Balosso, S., Antoine, D. J., & Vezzani, A. 2018. "High Mobility Group Box 1 is a novel pathogenic factor and a mechanistic biomarker for epilepsy." *Brain Behav Immun* 72:14-21. https://doi.org/10.1016/j.bbi.2017.10.008.

Rescigno, M., Martino, M., Sutherland, C. L., Gold, M. R., & Ricciardi-Castagnoli, P. 1998. "Dendritic cell survival and maturation are regulated by different signaling pathways." *J Exp Med* 188 (11):2175-80. https://doi.org/10.1084/jem.188.11.2175.

Riazi, K., M. A. Galic, and Q. J. Pittman. 2010. "Contributions of peripheral inflammation to seizure susceptibility: cytokines and brain excitability." *Epilepsy Res* 89 (1):34-42. https://doi.org/10.1016/j.eplepsyres.2009.09.004.

Richard, S. A., Min, W., Su, Z., & Xu, H. 2017. "High Mobility Group Box 1 and Traumatic Brain Injury." *Journal of Behavioral and Brain Science* 7:50-61.

Savill, J., Dransfield, I., Gregory, C., & Haslett, C. 2002. "A blast from the past: clearance of apoptotic cells regulates immune responses." *Nat Rev Immunol* 2 (12):965-75. https://doi.org/10.1038/nri957.

Scaffidi, P., T. Misteli, and M. E. Bianchi. 2002. "Release of chromatin protein HMGB1 by necrotic cells triggers inflammation." *Nature* 418 (6894):191-5. https://doi.org/10.1038/nature00858.

Scheffer, I. E., Berkovic, S., Capovilla, G., Connolly, M. B., French, J., Guilhoto, L., Hirsch, E., Jain, S., Mathern, G. W., Moshé, S. L., Nordli, D. R., Perucca, E., Tomson, T., Wiebe, S., Zhang, Y., & Zuberi, S. M. 2017. "ILAE classification of the epilepsies: Position paper of the ILAE Commission for Classification and Terminology." *Epilepsia* 58 (4):512-521. https://doi.org/10.1111/epi.13709.

Schmidt, A. M., and D. M. Stern. 2001. "Receptor for age (RAGE) is a gene within the major histocompatibility class III region: implications for host response mechanisms in homeostasis and chronic disease." *Front Biosci* 6:D1151-60. https://doi.org/10.2741/schmidt.

Shabab, T., Khanabdali, R., Moghadamtousi, S. Z., Kadir, H. A., & Mohan, G. 2017. "Neuroinflammation pathways: a general review." *Int J Neurosci* 127 (7):624-633. https://doi.org/10.1080/00207454.2016.1212854.

Shen, S., Qu, H., Huang, Y., Wu, D., Jiang, H., Yuan, C. 2015. "Expression level and distribution of HMGB1 in Sombati's cell model and kainic acid-induced epilepsy model." *Eur Rev Med Pharmacol Sci* 19 (15):2928-33.

Shi, Y., Zhang, L., Teng, J., & Miao, W. 2018. "HMGB1 mediates microglia activation via the TLR4/NF-kappaB pathway in coriaria lactone induced epilepsy." *Mol Med Rep* 17 (4):5125-5131. https://doi.org/10.3892/mmr. 2018.8485.

Shorvon, S. D. 2011. "The etiologic classification of epilepsy." *Epilepsia* 52 (6):1052-7. https://doi.org/10.1111/j.1528-1167.2011.03041.x.

Sloviter, R. S., and A. V. Bumanglag. 2013. "Defining "epileptogenesis" and identifying "antiepileptogenic targets" in animal models of acquired temporal lobe epilepsy is not as simple as it might seem." *Neuropharmacology* 69:3-15. https://doi.org/10.1016/j.neuropharm.2012.01.022.

Stros, M. 2010. "HMGB proteins: interactions with DNA and chromatin." *Biochim Biophys Acta* 1799 (1-2):101-13. https://doi.org/10.1016/j.bbagrm.2009. 09.008.

Tang, D., Kang, R., Zeh, H. J., & Lotze, M. T. 2011. "High-mobility group box 1, oxidative stress, and disease." *Antioxid Redox Signal* 14 (7):1315-35. https://doi.org/10.1089/ars.2010.3356.

Ugrinova, I., Pasheva, E. A., Armengaud, J., & Pashev, I. G. 2001. "In vivo acetylation of HMG1 protein enhances its binding affinity to distorted DNA structures." *Biochemistry* 40 (48):14655-60. https://doi.org/10.1021/bi0113364.

van Beijnum, J. R., W. A. Buurman, and A. W. Griffioen. 2008. "Convergence and amplification of toll-like receptor (TLR) and receptor for advanced glycation end products (RAGE) signaling pathways via high mobility group B1 (HMGB1)." *Angiogenesis* 11 (1):91-9. https://doi.org/10.1007/s10456-008-9093-5.

Venereau, E., De Leo, F., Mezzapelle, R., Careccia, G., Musco, G., & Bianchi, M. E. 2016. "HMGB1 as biomarker and drug target." *Pharmacol Res* 111:534-544. https://doi.org/10.1016/j.phrs.2016.06.031.

Vezzani, A., Auvin, S., Ravizza, T., Aronica, E. 2012. "Glia-neuronal interactions in ictogenesis and epileptogenesis: role of inflammatory mediators." In *Jasper's Basic Mechanisms of the Epilepsies*, edited by th, J. L. Noebels, M. Avoli, M. A. Rogawski, R. W. Olsen and A. V. Delgado-Escueta. Bethesda (MD).

Vezzani, A., Auvin, S., Ravizza, T., Aronica, E. 2011. "The role of inflammation in epilepsy." *Nat Rev Neurol* 7 (1):31-40. https://doi.org/10.1038/nrneurol.2010.178.

Vezzani, A., B. Lang, and E. Aronica. 2015. "Immunity and Inflammation in Epilepsy." *Cold Spring Harb Perspect Med* 6 (2):a022699. https://doi.org/10.1101/cshperspect.a022699.

Walker, L., and G. J. Sills. 2012. "Inflammation and epilepsy: the foundations for a new therapeutic approach in epilepsy?" *Epilepsy Curr* 12 (1):8-12. https://doi.org/10.5698/1535-7511-12.1.8.

Walker, L., Tse, K., Ricci, E., Thippeswamy, T., Sills, G. J., White, S. H., Antoine, D. J., Marson, A., & Pirmohamed, M. 2014. "High mobility group box 1 in the inflammatory pathogenesis of epilepsy: profiling circulating levels after experimental and clinical seizures." *The Lancet* 383:S105. https://doi.org/10.1016/s0140-6736(14)60368-8.

Walker, L. E., Griffiths, M. J., McGill, F., Lewthwaite, P., Sills, G. J., Jorgensen, A., Antoine, D. J., Solomon, T., Marson, A. G., & Pirmohamed, M. 2017. "A comparison of HMGB1 concentrations between cerebrospinal fluid and blood in patients with neurological disease." *Biomarkers* 22 (7):635-642. https://doi.org/10.1080/1354750X.2016.1265003.

Wang, H., Liao, H., Ochani, M., Justiniani, M., Lin, X., Yang, L., Al-Abed, Y., Wang, H., Metz, C., Miller, E. J., Tracey, K. J., & Ulloa, L. 2004. "Cholinergic agonists inhibit HMGB1 release and improve survival in experimental sepsis." *Nat Med* 10 (11) :1216-21. https ://doi.org/10.1038/nm1124.

Wang, N., Liu, H., Ma, B., Zhao, T., Chen, Y., Yang, Y., Zhao, P., & Han, X. 2021. "CSF high-mobility group box 1 is associated with drug-resistance and symptomatic etiology in adult patients with epilepsy." *Epilepsy Res* 177 :106767. https ://doi.org/ 10.1016/j.eplepsyres.2021.106767.

Willingham, S. B., Allen, I. C., Bergstralh, D. T., Brickey, W. J., Huang, M. T. H., Taxman, D. J., Duncan, J. A., & Ting, J. P. Y. 2009. "NLRP3 (NALP3, Cryopyrin) facilitates in vivo caspase-1 activation, necrosis, and HMGB1 release via inflammasome-dependent and -independent pathways." *J Immunol* 183 (3) :2008-15. https ://doi.org/ 10.4049/jimmunol.0900138.

Xie, Y., Yu, N., Chen, Y., Zhang, K., Ma, H.-Y., & Di, Q. 2017. "HMGB1 regulates P-glycoprotein expression in status epilepticus rat brains via the RAGE/NF-kappaB signaling pathway." *Mol Med Rep* 16 (2) :1691-1700. https ://doi.org/10.3892/mmr. 2017.6772.

Xu, Y. 2002. "Sphingosylphosphorylcholine and lysophosphatidylcholine : G protein-coupled receptors and receptor-mediated signal transduction." *Biochim Biophys Acta* 1582 (1-3) :81-8. https ://doi.org/10.1016/s1388-1981(02)00140-3.

Yang, H., Ochani, M., Li, J., Qiang, X., Tanovic, M., Harris, H. E., Susarla, S. M., Ulloa, L., Wang, H., DiRaimo, R., Czura, C. J., Wang, H., Roth, J., Warren, H. S., Fink, M. P., Fenton, M. J., Andersson, U., & Tracey, K. J. 2004. "Reversing established sepsis with antagonists of endogenous high-mobility group box 1." *Proc Natl Acad Sci USA* 101 (1) :296-301. https ://doi.org/10.1073/pnas.2434651100.

Yang, W., Li, J., Shang, Y., Zhao, L., Wang, M., Shi, J., & Li, S. 2017. "HMGB1-TLR4 Axis Plays a Regulatory ôle in the Pathogenesis of Mesial Temporal Lobe Epilepsy in Immature Rat Model and Children via the p38MAPK Signaling Pathway." *Neurochem Res* 42 (4) :1179-1190. https ://doi.org/10.1007/s11064-016-2153-0.

Zhang, J., Takahashi, H. K., Liu, K., Wake, H., Liu, R., Maruo, T., Date, I., Yoshino, T., Ohtsuka, A., Mori, S., & Nishibori, M. 2011. "Anti-high mobility group box-1 monoclonal antibody protects the blood-brain barrier from ischemia-induced disruption in rats." *Stroke* 42 (5) :1420-8. https ://doi.org/10.1161/STROKEAHA.110. 598334.

Zhang, Z., Liu, Q., Liu, M., Wang, H., Dong, Y., Ji, T., Liu, X., Jiang, Y., Cai, L., & Wu, Y. 2018. "Upregulation of HMGB1-TLR4 inflammatory pathway in focal cortical dysplasia type II." *J Neuroinflammation* 15 (1) :27. https ://doi.org/10.1186/s12974-018-1078-8.

Zhao, J., Wang, Y., Xu, C., Liu, K., Wang, Y., Chen, L., Wu, X., Gao, F., Guo, Y., Zhu, J., Wang, S., Nishibori, M., & Chen, Z. 2017. "Therapeutic potential of an anti-high mobility group box-1 monoclonal antibody in epilepsy." *Brain Behav Immun* 64 :308-319. https ://doi.org/10.1016/j.bbi.2017.02.002.

Zurolo, E., Iyer, A., Maroso, M., Carbonell, C., Anink, J. J., Ravizza, T., Fluiter, K., Spliet, W. G. M., van Rijen, P. C., Vezzani, A., & Aronica, E. 2011. "Activation of Toll-like receptor, RAGE and HMGB1 signalling in malformations of cortical development." *Brain* 134 (Pt 4) :1015-32. https ://doi.org/10.1093/brain/awr032.

Index

A

B

C

D

K

L

M

N

O

P

Q

R

S

T

V

W

About the Editors

Professor Eylem Taskin

Professor Eylem Taskin graduated from the Department of Physiology, Faculty of Medicine, Erciyes University, and Department of Biophysics, Faculty of Medicine, Cukurova University. She worked as a postdoctoral researcher at New York University for two years. Professor Taskin has published more than thirty articles and eight chapters and has edited two books as well. She has worked on the editorial boards of several scientific journals.

Associated Professor Celal Guven

Associated Professor Celal Guven graduated from the Department of Biophysics, Faculty of Medicine, Istanbul University. He has been working on cell cytoskeleton and protein-chemical interaction. Dr. Guven has published around twenty articles and eight chapters and has edited one book as well. He has worked on the editorial boards for different scientific journals.

Assistant Dr. Salih Tunc Kaya

Assistant Dr. Salih Tunc Kaya received his Ph.D. from the Department of Biology, Faculty of Science and Letters, Bolu University. He has been researching myocardial-ischemia reperfusion injury with arrhythmia. Dr. Kaya has published around fifteen articles, as well as five book chapters.